2006
INTERNATIONAL SYMPOSIUM ON
VLSI TECHNOLOGY, SYSTEMS, AND APPLICATIONS
(VLSI-TSA)

PROCEEDINGS OF TECHNICAL PAPERS

April 24-26, 2006
AMBASSADOR HOTEL
HSINCHU, TAIWAN
http://vlsitsa.itri.org.tw

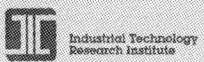

Copyright and Reprint Permission Statement

FOR PRINT VERSION

Copyright and Reprint Permission: Abstracting is permitted with credit to the source. Libraries are permitted to photocopy beyond the limit of U.S. copyright law for private use of patrons those articles in this volume that carry a code at the bottom of the first page, provided the per-copy fee indicated in the code is paid through Copyright Clearance Center, 222 Rosewood Drive, Danvers, MA 01923. For other copying, reprint or republication permission, write to IEEE Copyrights Manager, IEEE Operations Center, 445 Hoes Lane, P.O. Box 1331, Piscataway, NJ 08855-1331. All rights reserved. Copyright ©2006 by the Institute of Electrical and Electronics Engineers.

IEEE Catalog Number: 06TH8870

ISBN: 1-4244-0181-X

ISSN: 1930-885X

Publication Office
ERSO/ITRI
B300, Bldg. 11, 195-4, Chung Hsin Rd.
Chutung, Hsinchu, Taiwan 31015, R.O.C.
Tel: +886-3-591-3478
Fax: +886-3-582-0221
E-mail: vlsitsa@itri.org.tw

IEEE Conference Publications Managment Group
445 Hoes Lane
Piscataway, NJ 08854 U.S.A.
Tel: (732)562-3872
Fax: (732)981-1769
E-mail: confpubs@ieee.org

2006 International Symposium on VLSI Technology, Systems, and Applications

Hsinchu, Taiwan
24-26 April 2006

IEEE Catalog Number: CFP06846-POD
ISBN: 978-1-42440-181-9

FOREWORD

The objective of the International Symposium on VLSI Technology, Systems, and Applications (VLSI-TSA) is to bring together experts from all over the world to discuss current progress with local experts from Taiwan. Starting in 2005, the Symposium is splitting off its design segment to become the International Symposium on VLSI Design, Automation, and Test (VLSI-DAT). The VLSI-TSA Symposium will become an annual event and will extend its scope to include research advancements in the field of semiconductor manufacturing.

The symposium program includes three plenary talks, eight sessions, and two short courses. The contributed and invited papers are high quality, and will be presented by industrial and academic leaders and students from over 8 countries this year. The Plenary Session will feature the following invited talks: "Semiconductor Manufacturing Technology in the 21st Century", presented by professor Hiroshi Iwai from Tokyo Institute of Technology; "Challenges for Process and Product Integration at 45 nm", given by Dr. Hans Stork from Texas Instruments; and "3D System Integration Technologies", by Dr. Eric Beyne, IMEC. Two special sessions related to emerging technologies feature the progress in the areas of Flexible Electronics and Novel DRAM Technology. Eight experts were invited to present the cutting-edge results. The accepted papers cover the following areas: advanced memory, CMOS/ESD, novel devices (non-planar CMOS), metal gate/high-k stack, back-end technology, and simulation/characterization of advanced devices. Two sessions of short courses will cover Flash memory, DRAM memory, Advanced CMOS, and Manufacturing, as presented by four distinguished experts.

The main contents of VLSI-TSA are the contributed papers. This year, the conference has attracted about 97 submissions from all over the world, representing original pieces of work on the latest advances in the area of VLSI technology and manufacturing. The Technical Program Committee has selected 41 excellent papers for presentation.

We would like to thank the keynote, invited and contributed authors for their original and hard work. Appreciation is also extended to all the Technical Program Committee Members for their efforts in reviewing and selecting the papers. We would like to express special thanks to the Subcommittee Chairs for their great assistance in putting together an excellent technical program. Finally, we are most appreciative of the dedication and tireless efforts by Ms. Stacey Hsieh and the Local Organizing Committee in arranging all of the details of this Technical Digest as well as all aspects of 2006 VLSI-TSA.

Technical Program Committee Chair

Hsing-Huang Tseng
Chief Technologist
and CMOS Extension Program Manager
Front End Processes Division, SEMATECH
(Freescale Semiconductor assignee)

CONTENT

PLENARTY SESSION (Invited)

Session Chair: H. H. Tseng (SEMATECH/Freescale, USA)

K1 Semiconductor Manufacturing Technology in the 21st Century

Hiroshi Iwai..1

K2 Challenges for Process and Product Integration at 45nm

Hans Stork...18

K3 3D System Integration Technologies

Eric Beyne...19

SESSION 1: Memory

Session Chairs: Rama Divakaruni (IBM, USA), Rich Liu (Macronix, Taiwan)

T11 Future Memory Devices - from Stacked Memory, Gain Memory, Single-electron Memory to Molecular Memor (Invited)

Kazuo Nakazato 28

T12 Analysis of Electron and Hole Distributions on Scaled NBit Flash Cells

I. C. Yang, K. P. Chen, Y. W. Chang, and T. C. Lu...30

T13 P-channel SONOS Transient Current Modeling for Program and Erase

Pei-Ying Du, Jyh-Chyurn Guo, H.M. Lee, H. M. Chen, Rick Shen, and C.C.-H. Hsu......................................32

T14 Understanding of the Leakage Components and Its Correlation to the Oxide Scaling on the SONOS Cell Endurance and Retention

C. H. Chen, P. Y. Chiang, Steve S. Chung, Terry Chen, George C. W. Chou, and C. H. Chu............................34

T15 High-κ Hf-based Charge Trapping Layer with Al_2O_3 Blocking Oxide for High-density Flash Memory

S. Maikap, P. J. Tzeng, L. S. Lee, H. Y. Lee, C. C. Wang, P. H. Tsai, K. S. Chang-Liao,

W. -J. Chen, K. C. Liu, P. R. Jeng, and M.-J. Tsai..36

T16 V_{DD} Scaling for FinFET Logic and Memory Circuits: the Impact of Process Variations and SRAM Stability

C.-H. Lin, K. K. Das, L. Chang, R. Q. Williams, W. E. Haensch, and C. Hu....................................38

T17 Novel T Shape Structure PCM and Electrical-Thermal Characteristics

W. H. Wang, D. S. Chao, Y. C. Chen, C. M. Lee, H. H. Hsu, Y. Chuo, M. H. Tseng, M. H. Lee,

W. S. Chen, M. J. Kao, and M.-J. Tsai 40

T18 A New Read Method by Using DIBL Characteristics in Nitride Storage Device

L.P. Chiang, P. A. Chen, C. H. Hung , C. P. Tsao, H. H. Liao, and C. H. Lin....................................42

T19 Writing Architecture for Magnetic Random Access Memory with Negative Pulse Writing Scheme

C.P. Chang, C.C. Hung, Y.H. Wang, Y.J. Lee, K.L. Su, W.C. Chen, Y.H. Chen, C.S. Lin,

M.J. Kao, J.F. Huang, and M.-J. Tsai...44

SESSION 2: CMOS/ESD

Session Chairs: Clement Wann (IBM, USA), Jack Sun (TSMC, Taiwan)

T21 Strained Si and the Future Direction of CMOS (Invited)

Scott E. Thompson and Guangyu Sun...46

T22 Optimization of Source/Drain Extension for Robust Speed Performance to Process Variation in Undoped Double-Gate CMOS

Ji-Woon Yang, Daniel Pham, Peter Zeitzoff, Howard Huff, and George Brown 48

T23 Mo Gate Deformation Induced by Laser Annealing Process

Kentaro Shibahara, Akira Matsuno, Masaki Hino, and Ken-ichi Kurobe 50

T24 Low-Leakage Diode String Design without Extra Circuits for ESD Applications

Shao-Chang Huang, Yu-Hung Chu, Chen-Chi Kuo, T.Y. Huang, M.H. Song, and Mi-Chang Chang......................52

T25 Investigation on RF Characteristics of Stacked P-I-N Polysilicon Diodes for ESD Protection Design in 0.18-μm CMOS Technology

Yu-Da Shiu, Che-Hao Chuang, and Ming-Dou Ker...56

T26 65nm SOI CMOS Technology for High Performance Microprocessor Application

Samuel K.H.Fung, P.A.Grudowski, C.H.Wu, V. Kolagunta, N. Cave, C.T.Yang, S.J.Lian, V. Adams,

O. Zia, B. Min, N. Grove, K.H.Chen, W.J.Liang, D.H.Lee, H.T.Huang, J. Cheek, and H.C.Tuan......................58

T27 A Promising Planar Transistor with in-situ Doped Selective Si Epitaxy Technology (GORES MOSFET) for 32nm Node and Beyond

Y. Kikuchi, Y. Tateshita, T. Kataoka, J. Wang, Y. Miyanami, N. Yamagishi, T. Ikuta, Y. Yamamoto, S. Hiyama,

H. Ugajin, H. Ikeda, S. Fujita, R.Yamamoto, S. Kanda, T. Imoto, S. Kashiwadate, Y. Tagawa, H. Iwamoto,

T. Ohno, T. Kobayashi, M. Saito, S. Kadomura, and N. Nagashima......................................60

T28 Strain-Induced Channel Backscattering Modulation in Nanoscale CMOSFETs

Hung-Wei Chen, Hong-Nien Lin, Chih-Hsin Ko, Chung-Hu Ge, Horng-Chih Lin,

Tiao-Yuan Huang, and Wen-Chin Lee...62

T29 PSDG MOSFET

Deyuan Xiao, Gary Chen, Roger Lee, Daniel Lu, Leong Tan, Yung Liu, CC Shen, and Jong Woo Kim....................64

SESSION 3: Flexible Electronics (Invited)

Session Chairs: Roger De Keersmaecker (IMEC, Belgium), Yi-Jen Chan (ERSO/ITRI, Taiwan)

T31 Vapor and Solution Deposited Organic Thin Film Transistors

Thomas Jackson 66

T32 Printed Transistors and Passive Components for Low-cost Electronics Applications

Vivek Subramanian, Josephine B. Chang, Steven E. Molesa, Steven K. Volkman, and David R. Redinger 68

T33 Future Prospects of Flexible, Large-Area Sensors and Actuators with Organic Transistor ICs

Takao Someya, Takayasu Sakurai, and Tsuyoshi Sekitani...70

T34 Tubes, Ribbons and Wires for Flexible Electronics

Yugang Sun and John A. Rogers 72

T35 Printed Electronics for System Application

Z.Pei, C.P. Kung, P.Y.Lo, J.J.Chang, C.A.Chung. Stanley H. Huamg and Y.J.Chan...........................74

SESSION 4: Novel Devices

Session Chairs: Robin Van Den Nieuwenhuizen (AMD, USA), Mike Ma (UMC, Taiwan)

T41 **Advanced Substrate Engineering for the Nanotechnology Era (Invited)**

Carlos Mazure...78

T42 **A CMOS Bulk-Micromachined Thermal Imager**

L.-S. Zheng, D.-H. Liu, C.-Y. Hsu, D.-J. Yao, and M. S.-C. Lu.............................80

T43 **Impact of Back Gate Bias on Hot-Carrier Effects of n-channel Tri-Gate FinFETs (TGFET)**

Chia-Pin Lin and Bing-Yue Tsui..82

T44 **Ultra-Thin SOI CMOS Using Laser Spike Anneal**

Zhibin Ren, J. Sleight, J. M. Hergenrother, D. V. Singh, O. Gluschenkov, O. Dokumaci, L. Black,

J. Pan, K.-L. Lee, J. Ott, P. Ronsheim, J. Lee, W. Haensch, M. Ieong and C.Y. Sung.................... 84

T45 **Effect of Oxygen Absorption on Contact Resistance between Metal and Carbon Nano Tubes (CNTs)**

Bing-Yue Tsui, Chien-Li Weng, Chih-Lien Chang, Jeng-Hua Wei, and Ming-Jinn Tsai........................86

T46 **A Novel Deep Trench Isolation Featuring Airgaps for a High-Speed 0.13 μ m SiGe:C BiCMOS Technology**

L.J. Choi , E. Kunnen , S. Van Huylenbroeck , A. Piontek , A. Sibaja-Hernandez, F. Vleugels, T. Dupont,

P. Leray, K. Devriendt, X.P. Shi, R. Loo, S. Vanhaelemeersch and S. Decoutere.........................88

SESSION 5: DRAM

Session Chairs: Tak H. Ning (IBM, USA), J.P. Lin (Nanya Technology, Taiwan)

T51 **Beyond Scaling – Realizing Value Through the Integration of Memory**

and Autonomic Chip Features (Invited)

Subramanian Iyer...90

T52 **Trench DRAM Technologies for the 50nm Node and Beyond (Invited)**

W.Mueller, G.Aichmayr, W.Bergner, E.Erben, T.Hecht, A.Kersch, S.Kudelka, F.Lau,

J.Luetzen, A.Orth, J.Nuetzel, T.Schloesser, A.Scholz, U.Schroeder, A.Sieck, A.Spitzer,

M.Strasser, P-F.Wang, S.Wege, and R.Weis...92

T53 **Overview and Future Challenge of Floating Body Cell (FBC) Technology for**

Embedded Applications (Invited)

Akihiro Nitayama, Takashi Ohsawa, and Takeshi Hamamoto...94

T54 **Stack DRAM Technologies for the Future (Invited)**

Donggun Park, Wonshick Lee, and Byung-il Ryu..97

T55 **A Novel DRAM Cell Design and Process for 70NM Generation**

C.H. Chung, T. Chien, W.S. Kuo, J.S. Hsiao, CC Cheng, F. Li, S. Wu, B. Wang,

C. Wang, T. Hu, G. Hsiao, M. Che, R.Y. Hon, H.M. Chen, G. Chou, G. Chang,

H.C. Shu, K.Y. Huang, V. Tsai, L. Chou, and C.H. Chu 101

T56 **Reliability Assessment of the Embedded Dram Technology with Pmosfet Transfer Transistor and High-κ**

Dielectrics (Ta_2O_5) Mim Capacitor

R.F. Tsui, J.R. Shih, Kevin_Liu, Y.S. Tsai, H.W. Chin and Kenneth Wu..................................103

SESSION 6: Metal Gate / High-κ

Session Chais: Byoung-Hun Lee (SEMATECH/IBM, USA), Chih-Hsun Chu (ProMOS, Taiwan)

T61 Metal Gate Technology for 45nm and Beyond (Invited)

Kentaro Shibahara...105

T62 Electron Trapping Processes in High-κ Gate Dielectrics and Nature of Traps

G. Bersuker, J. Gavartin, J. Sim, C. S. Park, C. Young, S. Nadkarni, R. Choi, A. Shluger, and B. H. Lee...............107

T63 HfSiON Gate Dielectric for 45nm Node Low-Power Device

Tian-Choy Gan, Howard C.-H. Wang, Shang-Jr Chen, Ching-Wei Tsai,

Peng-Soon Lim, Huan-Just Lin, Ying Jin, Hun-Jan Tao, Shih-Chang Chen,

Ying Keung Leung, Carlos H. Diaz, Mong-Song Liang, and Yuh-Jier Mii...109

T64 Detection of Trap Generation in High-κ Gate Stacks due to Constant Voltage Stress

C.D. Young, D. Heh, R. Choi, J.J. Peterson, J. Barnett, B.H. Lee, P. Zeitzoff, G.A. Brown, and G. Bersuker...........111

T65 Relationship of HfO_2 Material Properties and Transistor Performance

P. D. Kirsch, M. A. Quevedo, G. Pant, S. Krishnan, S. C. Song, H. J. Li, J. J. Peterson, B. H. Lee,

R. W. Wallace, M. Kim and B. E. Gnade..113

T66 $Hf_xTa_yN_z$ Metal Gate Electrodes for Advanced MOS Devices Applications

Chin-Lung Cheng, Kuei-Shu Chang-Liao, Tzu-Chen Wang, Tien-Ko Wang, and Howard Chih-Hao Wang............115

T67 Impact of WSi_x Metal Gate Stoichiometry on Fully Depleted SOI MOSFETs Electrical Properties

J. Widiez, M. Vinet, B. Guillaumot, X. Garros, S. Minoret, T. Poiroux, O. Weber, L. Thevenod, P. Holliger,

B. Previtali, V. Barral, K. Sidi Ali Cherif, P. Grosgeorges, A. Toffoli, S. Maîtrejean, M. Cassé, F. Martin,

D. Lafond, O. Faynot, M. Mouis and S. Deleonibus...117

T68 NMOS and PMOS Metal Gate Transistors with Junctions Activated by Laser Annealing

S. Severi, E. Augendre, A. Falepin, C. Kerner, J. Ramos, P. Eyben, W. Vandervost, C. Curatola,

S. Felch, F. Nouri, P. Kraus, V. Parihar, T. Noda, R. Schreutelkamp, T. Y. Hoffmann, P. Absil,

K. De Meyer, M. Jurczak, and S. Biesemans..119

T69 Tunable Workfunction for Silicied Gates (FUSI) and Proposed Mechanisms

Y.H. Kim, C. Cabral. Jr. E. P. Gusev, L. Gignac, M. Gribelyuk, and M. Ieong.................................121

SESSION 7: BEOL

Session Chairs: Michel Brillouet (CEA-LETI, France) M. J. Tsai (ERSO/ITRI, Taiwan)

T71 Roadblocks and Critical Aspects of Cleaning for Sub-65nm Technologies (Invited)

Paul W. Mertens, G. Vereecke, R. Vos, S. Arnauts, F. Barbagini, T. Bearda, S. Degendt,

C. Demaco, A. Eitoku, M. Frank, W. Fyen, L. Hall, D. Hellin, F. Holsteyns, E. Kesters,

M. Claes, K. Kim, K. Kenis, H. Kraus, R. Hoyer, T.Q. Le, M. Lux, K-T. Lee, M. Kocsis,

T. Kotani, S. Malhouitre, A. Muscat, B. Onsia, S. Garaud, J. Rip, K. Sano, S. Sioncke,

J. Snow, J. Van Hoeymissen, K. Wostyn, K. Xu, V. Parachiev, and M. Heyns...................................123

T72 Effect of H_2 Addition during Cu Thin Film Sputtering

Masahiro Ooka and Shin Yokoyama...127

T73 Resistance Increase in Metal Nano-wires

Hsueh-Chung Chen, Hsien-Wei Chen, Shin-Puu Jeng, Chii-Ming M. Wu and Jack Y.-C. Sun.....................129

T74 Die-Based Electromigration Characterization For Copper / Low-K Dual Damascene Interconnects

Shou - Chung Lee, and Anthony S. Oates..131

T75 Integration of Cu Damascene with Pore-sealed PECVD Porogen Low-k (k=2.5)

Dielectrics for 65nm Generation

M.L. Yeh, C.C. Chou, T.I. Bao, K.C. Lin, I.I. Chen, K.P. Huang, Z.C. Wu, S.M. Jeng,

C.H. Yu, and M.S. Liang..133

T76 Wafer-Level Compliant Bump for 3D Chip-Stacking

Naoya Watanabe, Takeaki Kojima and Tanemasa Asano..135

SESSION 8: Simulation/ Modeling/ Characterization

Session Chairs: Subramanian Iyer (IBM, USA) Jenn-Gwo Hwu (NTU, Taiwan)

T81 Strained-SOI Technology for High-Speed CMOS Operation (Invited)

T. Mizuno, N. Sugiyama, T. Tezuka, Y. Moriyama, S. Nakaharai, and S. Takagi...........137

T82 RF Extrinsic Resistance Extraction Considering Neutral-Body Effect for

Partially-Depleted SOI MOSFETs

Sheng-Chun Wang, Pin Su, Kun-Ming Chen, Chien-Ting Lin, Victor Liang, and Guo-Wei Huang.....................139

T83 Hot Electrons Associated with the Long-Range Coulomb Interaction under the High-Density Regime

Tadayoshi Uechi and Nobuyuki Sano...141

T84 A New Series Resistance and Mobility Extraction Method by BSIM Model for Nano-Scale MOSFETs

William P.N. Chen, P. Su, J.S. Wang, C.H. Lien, C.H. Chang, K. Goto, and C.H. Diaz.........143

T85 Coupling Advanced Atomistic Process and Device Modeling for Optimizing Future CMOS Devices

B. Colombeau, S.H. Yeong, S.M. Pandey, F. Benistant, M. Jaraiz and S. Chu.................145

T86 An Asymmetrical Double-Gate VCO with Wide Frequency Range

Hung Ngo, Keunwoo Kim, Ching-Te Chuang, JB Kuang, Fadi Gebara, and Kevin Nowka............147

Industrial Sponsors... I

2006 International Symposium on VLSI Technology, Systems, and Applications................II

2006 International Symposium on VLSI Technology, Systems, and Applications

Semiconductor Manufacturing Technology in the 21st Century

Hiroshi Iwai

Frontier Collaborative Research Center, Tokyo Institute of Technology
4259, Nagatsuta-cho, Midori-ku, Yokohama, 226-8502,
Japan.

INTRODUCTION

The modern electronic circuits have now been evolved into ultra-large-scaled integrated (ULSI) circuits with extremely high performances. The silicon microchips, constituting with some silicon metal-oxide-semiconductor (MOS) transistors, have become indispensable key elements for our information society. For example, internet, mobile phones, video game players, digital cameras, and human-like robots could never be realized without the tremendous progress of the integrated circuit (IC) technology. The integrated circuits as well as their core device technology are expected to evolve further and with increasing importance in future intelligent society. Combining with artificial intelligent circuits and sensors, high intelligent robots may even perform better than human beings in some jobs such as elderly care and microsurgery. To realize such high intelligent systems, new integrated circuits with much higher performance and less power consumption are indispensable.

The electronic circuit development has been accomplished with the downscaling of component size since the replacement of vacuum tubes with transistors 40 years ago (see Fig.1) [1-4]. The circuit characteristics have benefited a lot from the downsizing. We are now able to integrate millions of transistors in a silicon chip with few centimeters square. The capacitance values are smaller in a smaller device. This leads to faster operating. speed and lesser power consumption. The size reduction in individual device makes higher integration density possible and allows parallel operations, which in turn further increases the circuit speed. Right now the operation speed of the latest microprocessor (MPU) has already reached 3 GHz and is expected to increase further [5] (see Fig.2), although recent trend indicates that the increase of the clock frequency may be gradually saturated .

Downsizing of the components has been the driving force for circuit evolution

1900	1950	1960	1970	2000
Vacuum Tube	Transistor	IC	LSI	ULSI
10 cm	cm	mm	10 μm	100 nm
10^{-1}m	10^{-2}m	10^{-3}m	10^{-5}m	10^{-7}m

In 100 years, the feature size was reduced to a millionth!

(a)

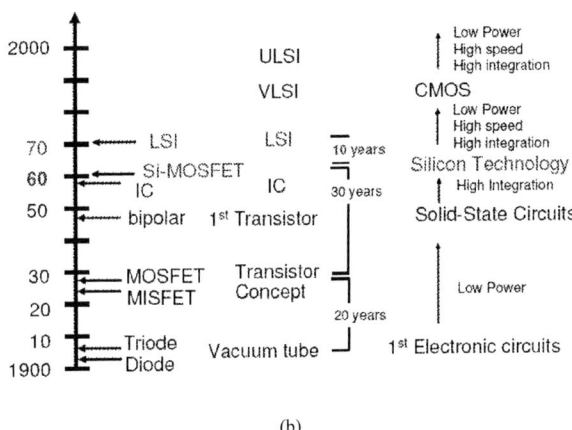

(b)

Fig.1 The downsizing trend of electronic components during the past century (a) and the major milestones of modern electronics development (b). The device feature size has been reduced to a millionth since 1900 and the circuit has been developed from discrete circuits to the ULSI circuits with millions of transistors.[1]

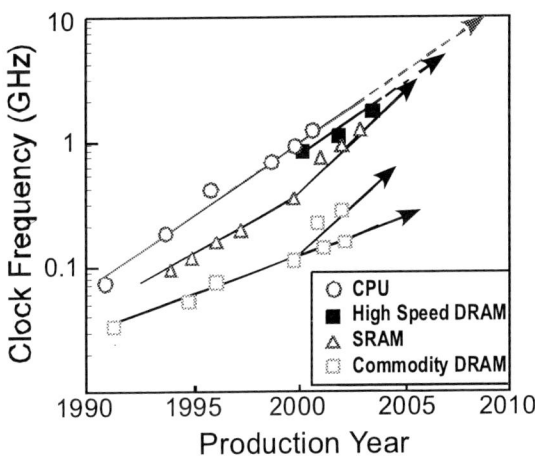

Fig.2 The device downsizing has a significant improvement in the speed of microprocessor and memory devices [5].

1-4244-0181-X/06/$20.00 ©2006 IEEE

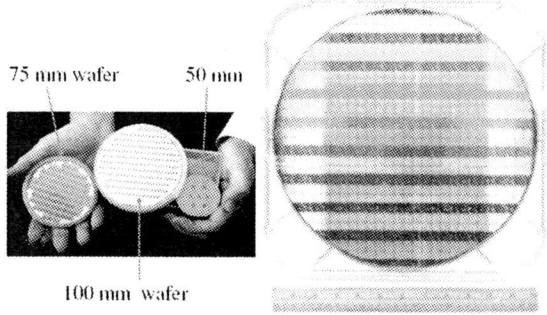

In addition to the device downsizing, the IC manufacturing methodologies have also been changed a lot during the past four decades. A readily visualizable change is the wafer size. The diameter of early wafer size was 50 mm and the latest one is 300 mm representing a 36 times increase in the chip area [1, 6] (see Fig.3). The throughput is further enhanced with the downsizing, improved yield [1, 7] (Fig.4) and the use of fully automatic high-precision machines and super clean environment [1, 8, 9] (see Fig.5). Thus the per-transistor and per-function cost has been reduced greatly.

Fig.3 The cost per chip has been reduced greatly by increasing the wafer size[1,6]

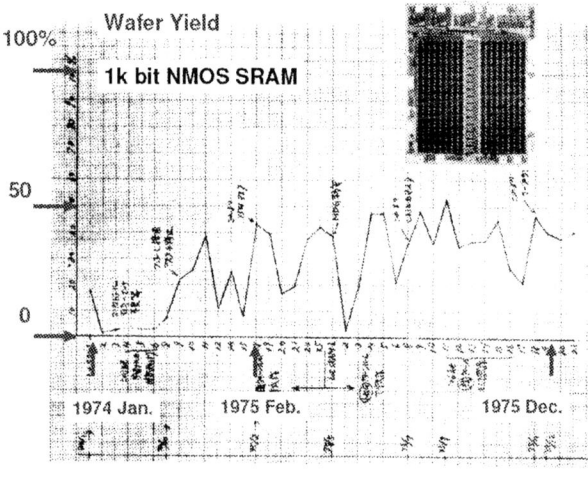

(a)

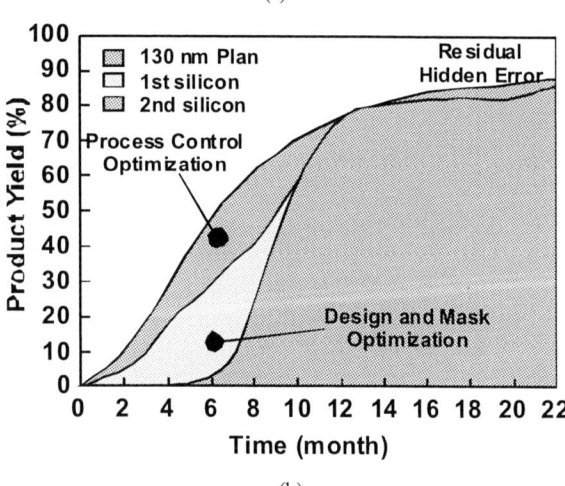

(b)

Fig.4 Yield was the major concern in large scale circuits. (a) In the middle of 70s, a 40% yield was accepted at the beginning stage of production. (b) Great improvement was made in recent technological nodes [1,7].

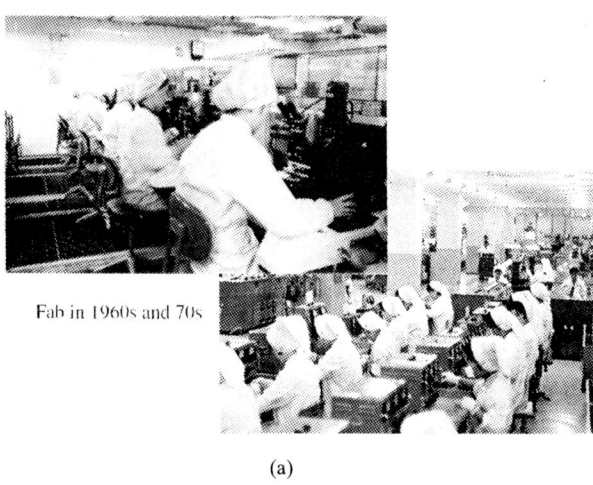

(a)

(b)

Fig. 5 The manufacturing plant has changed a lot during the past four decades. Early IC fabrication relied on manual operation (Toshiba Tama Works) (a). The latest 300 nm fabrication employed highly automatic machines running in a super clean room (b) [1,8,9]

With aggressively accelerated downsizing to the sub-100 nm range, we are facing a lot of serious problems in device, circuit as well as in manufacturing levels [1-18]. The skyrocketing increase in the production cost is a particular serious concern. Many people feel that we have to make some kind of drastic evolution or even a revolution in order to keep the scaling trend to continue towards the 10 nm range. This paper focuses on the future semiconductor manufacturing challenges. Some background information regarding the possible limits of scaling and the problems appeared in the sub-100 nm devices will be discussed, respectively, in section 2 and 3. The impacts of the future semiconductor manufacturing will be discussed in section 4. We shall also look forward to the possible geographical redistribution of the manufacturing centers and the new role playing of the present leaders in IC technology. Paradigm of post-downsizing era will be described in section 5.

TRENDS OF DOWNSIZING AND ITS LIMITS

The component scaling rate is really tremendous. The device dimensions have been reduced to a millionth at the production level in the past 100 years. Hundred years ago, no one could imagine that the mankind of our time is able to make some electronic circuits which consist of billions of electronic components with dimension smaller than bacteria and those circuits are controlling the operation of our society. Future scaling trends have been predicted by the International Technology Roadmap for Semiconductors (ITRS) [19] for 10 years up to 2018, when the physical gate length is expected to be 7 nm [1-4] (see Fig.6). Gate oxide thickness should be two orders of magnitude smaller than that of the gate length. Right now we have 1.2-nm thick oxynitride film being used in production and it is expected the silicon dioxide equivalent thickness (EOT) will be reduced to 0.5 nm in 10 years later. Although it is demonstrated that a MOS transistor with 1.5 nm [20-23] and recently even 0.8 nm [24] oxynitride gate insulator is still functioning, many serious problems are anticipated when realizing a large scale integrated circuit with such a thin gate insulator.

lithography at that time. In early 80's, 500 nm was predicted to be the limit because of the unacceptable large source/drain resistances in the scaled structure. Another predicted limit almost at the same time is 250 nm and the reason was the difficulty in managing the direct-tunneling leakage through the gate oxides and dopant fluctuations in the channel [26]. In the early 90's, 100 nm was thought to be the limit because of many expected difficulties in reducing the physical parameters of the MOSFETs. Fortunately, all those anticipated limits have been proven, with the available of shorter gate length commercial product, to be incorrect [1-25,27]. The latest prediction for the scaling limit is 10 nm because of several reasons such as direct-tunneling between source and drain. This was also proven not the case with the experimental confirmation of proper functioning 5 nm gate length MOSFETs [28] (see Fig.7).

It is very difficult to predict the limit of the scaling, although most of the people feel that we are approaching the limit and that it is somehow closer. The ultimate limit of the scaling is the distance of atoms in silicon crystals and that is about 0.3 nm [1-4] (see Fig.6). Note that the dimension of the 5 nm transistor is only 18 times of this limit. In the ultimate atomic dimension, we might expect some signal modulation effect through such a single atomic size gate electrode, but the modulated signal should be too weak to be transferred to another node. However, there is no practical solution at this moment for interconnects to contact such small atomic nodes. Thus the limit of the scaling should be considered from the viewpoint of integration of large scale manufacturable devices. There are many issues for the integration of ultra-small components. They are grouped into two categories: (1) performance and power, and (2) cost of design and manufacturing which also includes the yield and reliability issues and will be discussed in more detail in next section.

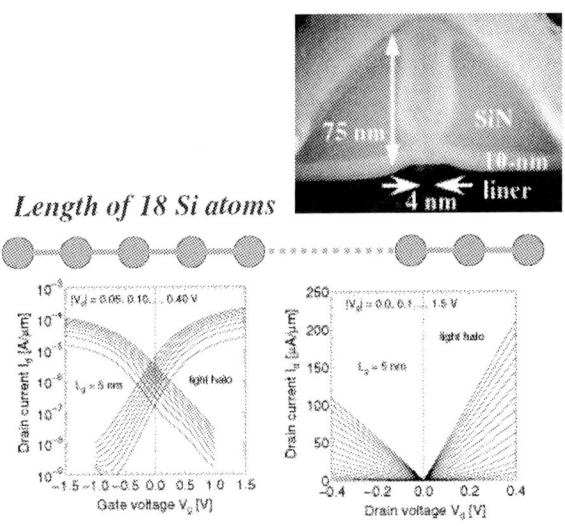

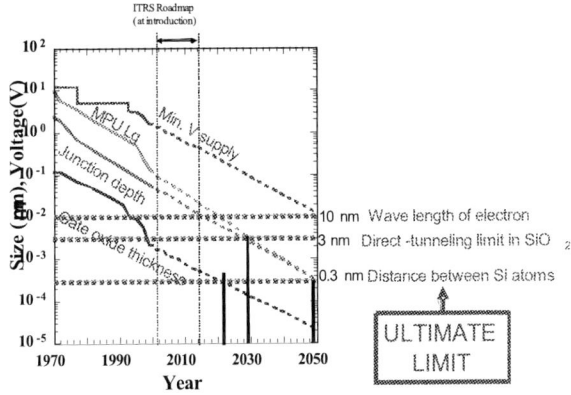

Fig.6 The downsizing trends on device parameters, possible constraints, and the ultimate limit of MOS transistors [1]

Fig.7 Cross-sectional view and current-voltage characteristics of a 5 nm gate length MOS transistor [28]

Looking back to the history of the CMOS technology developments, we have encountered so many expected and unexpected problems [1-25]. Fortunately, all those problems had been overcome. Historically, many possible limits for the downsizing were proposed. In the late 70's, one micrometer was thought to be the limit because of the forecasted difficulty in suppressing the short-channel effects as well as the expected resolution limit of optical

DEGRADATION IN PERFORMANCE AND POWER CONSUMPTION

The IC technology has benefited a lot in various aspects, such as speed, power, and cost, from downsizing of the MOS devices from 10 m to the 100-nm range. When approaching the atomic scales,

3

some performance degradations rather than improvements were reported. This section discusses the regimes for performance degradation in the sub-10 nm devices. Some possible technological options for conquering these issues will also be discussed.

Degradation of performance with downscaling

One of the major problems for performance degradation in the ultra-large scale circuits is the interconnect delay due to the increase in the resistance and the capacitance values of narrow and dense interconnection metal lines. For example, when the size of copper wire is reduced to less than 100 nm width, the resistivity of the conductor will increase pronouncedly because of the increased surface scattering effect [19] (see Fig. 8). Furthermore, the performance improvement is also questionable for the ultra-small MOSFET itself. According to the scaling theory, the drain current per unit gate width should stay constant. However, a significant reduction of the drain current value per unit gate width for sub-100 gate length MOSFETs was reported recently [1-4] (see Fig. 9). This phenomenon is due to the non-optimized MOSFET structure and process.

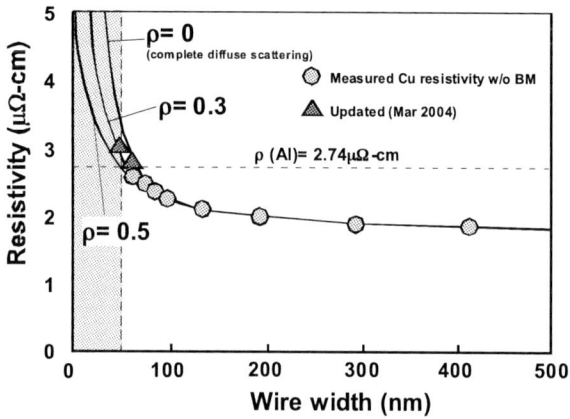

Fig.8 Increase of resistivity as a result of surface scattering is a major concern for the ultra-fine interconnects. Significant increase of resistivity was found for copper wire less than 100 nm width [19].

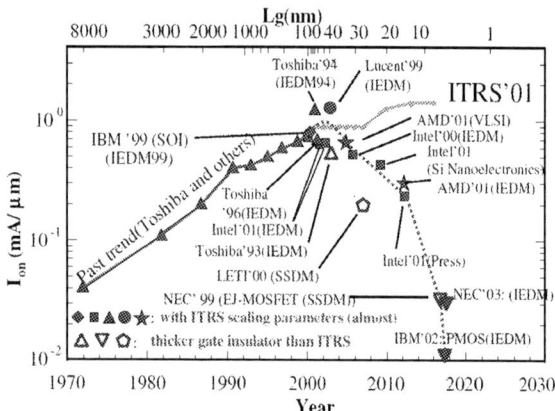

Fig.9 Significant reductions of the unit drain currents were reported in the sub-100 nm transistors which are diverged from the prediction of ITRS [1]

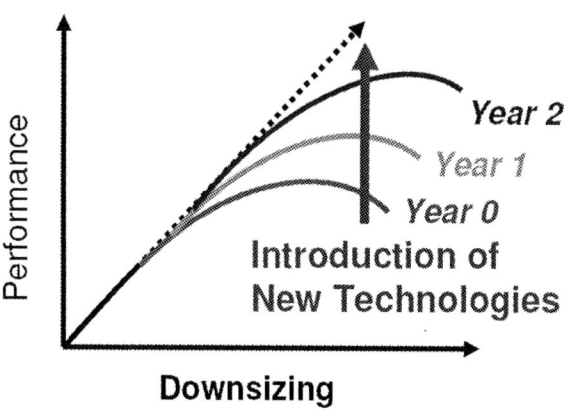

Fig.10 Possible scenarios of the influence of new technologies in transistor downsizing [1]

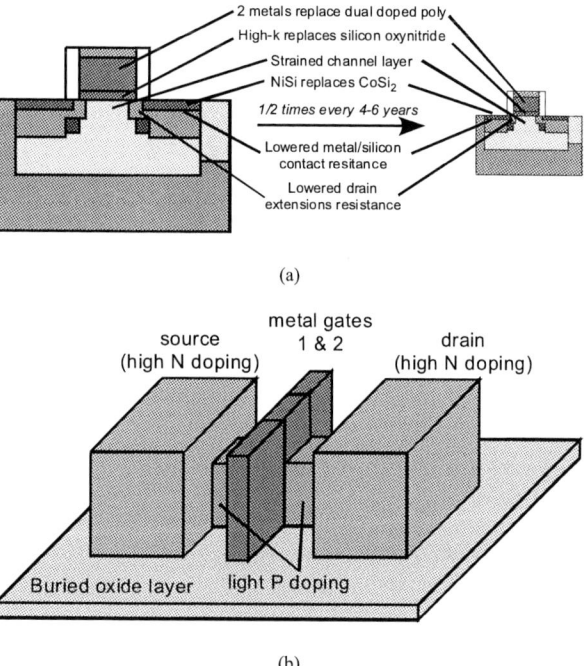

(a)

(b)

Fig.11 Scaling trend and challenges for some future technology node. The bulk MOS structure (a) can be further downscaled by introducing new materials. (b) Multiple gate or Fin-FET structure should be used to have a better control of the short-channel effects [19].

Without new technology, further downscaling may only result in performance degradation [1] (see Fig.10). Performance enhancements can be achieved and further downscaling may be proceeded with the introduction of new technologies and materials at least for another 10 years and furthermore with that of three dimensional structures [19] (Fig.11),. Several achievements in finding new materials and developing new process for sub-100 nm device manufacturing have been made recently. These processes or materials include the elevated source/drain [29-33], plasma doping with flash or laser annealing [34-43], NiSi silicide [44-56], strained Si channel for mobility enhancement [57-64], silicon on insulator (SOI) [65-67], three-dimensional structure [68-75] high dielectric constant (high-k) gate insulator [76-91], metal gate [90-96], and low dielectric constant (low-k) interlayer insulator for interconnects [99-101]. These measures are already on schedule for future technology nodes [102] (see Fig.12). However, some unexpected device parameter degradations were reported with the new materials. High-k gate insulator is an example.

Design Rule (nm)	180	130	90	65	45	32	22
Diameter of a wafer (mm)	200	200/300	300	300	300	300	300
Lithography	KrF		ArF	ArF Immersion		EUV	
Interconnect	Al						Cu
Interlayer	SiO$_2$						Low-k
Channel	Si						Strained Si
Gate electrode	poly-Si						Metal
Gate insulator	SiO$_2$						High-k

Fig.12 Predicted technology trend and the introduction or new materials or processes for continuous downsizing in future nodes [102]

In the sub-10 nm gate-length transistors, the small drain current (of several tens of micro-Ampere per micrometer) at the scaled supply voltage becomes a major concern. In addition, the fringing capacitance of the gate electrode, and the inversion layer capacitance will degrade the performance of such small MOSFETs [1-4] (see Fig.13). It is doubtful at this moment for such a small MOSFET be used for high-speed devices. It is possible that the roadmap for downsizing may be delayed unless there are some breakthroughs in certain aspects and the trends illustrated in Fig.14 [1] may be our future scenarios in downsizing.

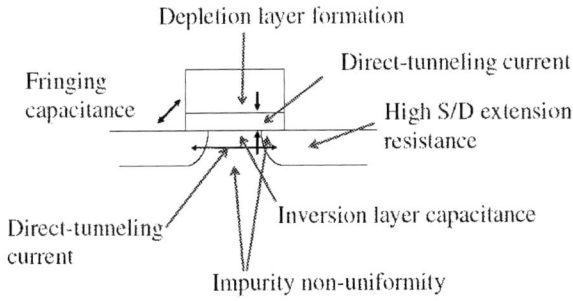

Fig.13 Challenging issues further downsizing of MOS transistor below sub-10 nm range [1].

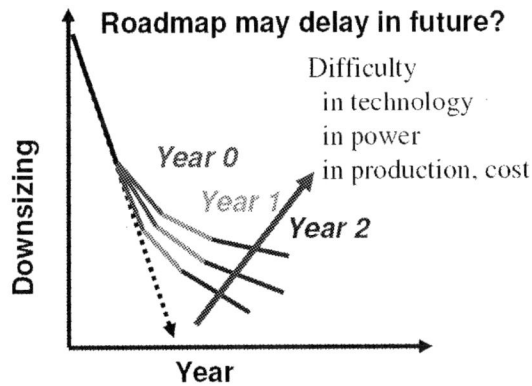

Fig.14 Roadmap of downsizing may be delayed in some future nodes because of the difficulties in the ultimate technology, large power dissipation or unaffordable manufacturing cost for smaller size transistor [1].

Fortunately, scaling is not the only solution for performance improvement. The improved circuit structures and system architectures can also make the integrated circuits to perform better. Parallel processing and optimized interconnect with the aid of routing tools are two best examples. As will be described later in more detail, system-on-chip (SOC) approach by embedding the DRAM in the logic unit will increase the transfer speed between logic and memory units. It is expected that the overall performance of our electronic system could be enhanced further at least down to the generation of 20 or 10 nm gate lengths as consequences of both the improved device technology and the new system structures. However, for sub-10 nm gate length transistors, there are still some uncertainties at least the drain current level should be improved

Increase in power consumption for high performance

Increasing power consumption for high-performance logic integrated circuits is another serious problem. If the trends of the clock frequency and the chip density continue to increase, the power consumption of a high-performance microprocessor (MPU) would reach 10 kW within several years and power density of the silicon chip surface may be as large as 1000 W/cm2 which is equivalent to the level of a rocket nozzle surface [103]. This situation is certainly not a realistic value and the implication is that we may not able to keep the rising trends of the clock frequency and the chip density at that time. The significant increase in power density is the consequences of the insufficient supply voltage reduction and the exponential increase in the transistor density. Low voltage technology (although is not easy) and appropriate control of the increase in chip density and chip size together with some new cooling technologies may partially solve this problem. It is noted that the growth rates of the chip area and the chip density were started to slow down in recent years. Innovative system power management will also help in this issue. Changeable clock frequency and supply voltage may be used in some functional blocks of some systems [103]. Power and performance of a chip can be traded off. For the low standby-power device such as cellular phones, subthreshold leakage current between source and drain [104], and gate dielectric leakage current significantly reduce the battery operation duration. The gate leakage current can be solved by using a physically thicker high-k gate dielectrics and the subthreshold leakage current can be suppressed by using a three-dimensional (3D) structure, such as fin-FET [19] (see Fig.11(b)). Those measures would be introduced for low standby-power devices even earlier than the high performance logic units as soon as the reliability of the process and fabricated devices become established and the fabrication cost becomes reasonable. Suppression of the leakage current is also important for the pass transistors or transfer gates of low power memories and fin-FETs and high-k dielectrics will be introduced to those memories [75] not too far.

MANUFACTURING

Cost reduction has been the largest driving force for the downscaling. The costs per bit of a memory chip or the cost per operation of a microprocessor have been reduced tremendously with the transistor downsizing. Market is another important driving force. The demand for higher performance image processing capability for video game player, and the requirement of lower power consumption for mobile PCs also speed up the IC technology development. In this section, we shall focus on the manufacturing issues of the IC technology.

Manufacturing challenges for '45nm' and below

In the so called '45 nm commercial' technology node to be in volume production (VP) in 2007, we need a lithography capability of 65 nm half pitch (HP) for the metal lines of DRAM. However, the recent custom for many chip makers is to use one generation size smaller for the node name than that of the half pitch specified by ITRS. We will still use the conventional half pitch in order to avoid any confusion. In that HP65 nm technology node, 35 nm physical gate length under the electrode (Lg) will be used for high performance MPU. The logic devices are supposed to be realized by resist thinning or slimming method [27,105,106]. It may be confusing again that the value of Lg for high performance MPU and logic is sometimes larger than that specified in ITRS.

To pattern such fine structure may require new photolithography tools [19] (see Fig.15). Several years ago, it was suggested that we had to use F2 laser for the 'commercial 45 nm' node (HP65nm) and an EUV (Extremely Ultra Violet) unit, whose wavelength is a tenth of the light source of the wavelength that we are currently using, for the 'commercial 32 nm' node (HP45 nm, Lg25nm, VP2010). Recently it was found that immersion with 193 nm wave length ArF laser can also be used for 'commercial 45 nm' node (HP65nm) and even for the 'commercial 32 nm' (HP45 nm) node [107,108]. The optical lithography resolution can be greatly enhanced by replacing the air gap between the last projection lens and the wafer with a larger refractive index liquid [7] (see inset of Fig.16). It is a tremendous achievement that one third to quarter wavelength resolution could be achieved with the combination of the immersion with some enhancement techniques [7] (Fig.16). Although there are still so many issues to be solved for volume production, those are good news for the chip makers. For 'commercial 22 nm' node (HP32 nm), possibilities of using EUV and other innovative techniques as well as the resist materials for those new systems are under investigation.

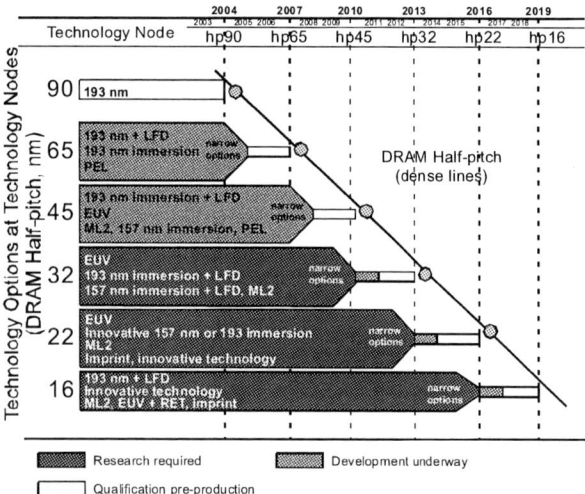

Fig.15 Potential lithography solutions for various technology nodes. With immersion technology, 193 nm wavelength ArF laser can still be used for HP45 nm node. Lithography with Extremely Ultra Violet (EUV) source may be needed for HP32 nm node and beyond [19].

Another big challenge is the limit of the gate oxide. In ultra small devices, it is crucial to suppress the short-channel effect or sub-threshold leakage current and to keep drive current high. To suppress the direct tunneling leakage current through the ultrathin gate insulator, high-k materials have to be introduced with the combination of metal gate electrodes. However, the interface between the gate dielectric and silicon substrate is the most sensitive and critical portion for MOS transistor. Several problems, such as thermal instability of the materials, poor interface properties with silicon, forming interface silicate layers, channel mobility degradation, high interface and oxide trap densities [88-91], turn out when replacing the silicon dioxide [23] or oxynitride [110-117] – which has been used for 40 years – with a high-k metal oxide. High-k will not be the real major gate dielectric materials at least until the '32 nm' node (HP 45nm, Lg=25 nm, VP2010), although some chip vendors may start to introduce for 65-45 nm node for low standby power application in a couple of years. However, in any case, the high-k gate must be used in some near technology nodes even for high-performance logic devices because the silicon dioxide or oxynitride is no longer able to provide sufficient large capacitance with the constraint of gate leakage current.

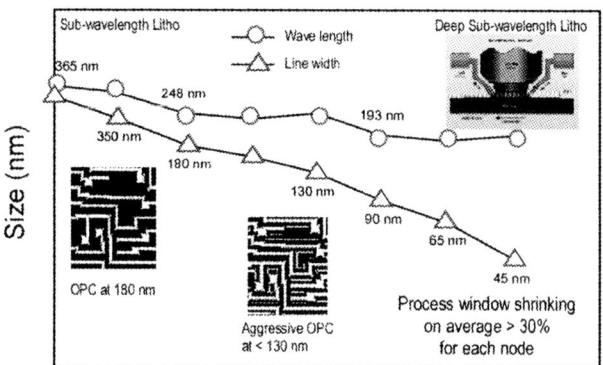

Fig.16 The sub-wavelength technology is able to make the photolithography with resolution better than quarter wavelength [7].

However, the resolution will be pushed to its limit in some future nodes. The processing window is shrinking on average of over 30% for each node since the 365 nm nodeIntroduction of lower-k materials for interlayers of interconnects to reduce the wiring capacitance also faces some severe technical difficulties. The fundamental issues are the fragility, chemical resistance and the pattenability of the low-k materials. Although there is a significant progress recently in achieving the effective-k value below 3 such as SiOC, polymer and porous films, the progress of low-k material application may lag behind the ITRS prediction. In addition to these issues, there are several new materials and processes, which may be used for the HP65 node and below. The increase of the contact resistance is governed by the source/drain junction extension regions and the parasitic resistance [118-121]. The source and drain contacts in the present MOS devices are normally made with the self-aligned silicide process. This process provides a good ohmic contact. Nickel silicide has advantages of less Si consumption, which is important for device applications on ultra-thin Si layers, and can be readily form a simple single-step annealing at 400 to 600oC[44,48,50-53]. Silicon consumption at the most conductive top layer (normally with the most heavily doped area) as well as the dopant segregation effect may still cause some fluctuation in the contact resistance of the ultra-small devices.

The sheet resistance and the abruptness of the source and drain extension can be improved with some new doping and annealing techniques such as low-energy plasma doping with flash or laser annealing. The recent ITRS roadmap for ULSI technology has predicted that the MOS transistor source/drain junction depths will be reduced down to some nm. The junction formation of today's IC fab is usually based on ion implantation technique. Although this process can provide precise control on the doping profile, the large penetration range prevents this technique for being used in the future technology nodes. Plasma doping (PD) has been demonstrated to make such ultra-shallow junction formation [34,35,42,43]. PD has several advantages such as high doping current, low implant energies, short process time, and low cost. In addition, the PD process is particularly suitable for ultra-shallow junction application.

Regarding the device structure, although SOI may be used for some specific applications, bulk wafer will keep its position as the major material at least down to the 'commercial 32 nm' (HP45 nm) nodes (see Fig.11(a)), where the gate length of MPU transistors is supposed to be 25 nm. Suppression of the sub-threshold leakage current would be the motivation for introducing SOI and double gate (DG) structures such as Fin- or Tri-gate FETs for the sub-20 nm gate length region [70-75]. DG-FET or Fin-FET is considered as the ultimately scalable MOS device structure. It minimizes the short-channel effects with 2D or 3D conformed electric field. It relieves the channel doping level for punch-through control. Because of the reduced channel doping the mobility in the DG-FET and Fin-FET can be theoretically enhanced thanks to the reduced coulomb scattering compared to the conventional device. However, there are some technical problems which degrade the mobility and conductance of the fin-FET, such as non-strait etching and surface roughness of the fin, and higher source/drain resistance because of the narrow fin itself.

Impact of 300 mm wafer and beyond on the economics

Since entering 'commercial 90 nm' node (HP130nm) era, there had been only several in building 300 mm wafer fab lines until a couple of years ago. By the end of 2005, however, there are so many thirty 300-mm fabs, which will start its operation. It is possible that the revenue of semiconductor companies might fall into another valley of the production cycle in 2006 because of the over-increased chip supply capacity. Nevertheless, the benefit of 300 mm production is huge. These mega fabs may product 10,000 to 30,000 wafers per month. Meanwhile, as the wafer area increases by 2.3 times when moving from the 200 mm to 300 mm line, while the production cost is estimated to increase by about 1.5 times only [8] (see Table 1). Significant increases in the costs are in the technology development, circuit design, mask and manufacturing sectors [9] (see Fig. 17).

Table 1. Significant merits are received from large scale production with larger wafer. Moving form 8" to 12", the operation cost increases about 30% while the wafer size or throughput increased by 2.25 times [8].

	12" Fab (per wafer)	8" Fab (per wafer)	12"/8" usage ratio	12"/8" wafer size ratio
Power (kW-hr)	1100	660	1.7	
				2.25
Water (m³)	6.1	4.7	1.3	
Waste Water (m³)	3.8	2.9	1.3	
Waste Gas (cm-hr)	20000	13000	1.5	

Low volume consumer chip production and/or shorter product lifetime may not be cost effective with these mega fabs. This becomes more significant when the chip size shrinks with scaling while the wafer size increases. Single wafer process and precise controlling of the process scheduling will rise the efficiency and decrease the wafer fabrication cycle time. The market share and the role of FPGAs (Field Programmable Gate Arrays) will increase more

significantly for smaller volume production in near future [122] (see Fig.18). Right now LSI chips with more than 5 million logic gates with 500 MHz operation can be still produced with FPGA. The performances of FPGA circuits are expected to improve by using smaller-sized devices and with improved structures.

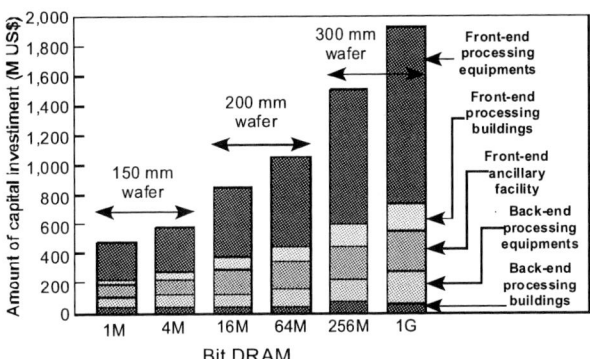

Fig.17 The capital investment for each DRAM generation is escalating. Significant investment was placed on the front-end processing equipment for DRAM [9].

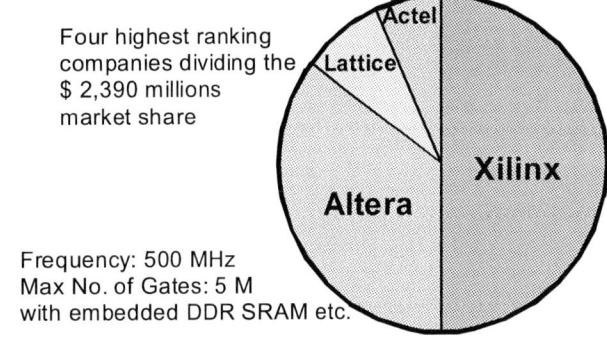

Fig.18 The market share of FPGA will be increased particularly for small volume production which cannot be accommodated with 300 mm mega fabs [122].

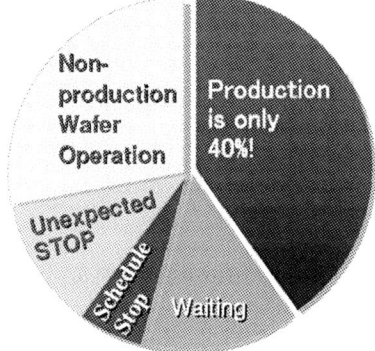

Fig.19 Low utilization of expensive equipment for wafer production is a big concern in modern semiconductor manufacturing. The equipment efficiency (production related operation) is only 40% [122]

Besides, manufacturing economics are still a major concern of 300 mm mega fabs. DRAM production with 300 mm fab, for example, the capital investment, particularly the investment in the front-end processing equipment is almost double the same cost for the 200 mm facility [9] (see Fig.17). In addition, the efficiency of such expensive equipment may not be very high. The equipment efficiency is only 40% [123] (see Fig.19). Cost and lifetime of mask is another important economical issue. In some cases for foundry, about 75% of mask sets were only used to produce less than 100 wafers [124] (see Fig.20). Better product development practices are expected for future designs development and product verification. Some computer tools such as TCAD, ECAD and CIM may help to solve some of the issues. Those tools can reduce the cost and the times of technology development and circuit design. CIM, tightly linked with TCAD and ECAD and providing real time intelligent feedback of the fabrication process, will significantly enhance the yield and reliability of production.

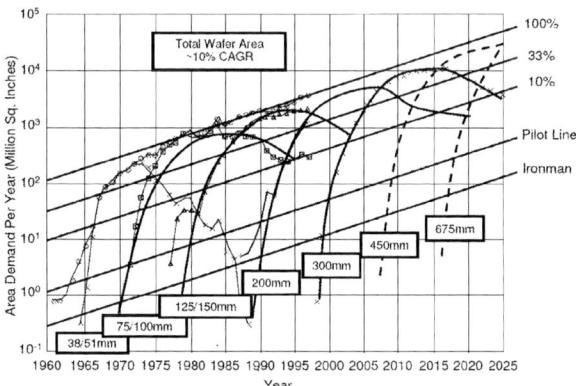

Fig.21 Wafer generation model. The recently introduced 300 mm fab will govern the IC productivity for at least 10 years from now [125].

What is the prospect of 450 mm wafer fab? According to the wafer generation model given in Fig. 21, the 450 mm wafer is expected to be introduced for the 'commercial 22 nm' (HP32 nm Lg15nm, VP2013) or 'commercial 15 nm' (HP22nm, Lg9nm, VP2016) node. Thus, it is almost right time to start the development. 450-nm fab will help the memory and MPU vendors, and probably some large Si foundries to save cost of in volume production. However, tremendous development cost is not only for the wafer production, but also for the equipment, factory and wafer processing. Many technical problems such as defects, thickness non-uniformity, wafer distortion and slip generation during the thermal process have to be solved. Further details as well as some other issues are listed in Table 2 [8]. The key point is who are willing to pay the development cost and take the risk for 10 years later. It is noted that some new 200 mm fabs are still under construction or planned even with available of 300 nm-fab lines. Based on the same philosophy, it is reasonable to anticipate that a significant amount of IC fabs will remain at 300 mm in the 450 mm era [125] (see Fig.21). Strong leadership and capability, for fund congregating and large-scale collaboration, are indispensable for initiating the 450 mm program

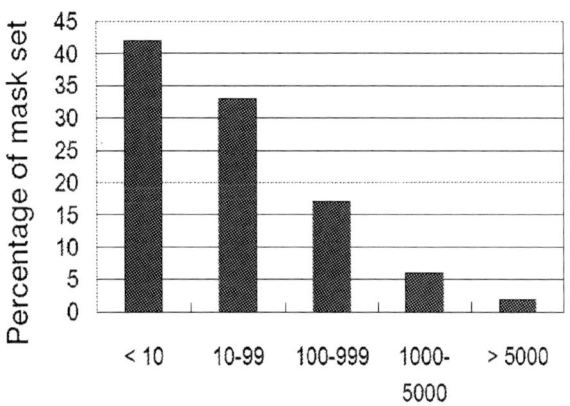

Fig.20 Cost and lifetime of mask is another big issue. According to a figure from TSMC, about 75% of masks can only produce 100 or less of wafers [124]

Table 2. Some technical issues of introducing 450 mm wafer [8].

450mm Attribute	Key Technology Decisions
Wafer	Material, Size, Thickness, ID, Registration, Edge Exclusion
Wafer Carrier	Number of Wafers, Size, Door Type, AMHS Strategy, ID
Production Equipment	Single wafer vs. mini-batch, cleanliness, interface standards, Productivity, targets relative to 300mm (NPW usage, etc.)
Factory	Factory Size, Egress, Cleanliness, Sub-Fab attributes, Clean-room height
Automated Material Handling Systems	Direct transport concepts, carrier/wafer delivery time, overall throughput, efficient storage concepts
Manufacturing Systems	Process Control & Yield Data Standards, Carrier delivery time, Decision Making Time, Data Flow

Role of East and Southeast Asia for manufacturing

Most of the new fabs being planned or under construction are in the East and Southeast Asia. In 10-year's time, the distribution of semiconductor manufacturing sites in Asia (including Japan) will be quite substantial, though a certain amount of the IC fabs will remain in North America and Europe [19] (see Fig.22). Mainland China, Taiwan, Korea and India have been the major sources of semiconductor technology researchers and engineers for US. Some of them started to return to their home countries with the increased job opportunities for semiconductor manufacturing in their countries.

With the global-wise population aging trends in next few decades, countries [126] (see Fig.23) with large population do not only have advantages in human resources and capitals for technology development, but also provide huge markets for industrial products. Market is not only a driving force, but also a necessary condition for sustaining the operation of mega fabs from the IC manufacturing point of view. Because of huge population for relatively cheap and high-quality labor, large markets, relatively stable political and economical situations, the Far East and some of Southeast Asia are the most suitable regions for semiconductor manufacturing.

Currently, Korea and Taiwan are in the first place for semiconductor memory manufacturing and semiconductor foundry, respectively. They also lead the technology development in Asia region. Singapore is now catching up with them. Mainland China will be another super power for semiconductor manufacturing. The share of China semiconductor manufacturing will keep fast growing with the support of booming IC design houses [127] (Fig.24), constructing new fabs with remarkable increase in industrial investment [128] (Fig.25), and will be the most important huge and rapidly expending market [129] (see Fig.26). As many other industries and other sectors of electronic products [130] (see Fig.27), Mainland China will eventually become "the factory of the world" in semiconductor manufacturing in long term. It would take at least several to ten of years when it may catch up with the leading-edge technology of Taiwan and Korea. Development of leading-edge semiconductor technology does not only require huge resource but also accumulation of lots of experienced engineers in various sectors.

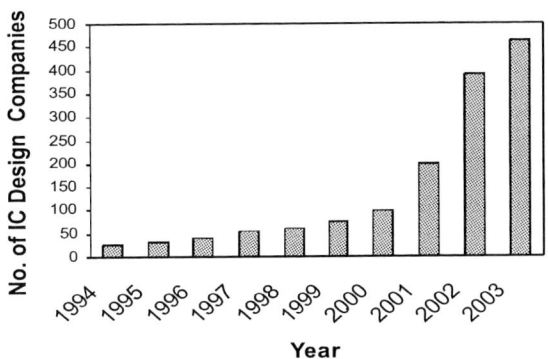

Fig.24 Rapid growth of IC design companies was registered in mainland China in recent years [127].

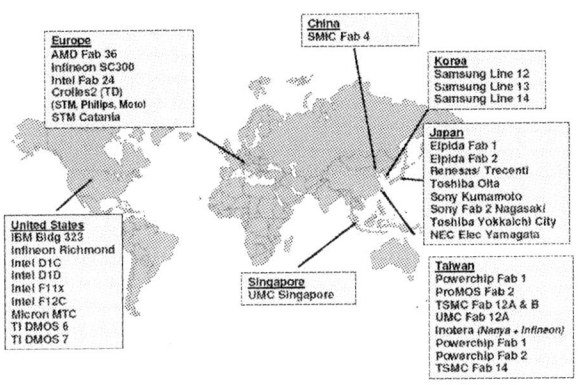

Fig.22 The global distribution of 300 mm Fabs. Number of Fabs is expected to increase in East and Southeast Asia and eventually dominates the world semiconductor manufacturing [19].

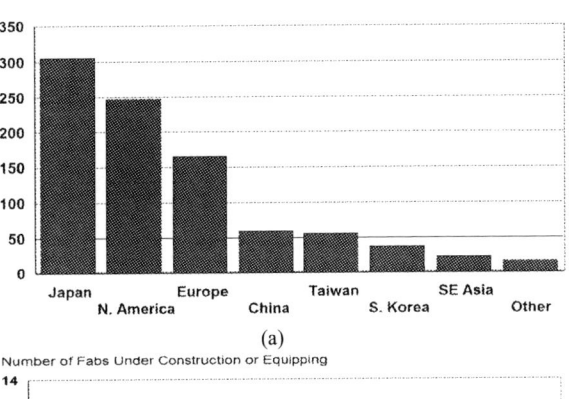

(a)

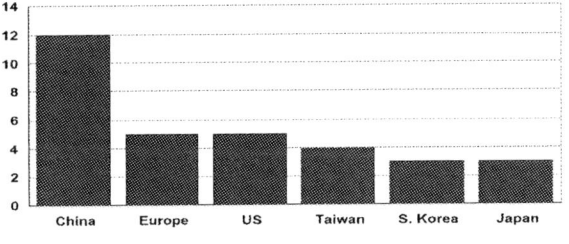

(b)

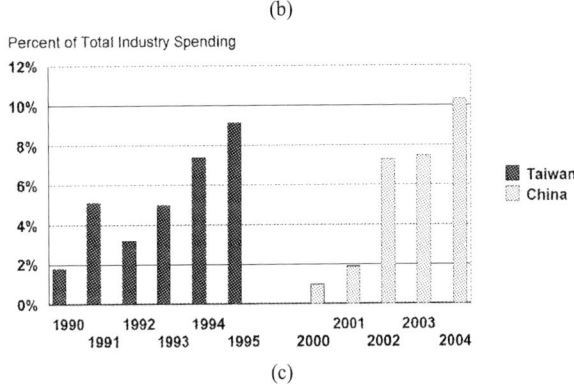

(c)

Fig.25 China will become the world factory for IC manufacturing. The recent trend in rapid growth of industrial investment and the new fabrication lines are strong rationales for supporting this prediction [128]

Population in million

Fig.23. Population is market and opportunity. China and India, being the largest and the second largest population countries, have greater opportunity for market growth and manufacturing [126].

9

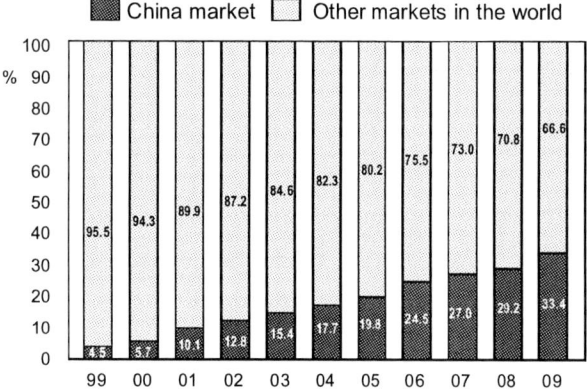

Fig.26 Rapid increase in the market share for IC demand in China and is expected to maintain a fast growth rate for several years. The huge market will be a strong ground for expending Chinese IC manufacturing [130].

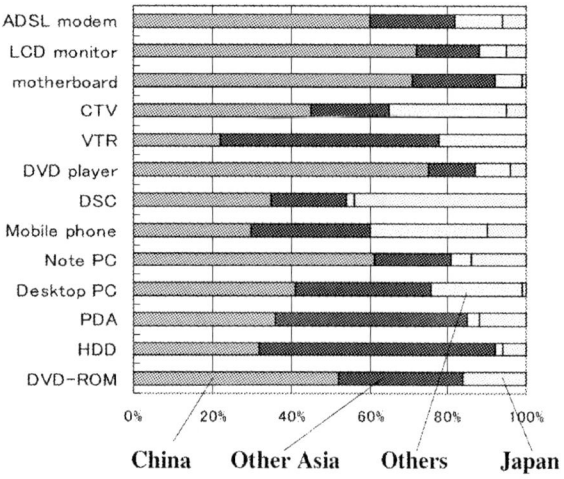

Fig.27 China is the world factory of many electronic products. Semiconductor manufacturing will be the next, may be in ten years' time [130].

What will be the roles of North America, Europe and Japan in future? We expect that those countries will move to some emerging applications and related technologies as the past. 'Emerging' application does not mean 'exotic' one but means a kind of applications more popular and related to our desire or dream for our near future society and life. Certainly North America, Europe and Japan will still keep leading semiconductor fabs in their countries as a strategic for future high technology development. It is noted that the major supplies for equipment, wafers and other materials for semiconductor manufacturing still remain in those countries [131,132] (see Fig. 28), even though the front-end manufacturing has moved to non-Japan Asia. In addition, Japan will serve a unique and different role. Japan has been extremely good at creating, nourishing and/or refining new applications for both consumers and manufacturing. For example, Japan has been the major force for developing pocket calculators, digital watches, digital still and movie cameras, video recorders, video games, various kinds of new functions of cellular phones and intelligent automobiles, various kinds of new electronic displays, and advanced high-intelligent robots for industrial and home applications. We expect that Japan will keep on contributing for new products and application ideas for

consumer, medical, automobile and some unknown future applications. It will help in creating new markets also. North America and Europe will serve the similar role. In addition, their strengths in the basic research will maintain for at least for some decades.

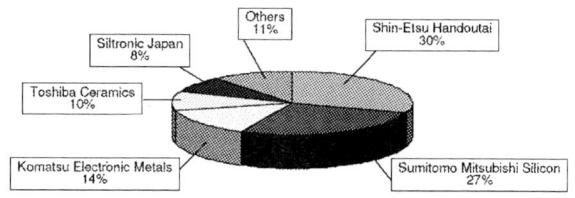

Total: 757,000 million dollars/149,700,000 sheets/5147 million square inches

(a)

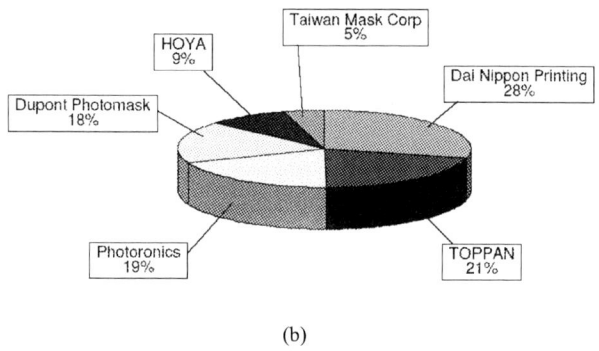

(b)

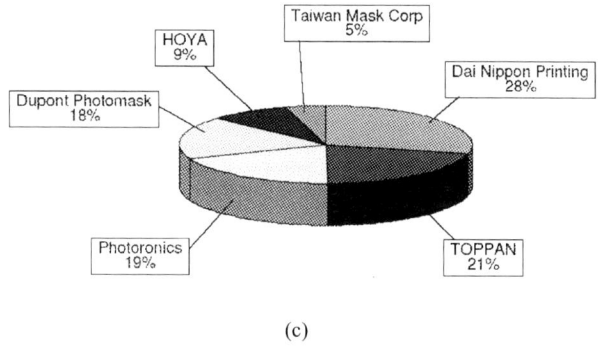

(c)

Fig.28 North America, Europe, and Japan are the major suppliers for equipment, silicon wafers and other materials for semiconductor manufacturing.. It is expected that they will still keep this advantages for decades [131, 132].

As shown in Fig.29 [19], 'Electronic End Equipment' is the base of all human political and economical activities which may be quantified by the world total GDP of about 30,000 billion US dollars. 'Semiconductor' is the key component of the 'Electronic End

10

Equipment.' Thus, 'Semiconductor' is a strategically important technology for all countries although its 300 billion US dollars revenue is only about one percent of the global GDP.

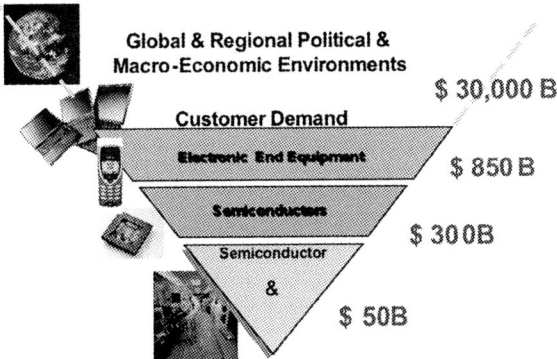

Fig.29 'Electronic End Equipment' is the base of all the human activities in the 'Global & Regional Political & Macro-Economic Environments' and 'Semiconductors' are strategically important for every country as the key component.[1.19]

FUTURE PARADIGM CHANGE FOR POST-DOWNSIZINGTRENDS OF DOWNSIZING AND ITS LIMITS

Silicon MOSFETs have been the smallest electronic device for several decades. Thirty five years ago, the gate oxide thickness was already in the nanoscale (120 nm) for commercial products. The gate oxide thickness is now 1.2 nm in production and 0.8 nm in research. The gate length used for high performance MPU and logic unit is 50 nm in production and 5 nm in research. Note that the 5-nm gate length is the distance of 18 atoms and 0.8-nm oxide thickness is two atomic layers only [24] (see Fig.30). Si technology is no doubt the most successful nano-devices. We do not see that there is any realistic replacement for silicon devices. Even the Si devices reach the downsizing limit no matter 10 nm, 5 nm, or 1 nm, other emerging devices such as molecular transistors will also reach their limit of downsizing in similar dimensions.

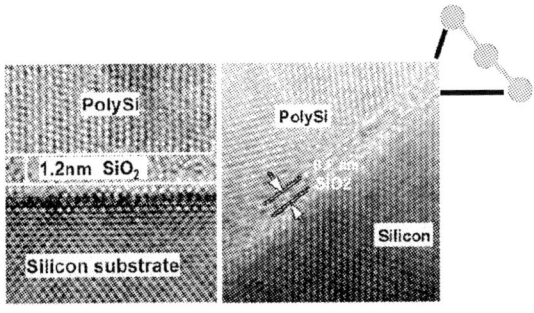

Fig.30 Transmission electron microscope pictures of a 1.2 nm thick and 0.8 nm thick gate oxide films. The 0.8 nm thick gate oxide is about two atomic layers thick

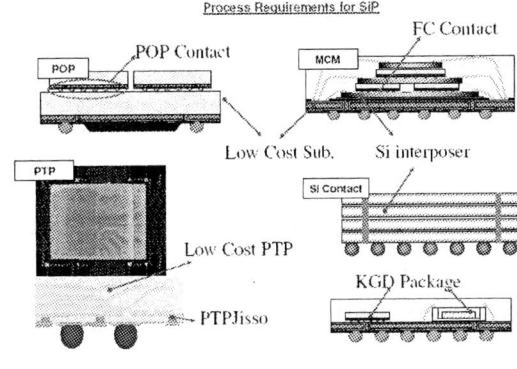

(a)

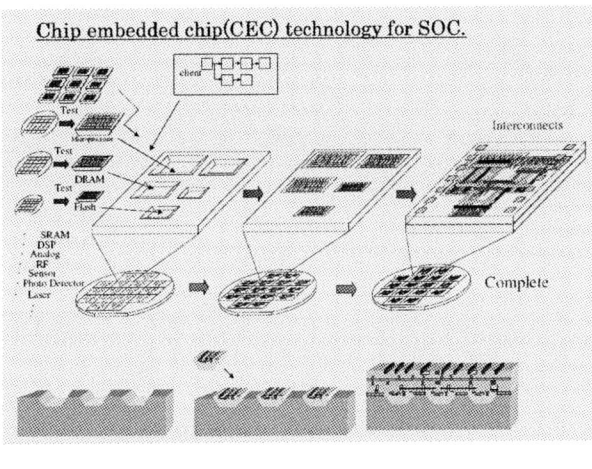

(b)

Fig.31 New packaging technologies such as system-in-package (a), chip-embedded-chip (b) will be the major driving force for performance booster for future large scale systems [1, 133].

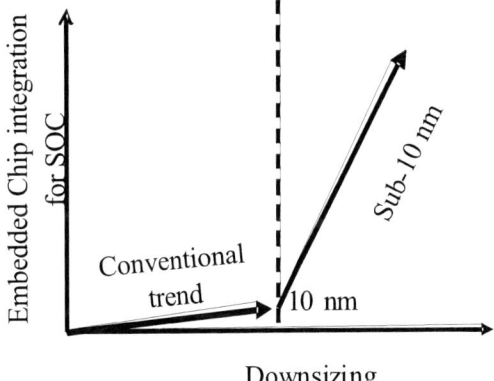

Fig.32 The introduction of system-on-chip is expected to have a great improvement in system performance for sub-10 nm technology [1].

There are still plenty of rooms for up scaling in the chip density. Interconnect complexity for billions of transistors will be one of the major constraints. This constraint can be overcome with new packaging methods such as SOC (System-on-Chip) and/or SIP (System-in-Package) [133] (see Fig.31(a)). These techniques will be introduced after or even before the ultimate scaling limit because of better yield control and more cost effective. For example, embedding functional sub-chips on every chip on a silicon wafer is a useful technique to reduce the cost and power, while keeping high performance [6]. The potential performance improvement with the introduction of this technology is illustrated in Fig.31 (b) [1-3]. The sub-chips such as MPU, DRAM, SRAM, RF-devices, lasers, sensors, biological chips, can be pre-built silicon and non-silicon. They are fabricated separately from the mother chips and then are fitted in the grooves made on each mother chip on a wafer. After the fitting of sub-chips, usual LSI multi-level interconnects process follows. This approach will significantly reduce the signal delay between the chips because it can greatly reduce the large parasitic capacitances from bonding pads, bonding wires, and package pins. It saves power, reduces the wiring cost and increases the system performance. It is noted that the advantages of scaling, such as improving performance, cutting power and cost, starts to decrease because of the aforementioned constraints. Other techniques, such as 'chip embedded chip' or hybrid integration of different functional sub-chips on silicon is shown in Fig.32 [1-3]. Combination of SOC and SIP would also improve the system performance and reduce the manufacturing cost for ultra large-scale systems.

Other ways to further improve the system performance after the downsizing limit is the system architecture optimization and algorithm evolution. Compared with biological systems, such as human brain or even as tiny as a mosquito, the performance per unit volume or per unit power is much lower than the modern electronic computer (see Fig.33) [1-4]. However, no one knows how to realize such efficient systems today. It would take at least another 50 years when such an efficient biological system may be realized in the electronic way.

New materials and technologies are required for further downscaling the device to the limit. Immersion lithography for ultra fine patterning, strained channels, nickel salicide, high-k gate dielectric, low-k interlayer for interconnect, plasma doping, flash and laser annealing for source and drain doping, elevated source and drain and three-dimensional MOSFETs for controlling short-channel effects, would help to overcome the materials and technological constraints and improve the device performance in the ultra-small scale.

On the manufacturing side, the choice of 300 mm wafers should an economical solution for mega-fabs though a substantial increase in running cost and some kinds of inefficiency were reported. Computer-aided tools such as TCAD, ECAD and CIM will help to save the cost and the cycle time of fabrication. For smaller volume production, the market share of FPGA type chips will increase significantly in near future. It is still questionable if the wafer size will move from 300 to 450 mm in coming decade but 300-mm fabs will definitely remain as important fabs even in the 450 mm era. Launching a 450-mm plant does not only require a huge resource in money wise, human resource, tremendous of intellectual properties, good team work, it also requires a huge product development teams and a huge market to sustain the non-stopped operation.

East Asia including some of the Southeast Asia regions is the most suitable place for semiconductor manufacturing because of relatively cheaper and higher quality human resources as well as high political and economical stability. East Asia will become the "world factory" for semiconductor manufacturing in a decade. North America, Europe, and Japan will move to some emerging technologies but may still keep some leading semiconductor fabs as a strategic for higher technology development. Those countries will still be the major suppliers for equipment, silicon wafers and other materials for semiconductor manufacturing. In addition, those countries will contribute for creating new applications and markets for new consumer products. When scaling towards 10 nm, the benefits in terms of performance, power consumption and cost, will fall. The system on chip (SOC) approach such as hybrid integration of multi-functional sub-chips on silicon as well as system in package (SIP) will be the important measures for system integration and performance improvement in that era.

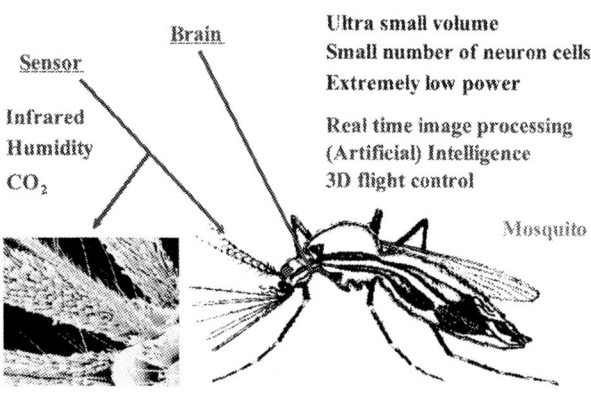

Fig.33 In terms of individual element response time, the MOS device is several orders of magnitude faster than a biological system. The synapse interaction is normally in the range of some millisecond. In terms of scale, the ULSI circuits will soon approach the scale of human brain (10^{10} neural cells and 10^{14} synapses). However, in term of intelligence and power consumption, the performances of the modern computer are even far below an inset such as mosquito. A revolution in system architecture and computing method may happen in future [1].

Past innovations for semiconductor technologies I

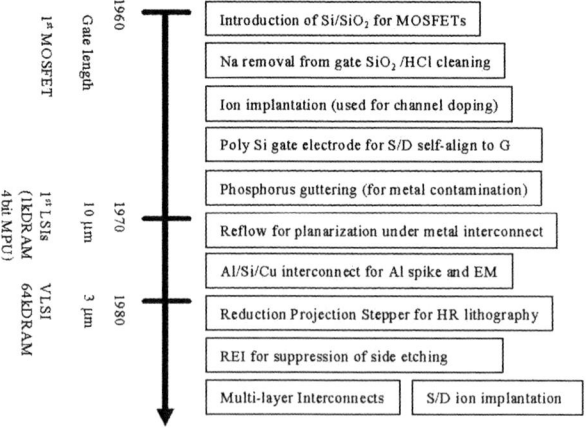

(a)

Past innovations for semiconductor technologies II

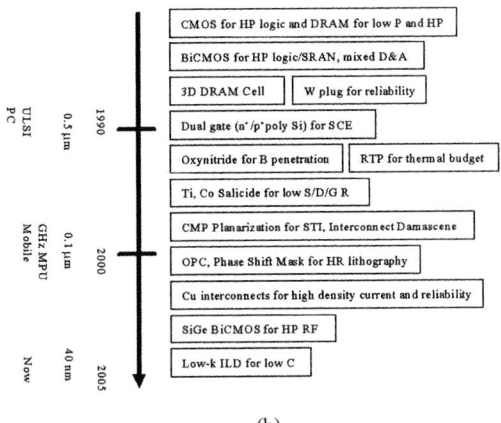

(b)

Future innovations for semiconductor technologies I

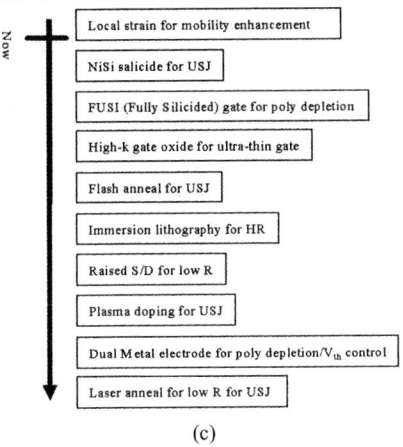

(c)

Future innovations for semiconductor technologies II

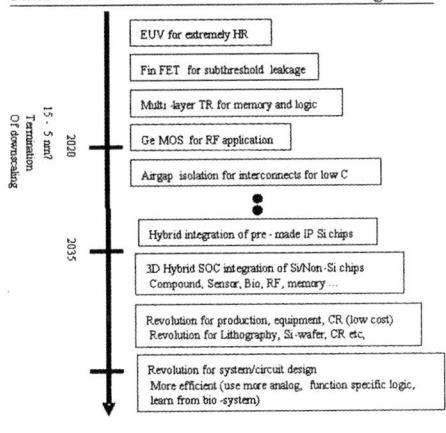

(d)

Fig.34 Past and future integrated circuits technology evolutions and revolutions

Finally, past and future technology evolutions and revolutions expected are shown in Fig.34. In the post scaling era when downsizing reaches its limits, the technology developments will concentrate for the manufacturing items, regarding how to make the chip chaper. Thus, revolution in the manufacturing will occur and big change from the current style clean room, wafers, and equipments will proceed.

ACKNOWLEDGEMENT

The authors would like to their express deep gratitude to so many people for their useful materials and information and constructive discussions. Particular appreciations are Hei Wong City University of Hong Kong, T. Masuhara, M. Ishino and S. Okazaki of ASET, S. Komatsu, H. Ishiuchi, K. Ishimaru, M. Ihara, S. Inada, R. Kuwae, and Y. Katsumata of Toshiba, M. Honma of NEC, T. Osada of Fujitsu, T. Hokari of JEITA, D. Uchida, Y. Ando, D. Tracy of Semi, F.C. Tseng, J. Sun, C.H. Diaz, C. Wang, and J.J. Lin of TSMC, C.G. Hwang, and K. Kim of Samsung, R. Chau of Intel, Qi Xiang of AMD, M. Kameyama of Nikon, A, Koike, S. Ikeda, and K. Uchino of Trecenti, T. Shigematsu, M. Nakamae and S. Sumita of SUMCO, Y. Ishikawa, M. Yamaguchi, and Y. Tada of TEL, N. Hayashi, S. Noguchi, and Y. Ohta of DHP, F. Sato and A. Izumi of Deuche Bank, Y. Mochizuki of Nikkei BP, Y. Nishi of Stanford, C. Y. Chang, S. Chung, H.C. Lin, and W. Hwang of NCTU, S.W. Sun and M. Ma of UMC, G. Chen of TSIA, B.T. Dai of NDL, M.J. Tsai and P.P. Tsai of ITRI, M. Yang of SIC R&D Center, K. Monahan of KLA Tencor, K. Inoue of ERI-JSPMI, and J. Candelaria of Freescale, K. Kakusima of TIT.

REFERENCES

1. H. Iwai, Future semiconductor manufacturing-challenges and opportunities, *IEDM Tech. Dig.,* 11-16 (2004)
2. H. Iwai, CMOS downsizing toward sub-100 nm, *Solid-State Electron.,* **48**, 497-503 (2003).
3. H. Iwai and S. Ohmi, Silicon integrated circuit technology from past to future, *Microelectron. Reliab.,* **42**, 465-491 (2002).
4. H. Iwai, H. S. Momose, M. Saito, M. Ono, and Y. Katsumata, The future of ultra-small-geometry MOSFETs beyond 0.1 micron, *Microelectronic Engineering,* **28**, 147-154 (1995)
5. K. Kim, Memory technology in nano-era, *2005 IEEE EDS Colloquium (WIMNACT-7),* 231-261 (2005)
6. M. Bohr, Intel's 90 nm Technology: Moore's Law and More, *Intel Developer Forum, Fall 2002,* San Jose, September (2002)
7. K. Monahan, Yield Challenges at the 90nm Technology Node and Beyond, *The Electrochemical Society International Semiconductor Technology Conference,* Shanghai, Key note session (2004)
8. J. J. Lin, 300 mm Manufacturing, *IEDM Short Course 'The Future of Semiconductor Manufacturing'* (2002)
9. JEITA IC Guide Book Edit Committee, *IC Guide Book (in Japanese)* (2003)
10. H. Iwai, CMOS Device Architecture and Technology for the 0.25 Micron to 0.025 Micron Generation, *ESSDERC 93,* Grenoble, France, 513-520 (1993)
11. C. Fiegna, H. Iwai, T. Wada, M. Saito, E. Sangiorgi, and B. Ricco, Scaling the MOS transistor below 0.1 μm: methodology, device structures and technology requirements, *IEEE Trans. Electron Devices,* **41**, 941-951 (1994)
12. H. Iwai, CMOS Technology – Year 2010 and beyond, *IEEE Journal of Solid-State Circuits,* **34**, 357-366 (1999)

13. D. Frank, R. Dennard, E. Nowak, P. Solomon, Y. Taur, and H.-S. Wong, Device scaling limits of Si MOSFETs and their application dependencies, *Proc. IEEE,* **89**, 259-288 (2001)

14. H. Wong, On the scaling issues of ultrathin MOS gate dielectrics, *Proc. 2004 Int'l Conf. on Communications, Devices, and Intelligent Systems*, 393-396, Kolkata (2004)

15. J. Plummer and P. Griffin, Material and process limits in silicon VLSI technology, *Proc. IEEE* **89**, 240-258 (2001).

16. S. Asai and Y. Wada, Technology challenges for integration near and below 0.1 μm, *Proc. IEEE* **85**, 505, (1997)

17. M. Norishima, H. Iwai, Y. Niitsu, and K. Maeguchi, Impurity diffusion behaviors of bipolar transistor under low-temperature furnace annealing and high-temperature RTA and its optimization for 0.5 μm Bi-CMOS process, *IEEE Trans. Electron Devices*, **39**, 33-40 (1992)

18. H. Iwai and H. S. Momose, Technology toward low power/low voltage and scaling of MOSFETs, *Microelectronic Engineering*, **39**, 7-30 (1997)

19. *International Technology Roadmap for Semiconductors*, 2003 Edition, Semiconductor Industry Association (SIA), Austin, Texas: SEMATECH, USA, 2706 Montopolis Drive, Austin, Texas 78741;

20. H. S. Momose, M. Ono, T. Yoshitomi, T. Ohguro, S. Nakamura, M. Saito, and H. Iwai, Tunneling gate oxide approach to ultra-high current drive in small-geometry MOSFETs, *IEDM Tech. Dig.*, 593-596 (1994)

21. H. S. Momose, M. Ono, T. Yoshitomi, T. Ohguro, S. Nakamura, M. Saito, and H. Iwai, 1.5 nm Direct-Tunneling Gate Oxide Si MOSFET's, *IEEE Trans. Electron Devices*, **43**, 1233-1242, (1996)

22. H. S. Momose, S. Nakamura, T. Ohguro, T. Yoshitomi, E. Morifuji, T. Morimoto and H. Iwai, Study of the manufacturing feasibility of 1.5-nm direct-tunneling gate oxide MOSFET's: Uniformity, reliability, and dopant penetration of the gate oxide, *IEEE Trans. Electron Devices*, **45**, 691-700 (1998)

23. H. Iwai and H. S. Momose, Ultra-thin gate oxide – performance and reliability, *IEDM Tech. Dig.,* 163-166 (1998)

24. R. Chau, J. Kavalieros, B. Roberds, B. Schenker, D. Lionberger, D. Barlage, B. Doyle, R. Arghavani, A. Murthy, G. Dewey, 30 nm physical gate length CMOS transistors with 1.0 ps n-MOS and 1.7 ps p-MOS gate delays, *IEDM Tech. Dig.*, 45-48 (2000)

25. H. S. Momose, M. Ono, T. Yoshitomi, T. Ohguro, S. Nakamura, M. Saito, and H. Iwai, Prospects for Low-power, High-speed MPUs Using 1.5 nm Direct-tunneling Gate Oxide MOSFETs, *Solid-State Electron,* **41**, 707-714 (1997)

26. C. Mead and L. Conway, 'Introduction to VLSI systems,' Addison Wesley, 37 (1979)

27. M. Ono, M.Saito, T. Yoshitomi, C. Fiegna, T. Ohguro, and H. Iwai, Sub-50 nm Gate Length N-MOSFETs with 10nm phosphorus Source and Drain Junction, *IEDM Tech. Dig.*, 119-122, (1993)

28. H. Wakabayashi, S. Yamagami, N. Ikezawa, A. Ogura, M. Narihiro, K. Arai, Y. Ochiai, K. Takeuchi, T. Yamamoto, and T. Mogami, Sub-10-nm planar-bulk-CMOS devices using lateral junction control, *IEDM Tech. Dig.*, 989-991 (2003)

29. T. Yoshitomi, M. Saito, T. Ohguro, M. Ono, H. S. Momose, and H. Iwai, Silicided Silicon-Sidewall Source and Drain (S^{4}D) structure for high-performance 75-nm gate length pMOSFETs, *Symp. on VLSI Technology*, 11-12 (1995)

30. T. Yoshitomi, M. Saito, T. Ohguro, M. Ono, H. S. Momose, and H. Iwai, Hot-Carrier Reliability of S^{4}D n-MOSFETs, *ESSDERC '96*, 65-68 (1996)

31. R. Chau, J. Kavalieros, B. Doyle, A. Murthy, N. Paulsen, D. Lionberger, D. Barlage, R. Arghavani, B. Roberds, and M. Doczy, A 50nm depleted-substrate CMOS transistor, *IEDM Tech. Dig.*, 621-624 (2001)

32. C. B. Oh, M. H. Oh, H. S. Kang, C. H. Park, B. J. Oh, Y. H. Kim, H. S. Rhee, Y. W. Kim, and K. P. Suh, Double raised source/drain transistor with 50 nm gate length on 17 nm UTF-SOI for 1.1 μm2 embedded SRAM technology, *IEDM Tech. Dig.*, 31-34 (2003)

33. H. Park, W. Rausch, H. Utomo, K. Matsumoto, H. Nii, S. Kawanaka, P. Fisher, S. Oh, J. Snare, W. Clark, A. C. Mocuta, J. Holt, R. Mo, T. Sato, D. Mocuta, B. H. Lee, O. Documaci, P. O'Neil, D. Brown, J. Suenaga, Y. Li, L. Brown, J. Nakos, K. Hathorn, P. Ronsheim, H. Kimura, B. Doris, G. Sudo, K. Scheer, S. Mittl, T. Wagner, T. Umebayashi, M. Tsukamoto, Y. Kohyama, J. Cheek, I. Yang, H. Kuroda, Y. Toyoshima, J. Pellerin, D. Schepis, Y. Li, P. Agnello, and J. Welser, High performance CMOS devices on SOI for 90nm technology enhanced by RSD (raised source/drain) and thermal cycle/Spacer engineering, *IEDM Tech. Dig.*, 635-638 (2003)

34. Y. Sasaki, C. G. Jin, H. Tamura, B. Mizuno, R. Higaki, T. Satoh, K. Majima, H. Sauddin, K. Takagi, S. Ohmi, K. Tsutsui, H. Iwai, B$_2$H$_6$ Plasma Dopong with In-situ He Pre-amorphization *Symp. on VLSI Technology*, 180-181 (2004)

35. B. Mizuno, Y. Sasaki, C. Jin, H. Tamura, K. Okashita, H. Ito, K.Tsutsui, H. Iwai, Plasama Doping, *7th International Conference on Solid-State and Integrated Circuits Technology, Proceedings (ICSICT)*, 423-427 (2004)

36. K. Tsutsui, R. Higaki, T. Sato, Y. Sasaki, H. Tamura, B. Mizuno, H. Iwai, Effects of Surface Conditions on Dose Controllability of Plasma Doping Process, *7th International Conference on Solid-State and Integrated Circuits Technology (ICSICT)*, 439-444 (2004)

37. I. Aiba, Y. Sasaki, K. Okashita, H. Tamura, Y. Fukagawa, K. Tsutsui, H. Ito, K. Kakushima, B. Mizuno, H. Iwai, Feasibility Study of Plasma Doping on Si Substrates with Photo-Resist Patterns *International Workshop on Junction Technology(IWJT)*, 71-72 (2005)

38. K. Tsutsui, K. Majima, Y. Fukagawa, Y. Sasaki, K. Okashita, H. Tamura, K. Kakushima, H. Ito, B. Mizuno, H. Iwai, Analysis of Conductivity in Ultra-shallow p$^+$ Layers Formed by Plasma Doping, *International Workshop on Junction Technology(IWJT)*, 73-74 (2005)

39. H. Sauddin, H. Tamura, K. Okashita, Y. Sasaki, H. Ito, B. Mizuno, K. Kakushima, K. Tsutsui, H. Iwai, Reverse Current of Plasma Doped p$^+$/n Ultra-Shallow Junction, *International Workshop on Junction Technology(IWJT)*, 75-76 (2005)

40. N. W. Cheung, Plasma immersion ion implantation for semiconductor processing, *Mater. Chem. Phys.,* **46**, 132-139 (1996).

41. S. Qin, N.E. McGruer, C. Chan and K. Warner, Plasma immersion ion implantation doping using a microwave multipolar buceket plasma, *IEEE Trans. Electron Devices*, **39**, 2354-2358 (1992).

42. B. L. Yang, E. C. Jones, N. W. Cheung, J. Shao, H. Wong and Y. C. Cheng, N$^+$P ultra-shallow junction on silicon by immersion ion implantation, *Microelectron. Reliab.*, **38**, 1489-1494 (1998).

43. B. L. Yang and N. W. Cheung, S. Denholm and J. Shao, H, Wong, P. T. Lai and Y. C. Cheng, Ultra-shallow n$^+$p junction formed by PH$_3$ and AsH$_3$ plasma immersion ion implantation, *Microelectron. Reliab.*, **42**, 1985-1989 (2002).

44. T. Morimoto, H. S. Momose, T. Iinuma, I. Kunishima, K. Suguro, H. Okano, I. Katakabe, H. Nakajima, M. Tsuchiaki, M. Ono, Y. Katsumata, and H. Iwai, A NiSi salicide technology for advanced logic devices, *IEDM Tech. Dig.*, 653-656 (1991)

45. T. Iizima, A. Nishiyama, Y. Ushiku, T. Ohguro, I. Kunishima, K. Suguro, and H. Iwai, A novel selective Ni₃Si contact plug technique for deep-submicron ULSIs, *Symp. on VLSI Technology*, 70-71 (1992)

46. T. Iinuma, K. Inou, H. Nakajima, S. Matsuda, I. Kunishima, K. Suguro, Y. Katsumata, and H. Iwai, A self-aligned emitter base NiSi electrode technology for advanced high speed bipolar LSIs, *IEEE Bipolar/BiCMOS Circuits and Technology Meeting (BCTM)*, 92-95 (1992)

47. Q. Wang, C.M. Osburn, and C.A. Canovai, Ultra-shallow junction formation using silicide as diffusion source and low thermal budget, *IEEE Trans. Electron Devices*; **39**, 2486-2496 (1992)

48. T. Ohguro, S. Nakamura, M. Koike, T. Morimoto, A. Nishiyama, Y. Ushiku, T. Yoshitomi, M. Ono, M. Saito, and H. Iwai, Analysis of Resistance Behavior in Ti- and Ni- Salicided Polysilicon Films, *IEEE Trans. Electron Devices*, **41**, 2305-2317 (1994)

49. T. Ohguro, S. Nakamura, E. Morifuji, M. Ono, T. Yoshitomi, M. Saito, H. S. Momose, and H. Iwai, Nitrogen-doped nickel monosilicide technique for deep submicron CMOS salicide, *IEDM Tech. Dig.*, 453-546 (1995)

50. T. Morimoto, T. Ohguro, H. S. Momose, T. Iinuma, I. Kunishima, K. Suguro, I. Katakabe, H. Nakajima, M. Tsuchiaki, M. Ono, Y. Katsumata, and H. Iwai, Self-Aligned Nickel-Mono-Silicide Technology for High-Speed Deep Submicrometer Logic CMOS ULSI, *IEEE Trans. Electron Devices*, **42**, 915-922 (1995)

51. H. Iwai, T. Ohguro, and H. Ohmi, NiSi salicide technology for scaled CMOS, *Microelectronic Engineering*, **60**, 157-169 (2002)

52. C. M. Osburn, J. Y. Tsai, and J. Sun, Metal silicides: Active elements of ULSI contacts, *J. Electron Mater.*, **25**, 1725 (1996)

53. M. C. Poon, M. Wong, F. Deng, S. S. Lau, H. Wong, Thermal stability of cobalt and nickel silicides, *Microelectron. Reliab.*, **38**, 1495-1498 (1998)

54. Q. Xiang, C. Woo, E. Paton, J. Foster, B. Yu and M.-R. Lin, Deep Sub-100nm CMOS with Ultra Low Gate Sheet Resistance by NiSi, *Symp. on VLSI Tech.*, 76-77 (2000)

55. M. Sun, M. Kim, J.-H. Ku, K. Roh, C. Kim, S. Youn, S.-W. Jung, S. Choi, N. Lee, H.-K. Kang, and K. Suh, Thermally Robust Ta-Doped Ni SALICIDE Process Promising for Sub-50nm CMOSFETs, *Symp. VLSI Technology*, 81-82 (2003)

56. K. Rim, J. Chu, , H. Chen, K.A. Jenkins, T. Kanarsky, K. Lee, A. Mocuta, H. Zhu, R. Roy, J. Newbury, , J. Ott, , K.Petarca, P. Mooney, D. Lacey, S. Koester, K. Chan, D. Boyd, M. Ieong, and H.-S.Wong, Characteristics and device design of sub-100nm strained Si n-and pMOSFETs', *Symp. VLSI Technology*, 98–99 (2002)

57. K. Ismail, S. F. Nelson, J. O. Chu, and B. S. Meyerson, Electoron transport properties of Si/SiGe heterostructures: Measurements and device implications, *Appl. Phys. Lett.*, **63**, 660-662 (1993)

58. J. Welser, J. L. Hoyt, and J. F. Gibbons, NMOS and PMOS transistors fabricated in strained silicon/relaxed silicon-germanium structures, *IEDM Tech. Dig.*, 1000-1002 (1992)

59. H. Iwai, H. S. Momose, S. Takagi, T. Morimoto, S. Kitagawa, S. Kambayashi, K. Yamabe, and S. Onga, Analysis of an ONO gate film effect on n- and p-MOSFET motilities, *Symp. on VLSI Technology*, 131-132 (1990)

60. H. S. Momose, T. Morimoto, K. Yamabe, and H. Iwai, Relationship between mobility and residual-mechanical-stress as measured by Raman spectroscopy for nitrided-oxide-gate MOSFETs, *IEDM Tech. Dig.*, 65-68 (1990)

61. T. Ghani, M. Armstrong, C. Auth, M. Bost, P. Charvat, G. Glass, T. Hoffmann, K. Johnson, C. Kenyon, J. Klaus, B. McIntyre, K. Mistry, A. Murthy, J. Sandford, M. Silberstein, S. Sivakumar, P. Smith, K. Zawadzki, S. Thompson, and M. Bohr, A 90 nm high volume manufacturing logic technology featuring novel 45nm gate length strained silicon CMOS transistors, *IEDM Tech. Dig.*, 978-980 (2003)

62. S. Takagi, T. Mizuno, T. Tezuka, N. Sugiyama, T. Numata, K. Usuda, Y. Moriyama, S. Nakaharai, J. Koga, A. Tanabe, N. Hirashita, and T. Maeda, Channel structure design, fabrication and carrier transport properties of strained-Si/SiGe-on-insulator (strained-SOI) MOSFETs, *IEDM Tech. Dig.*, 57-60 (2003)

63. D. Chanemougame, S. Monfray, F. Boeuf, A. Talbot, N. Loubet, F. Payet, V. Fiori, S. Orain, F. Leverd, D. Delille, B. Duriez, A. Souifi, D. Dutartre, and T. Skotnicki, Peformance boost of scaled Si PMOS through Novel SiGe Stressor for HP CMOS, *Symp. on VLSI Technology*, 180-181 (2005)

64. J. Cai, K. Rim, A. Bryant, K. Jenkins, C. Ouyang, D. Singh, Z. Ren, K. Lee, H. Yin, J. Hergenrother, T. Kanarsky, A. Kumar, X. Wang, S. Bedell, A. Reznicek, H. Hovel, D. Sadana, D. Uriarte, R. Mitchell, J. Ott, D. Mocuta, P. O'Neil, A. Mocuta, E. Leobandung, R. Miller, W. Haensch, and M. Leong, Performance comparison and channel length scaling of strained Si FETs on SiGe-on-insulator (SGOI), *IEDM Tech. Dig.*, 165-168, (2004)

65. R. Tsuchiya, M. Horiuchi, S. Kimura, M. Yamaoka, T. Kawahara, S. Maegawa, T. Ipposhi, Y. Ohji, and H. Matsuoka, Silicon on thin BOX: A new paradigm of the CMOSFET for low-power and high-performance application featuring wide-range back-bias control, *IEDM Tech. Dig.*, 631-634 (2004)

66. S. Monfray, D. Chanemougame, S. Borel, A. Talbot, F. Leverd, N. Planes, D. Delille, D. Dutartre, R. Palla, Y. Morand, S. Descombes, M. Samson, N. Vulliet, T. Sparks, A. Vandooren, and T. Skotnicki, SON (Silicon-on-nothing) technological CMOS platform: Highly performant devices and SRAM Cells, *IEDM Tech. Dig.*, 635-638 (2004)

67. F.-L. Yang, Che.-C. Huang, Chi.-C. Huang, T.-X. Chung, H.-Y. Chen, C.-Y. Chang, H.-W.Chen, D.-H. Lee, S.-D. Liu, K.-H. Chen, C.-K. Wen, S.-M. Cheng, C.-T. Yang, L.-W.Kung, C.-L. Lee, Y.-J. Chou, F.-J. Liang, L.-H. Shiu, J.-W. You, K.-C. Shu, B.-C. Chang,J.-J. Shin, C.-K. Chen, T.-S. Gau, P.-W. Wang, B.-W. Chan, P.-F. Hsu, J.-H. Shieh, S.K.-H. Fung, C.H. Diaz, C.-M. Wu, Y.-C. See, B.J. Lin, M.-S. Liang, J.Y.-C. Sun and C. Hu, 45 nm Node Planar-SOI Technology with 0.296μm² 6T-SRAM Cell, *Symp. on VLSI Technology*, 8-9 (2004)

68. D. Hisamoto, T. Kaga, E. Takeda, Impact of Vertical SOI "DELTA" Structure on Planar Device Technology, *IEEE Trans. Electron Devices*, **38**, 1419-1424 (1991)

69. K. Suzuki, T. Tanaka, Y. Tosaka, H. Horie, and Y. Arimoto, Scaling theory for double-gate SOI MOSFET's, *IEEE Trans. Electron Devices*, **40**, 2326 (1993)

70. H.-S. Wong, D. Frank, and P. Solomon, Device design considerations for double-gate, ground-plane, and single-gated ultra-thin SOI MOSFET's at the 25 nm channel length generation, *IEDM Tech. Dig.*, 407-410 (1998)

71. D. Hisamoto, W.-C. Lee, J. Kedzierski, H. Takeuchi, K. Asano, C. Kuo, E. Anderson, T.-J. King, J. Bokor, C. Hu, FinFET-A

Self-Aligned Double-Gate MOSFET Scalable to 20 nm, *IEEE Trans. Electron devices*, **47**, 2320-2325 (2000)

72. J. Kedzierski, D. M. Fried, E. J. Nowak, T. Kanarsky, J. H. Rankin, H. Hanafi, W. Natzle, D. Boyd, Y. Zhang, R. A. Roy, J. Newbury, C. Yu, Q. Yang, P. Saunders, C. P. Willets, A. Johnson, S. P. Cole, H. E. Young, N. Carpenter, D. Rakowski, B. A. Rainey, P. E. Cottrell, M. Ieong, and H. P. Wong, High-performance symmetric-gate and CMOS-compatible V_t asymmetric-gate FinFET devices, *IEDM Tech. Dig.*, 437-440 (2001)

73. B. Doyle, B. Boyanov, S. Datta, M. Doczy, S. Hareland, B. Jin, J. Kavalieros, T. Linton, R. Rios, and R. Chau, Tri-Gate Fully-Depleted CMOS Transistors: Fabrication, Design and layout, *Symp. on VLSI Technology*, 133-134 (2003)

74. D. Ha, H. Takeuchi, Y. Choi, T. King, W. P. Bai, D. Kwong, A. Agarwal, and M. Ameen, Molybdenum-gate HfO_2 CMOS finFET technology, *IEDM Tech. Dig.*, 643-646 (2004)

75. N. Collaert, M. Demand, I. Ferain, J. Lisoni, R. Singanamalla, P. Zimmerman, Y. S. Yim, T. Schram, G. Mannaert, M. Goodwin, J.C. Hooker, F. Neuilly, M.C. Kim, K. De Meyer, S. De Gendt, W. Boullart, M. Jurczak and S. Biesemans, Intergration of Tall Triple-Gate Devices with Inserted-Ta_xN_y Gate in a 0.274mm^2 6T-SRAM Cell and Advanced CMOS Logic Circuits, *Symp.on VLSI Technology*, 106-107 (2003)

76. B. H. Lee, L. Kang, W. Qi, R. Nieh, Y. Jeon, K. Onishi, and J. C. Lee, Ultrathin Hafnium Oxide with low leakage and excellent reliability for alternative gate dielectric application, *IEDM Tech. Dig.*, 133-136 (1999)

77. S. J. Lee, H. F. Luan, W. P. Bai, C. H. Lee, T. S. Jeon, Y. Senzaki, D. Roberts, and D. L. Kwong, High quality ultra thin CVD HfO_2 gate stack with poly-Si gate electrode, *IEDM Tech. Dig.*, 31-34 (2000)

78. T. Aoyama, T. Maeda, K. Torii, K. Yamashita, Y. Kobayashi, S. Kamiyama, T. Miura, H. Kitajima, and T. Arikado, Proposal of new HfSiON CMOS fabrication process (HAMDAMA) for low standby power device, *IEDM Tech. Dig.*, 95-98 (2004)

79. H. Iwai, S. Ohmi, S. Akama, C. Ohshima, A. Kikuchi, I. Kashiwagi, J. Taguchi, H. Yamamoto, J. Tonotani, Y. Kim, I. Ueda, A. Kuriyama, and Y. Yoshihara, Advanced gate dielectric materials for sub-100nm CMOS, *IEDM Tech. Dig.*, 625-628 (2002)

80. S. Ohmi, C. Kobayashi, E. Tokumitsu, H. Ishiwara and H. Iwai, Low Leakage La_2O_3 Gate Insulator Film with EOTs of 0.8-1.2 nm, *the 2001 International Conference on Solid State Devices and Materials (SSDM)*, 496-497 (2001)

81. Y Kim, S.-I Ohmi, K. Tsutsui, H. Iwai, Analysis of variation in leakage currents of Lanthana thin films, *Solid-State Electronics* **49**, 825-833 (2005)

82. A. Kuriyama, S. Ohmi, K. Tsutsui and H. Iwai, Effect of Post-Metallization Annealing on Electrical Characteristics of La_2O_3 Gate Thin Films, *Jpn. J. Appl. Phys.*, **44**, 1045-1051, (2005)

83. T. Watanabe, M. Takayanagi, K. Kojima, K. Sekine, H. Yamasaki, K. Eguchi, K. Ishimaru, and H. Ishiuchi, Impact of Hf concentration on performance and reliability for HfSiON-CMOSFET, *IEDM Tech. Dig.*, 507-510 (2004)

84. C. Choi, C.Y. Kang, S.J. Rhee, M.S. Abkar, S.A. Krishna, M. Zhang, H. Kim, T. Lee, F. Zhu, I. Ok, S. Koveshnikov and J.C. Lee, Fabrication of TaN-gated Ultra-Thin MOSFETs (EOT <1.0nm) with HfO_2 using a Novel Oxygen Scavenging Process for Sub 65nm Application, *Symp. VLSI Technoogy*, 226-227 (2005)

85. S. J. Rhee, H.-S. Kim, C.Y. Kang, C. H. Choi, M. Zhang, F. Zhu, T. Lee, I. Ok, M.S. Akbar, S.A. Krishnan and J.C. Lee, Optimization and Reliability Characteristics of TiO_2/HfO_2 Multi-metal Dielectric MOSFETs, *Symp. on VLSI Technology*, 168-169 (2005)

86. H.C .-H. Wang, C.-W. Tsai, S.-J. Chen, C.-T. Chan, H.-J. Lin, Y. Jin, H.-J. Tao, S.-C. Chen, C. H. Diaz, T. Ong, A.S. Oates, M.-S. Liang and M.-H. Chi, Reliability of HfSiON as Gate Dielectric for Advanced CMOS Technology, *Symp. on VLSI Technology*, 170-171 (2005)

87. G. D. Wilk, R. M. Wallace, and J. M. Anthony, High-k gate dielectrics: current status and materials Properties Considerations, *J. Appl. Phys.*, **89**, 5243-7275 (2001).

88. N. Zhan, M. C. Poon, C. W. Kok, K. L. Ng, H. Wong, XPS Study of the thermal instability of hafnium oxide prepared by HF sputtering in oxygen with rapid thermal annealing, *J. Electrochem. Soc.*, **150**, F200-202 (2003)

89. E. P. Gusev, E. Cartier D. A. Buchanan, M. Gribelyuk, M. Copel, H. Okorn-Schmidt, and C. D'Emic, Ultrathin high-k metal oxide on silicon: processing, characterization and integration issues, *Microelectron. Eng.*, **59**, 341-349 (2001)

90. H. Wong, K. L. Ng, N. Zhan, M. C. Poon, C. W. Kok, Interface bonding structure of hafnium oxide prepared by direct sputtering of hafnium in oxygen, *J. Vac. Sci. Techno. B*, **22**, 1094-1100 (2004).

91. N. Zhan, M.C. Poon, H. Wong, K.L. Ng and C.W. Kok, Dielectric breakdown characteristics and interface trapping of hafnium oxide films, *Microelectron J.*, **36**, 29-33 (2005)

92. C. Ren, D.S.H. Chan, Faizhal B. B., M.-F. Li, Y.-C. Yeo, A.D. Trigg, A. Agarwal, N. Balasubramanian, J.S. Pan, P.C. Lim and D.-L. Kwong, 4A-1 Lanthanide-Incorporated Metal Nitrides with Tunable Work Function and Good Thermal Stability for NMOS Devices, *Symp. on VLSI Technology*, 42-43 (2005)

93. Z.B. Zhang, S.C. Song, C. Huffman, J. Barnett, N. Moumen, H. Alshareef, P. Majhi, M. Hussain, M.S. Akbar, J.H. Sim, S.H. Bae, B. Sassman and B.H. Lee, Integration of Dual Metal Gate CMOS with TaSiN (NMOS) and Ru (PMOS) Gate Electrodes on HfO_2 Gate Dielectric, in *Symp. on VLSI Technology*, 50-51 (2005)

94. D.-G. Park, Z.J. Luo, N. Edleman, W. Zhu, P. Nguyen, K. Wong, C. Cabral, P. Jamison,B.H. Lee, A. Chou, M. Chudzik, J. Bruley, O. Gluschenkov, P. Ronsheim, A. Chakravarti, R. Mitchell, V. Ku, H. Kim, E. Duch, P. Kozlowski, C. D'Emic, V. Narayanan, A. Steegen, R. Wise, R. Jammy, R. Rengarajan, H. Ng, A. Sekiguchi and C.H. Wann, Thermally Robust Dual-Work Function ALD-MNx MOSFETs Using Conventional CMOS Process Flow, *Symp. on VLSI Technology*, 186-187 (2004)

95. S.H. Bae, W.P. Bai, H.C. Wen, S. Mathew, L.K. Bera, N. Balasubramanian, N. Yamada, M.F. Li and D.L. Kwong, Laminated Metal Gate Electrode with Tunable Work Function for Advanced CMOS, *Symp. VLSI Technology*, 188-189 (2004)

96. J. K. Schaeffer, C. Capasso, L. R. C. Fonseca, S. Samavedam, D. C. Gilmer, Y. Liang, S. Kalpat, B. Adetutu, H. Tseng, Y. Shiho, A. Demkov, R. Hegde, W. J. Taylor, R. Gregory, J. Jiang, E. Luckowski, M. V. Raymond, K. Moore, D. Triyoso, D. Roan, B. E. White Jr., and P. J. Tobin, Challenges for the integration of metal gate electrodes, *IEDM Tech. Dig.*, 287-290 (2004)

97. R. Jha, J. Lee, B. Chen, H. Lazar, J. Gurganus, N. Biswas, P. Majhi, G. Brown, and V. Misra, Evaluation of Fermi level pinning in low, midgap and high workfunction metal gate electrodes on ALD and MOCVD HfO_2 under high temperature exposure, *IEDM Tech. Dig.*, 295-298 (2004)

98. I. S. Jeon, J. Lee, P. Zhao, P. Sivasubramani, T. Oh, H. J. Kim, D. Cha, J. Huang, M. J. Kim, B. E. Gnade, J. Kim, and R. M. Wallace, A novel methodology on tuning work function of metal gate using stacking bi-metal layers, *IEDM Tech. Dig.*, 303-306 (2004)

99. S. Nitta, S. Purushothaman, S. Smith, M. Krishnan, D. Canaperi, T. Dalton, W. Volksen, R. D. Miller, B. Herbst, C. Hu, E. Liniger, J. Lloyd, M. Lane, D. L. Rath, M. Colburn, and L. Gignac, Successful dual damascene integration of extreme low k materials (k < 2.0) using a novel gap fill based integration scheme, *IEDM Tech. Dig.*, 321-324 (2004)

100. H. Miyajima, K. Watanabe, K. Fujita, S. Ito, K. Tabuchi, T. Shimayama, K. Akiyama, T. Hachiya, K. Higashi, N. Nakamura, A. Kajita, N. Matsunaga, Y. Enomoto, R. Kanamura, M. Inohara, K. Honda, H. Kamijio, R. Nakata, H. Yano, N. Hayasaka, T. Hasegawa, S. Kadomura, H. Shibata, and T. Yoda, Challenge of low-k materials for 130, 90, 65 nm node interconnect technology and beyond, *IEDM Tech. Dig.*, 329-332 (2004)

101. D. Ryuzaki, H. Sakurai, K. Abe, K. Takeda, and H. Fukuda, Enhanced dielectric-constant reliability of low-k porous organosilicate glass (k=2.3) for 45-nm-generation Cu interconnects, *IEDM Tech. Dig.*, 949-952 (2004)

102. F. Sato and A. Izumi, Moore's law has collapsed, *Deutsche Bank Report (*in Japanese) March 1 (2004)

103. P. P. Gelsinger, Microprocessor for the new millennium: Challenges, opportunities, and new frontiers, *IEEE Int'l Solid-State Circuits Conf. Tech. Digest (ISSCC)*, 22-23 (2001)

104. H. Wong, Drain breakdown in submicron MOSFET's: A review, *Microelectron. Reliab.*, **40**, 3-15, (2000)

105. J. Chung, M. C. Jeng, J. E. Moon, A. T. Wu, T. Y. Chan, p. K. Ko, and C. Hu, Deep-submicrometer MOS device fabrication using a photoresist-ashing technique, *IEEE Electron Devices Lett.*, **9**, 186-188, (1988)

106. M. Ono, M. Saito, T. Yoshitomi, C. Fiegna, T. Ohguro, H. S. Momose, and H. Iwai, Fabrication of sub-50-nm gate length n-metal-oxide semiconductor field effect transistors and their electrical characteristics, *J. Vac. Sci. Technol*, **B13**, 1740-1743 (1995)

107. M. Switkes and M. Rothschild, Resolution Enhancement of 157 nm Lithography by Liquid Immersion, Journal of Microlithography, Microfabrication, and Microsystems, **1**, 225-228, (2002)

108. B.J. Lin, Semiconductor foundry, lithography, and partners, *Proc. of SPIE*, **4688**, 11-24 (2002)

109. H. Iwai, H. S. Momose Ultra-thin gate oxide performance and reliability. *IEDM Tech. Dig.*, 162-166 (1998)

110. H. S. Momose, S. Kitagawa, K. Yamabe, and H. Iwai, Hot carrier related phenomena for n- and p-channel MOSFETs with nitrided gate oxide by RTA, *IEDM Tech. Dig.* 267-270 (1989)

111. T. Morimoto, H. S. Momose, K. Yamabe, and H. Iwai, Prevention of boron penetration from pt poly gate by RTP produced thin gate oxide, *ESSDERC 90, Nottingham*, 73-76 (1990)

112. T. Morimoto, H. S. Momose, K. Yamabe, and H. Iwai, Prevention of boron penetration from pt poly gate by RTP produced thin gate oxide, ESSDERC 90, Nottingham, 73-76, (1990)

113. H. S. Momose, T. Morimoto, Y. Ozawa, M. Tsuchiaki, M. Ono, K. Yamabe, and H. Iwai, Very lightly nitrided oxide gate MOSFETs for deep-sub-micron CMOS devices, *IEDM Tech. Dig.* 359-362 (1991)

114. T. Morimoto, H. S. Momose, Y. Ozawa, K. Yamabe, and H. Iwai, Effects of boron penetration and resultant limitations in ultra thin pure-oxide and nitrided-oxide gate-films, *IEDM Tech. Dig.*, 429-432 (1990)

115. H. Wong, M. C. Poon, C. W. Kok, P. J. Chan, and V. A. Gritsenko, Interface structure of ultra thin oxide prepared by N_2O oxidation, *IEEE Trans. Electron Devices*, **50**, 1941-1945 (2003)

116. H. Iwai, H. S. Momose, T. Morimoto, Y. Ozawa, K. Yamabe, Stacked-nitrided oxide gate MISFET with high hot-carrier immunity, *IEDM Tech. Dig*, 235–238. (1990)

117. H. Wong and V. A. Gritsenko, Defects in silicon oxynitride gate dielectric films, *Microelectron. Reliab.*, **42**, 597-605 (2002).

118. C. M. Osburn and K. R. Bellur, Low parasitic resistance contacts for scaled ULSI devices, *Thin Solid Films*, **332**, 428-436 (1998)

119. S. Thompson, P. Packan, T. Ghani, M. Stettler, M. Alavi, I. Post, S. Tyagi, S. Ahmed, S. Yang, and M. Bohr, Source/drain extension scaling for 0.1 μm and below channel length MOSFETs, *Symp. on VLSI Technology*, 132-133 (1998).

120. S. D. Kim, J. Yuan, and J. Woo, Source/Drain resistance modeling in bulk and ultra-thin bogy SOI MOSFETs, *IEEE Trans. on Electron Devices*, **49**, 457-466 (2002)

121. S. D. Kim, C.-M. Park, and J. Woo, Advanced model and analysis of series resistance for CMOS scaling into nanometer regime. I. Theoretical derivation, *IEEE Trans. on Electron Devices*, **49**, 457-466 (2002)

122. Sangyo Times, *Semiconductor Industry Review (in Japanese)* (2004)

123. Selete, *Equipment Engineering System Data collection Capability Requirement Document (DCRD) Version 2,* Data Source Sematech March (2003)

124. M. Liu, APC from the Foundry Perspective-Now and Beyond, *Proceedings XV AEC/APC Symposium, Colorado Springs,* 13-18 (2003)

125. A. Alan, Wafer generation model, Historical/Strategic Validation, *Global Economic Symposium III*, LSI Research, Sematech I3001 (2000)

126. Yano Kotaro Memorial Foundation, *World Census Diagrams (in Japanese),* Source UN (2003)

127. Shanghai Integrated Circuit R&D Center, *Chinese IC Industry: Present and Future (in Chinese)* (2004)

128. *Strategic Marketing Associates Report*, March (2004)

129. Sangyo Times, *Asia Semiconductor/LCD Handbook 2004 (in Japanese)* (2004)

130. A. Minamikawa, Who will drive the worldwide electronics market?, *SEMI Market Seminar, SEMICON Japan 2004*, Makuhari (2004)

131. Electronic Journal, *2004 Semiconductor Equipment Data Book (in Japanese)* (2004)

132. Electronic Journal, *2004 Semiconductor Materials Data Book (in Japanese)* (2004)

133. K. Takahashi and M. Ishino, Development of Three- dimensional Chip Stacking Technology *The Electrochemical Society International Semiconductor Technology Conference,* Shanghai, Session 4 (2004)

2006 International Symposium on VLSI Technology, Systems, and Applications

Challenges for process and product integration at 45nm

Hans Stork

SVP and CTO, Texas Instruments Inc., Dallas, TX, USA

To fully realize the benefits of CMOS scaling, new process generations must reach high yield at low cost, rapidly. Product design faces both higher complexity of process integration, including geometrical effects like strain engineering, as well as a larger need to accurately predict many systematic and random variations. These design-for-manufacturing challenges force an unprecedented interaction of physical design and process optimization for the first generation of 45nm CMOS systems.

2006 International Symposium on VLSI Technology, Systems, and Applications

3D System Integration Technologies

Eric Beyne
IMEC
Kapeldreef 75, B-3001 Leuven, Belgium
ERIC.BEYNE@IMEC.BE

Abstract

Electronic interconnection and packaging is mainly performed in a planar, 2D design style. Further miniaturization and performance enhancement of electronic systems will more and more require the use of 3D interconnection schemes. Key technologies for realizing true 3D interconnect schemes are the realization of vertical connections, either through the Si-die or through the multilayer interconnect with embedded die.

Different applications require different complexities of 3D-interconnectivity. Therefore, different technologies may be used. These can be categorized as a more traditional packaging approach, a wafer-level-packaging, WLP ('above' passivation), approach and a foundry level ('below' passivation) approach. We define these technologies as respectively 3D-SIP, 3D-WLP and 3D-SIC. In this paper, these technologies are discussed in more detail.

Introduction: Why 3D?

As the semiconductor roadmap strides on, packaging and interconnection technologies are required to follow. In order to stay in pace with system demands on scaling, performance and functionality 3D integration is gaining a lot of interest as a solution to this demand.[1] The reasons and requirements for 3D integration are however very diverse and often application specific.[2,3,4,5,6,10]

A basic reason for 3D-integration is system-size reduction. Traditional assembly technologies are based on 2D planar architectures. Die are individually packaged and interconnected on a planar interconnect substrate, mainly printed circuit boards. The area-packaging efficiency (ratio of die to package area) of individually packaged die is generally rather low (e.g. 5x5mm die in 7x7mm package: 50% area efficiency) and an additional spacing between components on the board is typically required, further reducing the area efficiency (for example above e.g. 1mm clearance: 30% area efficiency). If we consider the volumetric packaging density, the packaging efficiency drops to very low levels. If in the previous example, we consider the active area of a die to be about 10 μm, and the combined package and board thickness to be 2 mm, the volumetric packaging density is only 0.15%. There is clearly room for improvement of the packaging density.

A different reason for looking at 3D integration is performance driven. Interconnects in a 3D assembly are potentially much shorter than in a 2D configuration, allowing for a higher operating speed and smaller power consumption. This is of particular interest for advanced computing applications. Due to the rising on-chip clock speeds, only a limited distance may be traveled by a signal in a synchronous operating mode. Using 3D-IC stacking techniques, more circuits may be packed in a single synchronous region. This requires a technology with 3D interconnects with low parasitics;

in particular low capacitance and inductance are needed to avoid additional signal delay. The interconnection of circuit elements can be performed at several levels of the on-chip hierarchy. Of particular interest is the 3D stacking at the so-called "tile-level". As shown in figure 1, typical system-on-chip, SOC, devices are constructed of a number of functional blocks. The longest on-chip lines are those that are used to interconnect these tiles. These lines are typically in the top-on-chip interconnect layers and are referred to as 'global' interconnects in the on-chip wiring hierarchy. Within the tiles, 'local' and 'intermediate' wiring hierarchy levels are mainly used. In a 3D approach, the large die is split in a number of smaller die, using the 3D interconnects as 'global' interconnects between the tiles on both die. As this interconnect goes one or more levels down the traditional IC-pad level, a very high 3D interconnect density is required for such an application.

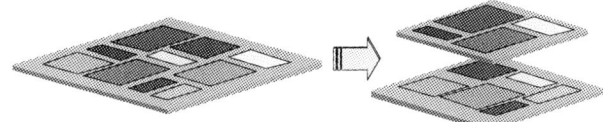

Figure 1: Conceptual view a 3D stacked SOC. Functional 'tiles' on the die are rearranged in multiple die that are vertically interconnected, resulting in much shorter global interconnect lines.

A third, and maybe most important, reason to consider 3D integration is so-called hetero-integration. As silicon semiconductor technologies continue to scale (vertical scaling), the realization of true SOC devices with a large variety of functional blocks becomes very difficult to achieve. Technologies need specific optimization for logic, analog, memory etc. to reach the desired performance levels and circuit density. Furthermore, the substrates used to build active devices may vary significantly between technologies, including non-silicon substrates, e.g. compound semiconductors. Also systems may contain other planar components, such as MEMS and integrated passive devices. Besides the 'vertical' scaling we are also experiencing a 'horizontal' scaling. Realizing the full system on a single SOC die is becoming increasingly difficult and often not economically justified. If however a high-density 3D technology is available, a "3D-SOC" device could be manufactured, consisting of a stack of heterogeneous devices. This device would be smaller, lower power and higher performance than a monolithical SOC approach.

Such an approach is the obvious choice for many sensor-array applications. Many sensor applications use particular substrate materials, such as IR and X-ray sensing, that are incompatible with Si-CMOS processing. These applications require however high-density circuits to read-out the signals from individual sensor pixels, a requirement best met with advanced CMOS technologies. The solution therefore consists in flip-chip (3D) mounting the sensor-array on a read-out electronics chip.

1-4244-0181-X/06/$20.00 ©2006 IEEE   19

Another possible application for this approach is the combination of logic and memory (see figure 2).

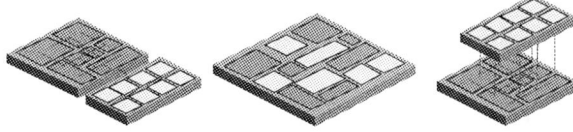

Figure 2: Different approaches for combining logic and memory circuits. Left: 2D interconnect between logic and memory die; Center: (2D-SOC) combined logic and memory device; Right: "heterogeneous 3D-SOC" stacking of a memory and logic device with 3D interconnects between individual logic tiles and memory banks.

Most applications require a combination of logic and memory. When large amounts of memory are needed, the memory is realized as a separate die, using a high density, optimized memory technology. Due to the use of large busses on the logic and memory die and the use of off-chip interconnects, only a relatively slow and power-hungry interconnect between memory and logic is possible. To overcome these limitations, e.g. for real-time data processing applications, a SOC approach is typically used. Although not optimal for the integration of high-density memory, the IC logic technology is used for integrating large amounts of memory. This allows for allocating smaller pieces of memory (memory-banks) to specific logic blocks. Distance between logic and memory is short, resulting in the required performance. The integrated memory is however of the same performance as dedicated memory technologies would offer. In particular, a much larger die area is consumed by the memory cells, resulting in a die are that is significantly larger than the case with 2 die solutions. 3D interconnect technology may solve this problem, by allowing for logic 'tiles' on a first die to directly access memory banks on a memory chip. In this case the number of 3D connections required from the memory die to the logic die will increase by an order of magnitude compared to the I/O count of standard memory devices. Similarly as for the example shown in figure 1, this approach uses 3D interconnects as "global-on chip" interconnect layers to realize a "heterogeneous 3D-SOC" structure.

3D TODAY

Currently, one particular type of 3D-IC packaging is highly popular, the so-called stacked die package, shown schematically in figure 3. [7] In this technology, standard BGA package technology is used to create a 3D-die stack. The individual die are thinned down aggressively – down to about 50 µm – and are glued on top of each other. The die are connected to the base interposer substrate by wire bonding. This method is used to stack as many as five functional die in a single package. It is particularly popular for portable applications where a processor die is combined with several types of memory die (ROM/RAM/Flash…) in a very small volume. Packaging area efficiencies of 100 to 300% may be obtained. The volumetric packaging density (considering 10 µm of "active" Si-thickness) is 0.5 to 1.5% (up from only 0.15% for CSP packages), still relatively low.

The BGA interposer substrate has typically only a very limited capability for routing signal lines, other than the traditional bond pad-to-solder ball connections. Increasing the interposer wiring capabilities requires the use of additional wiring layers and finer line/space board technologies, both resulting in a higher cost package. In fact, only in the case of memory die a relatively simple interposer can be used.

Stacking many die on each other also results in very long wire bonds at the top layers, typically several mm in length. This results in connections with several nH of inductance. Furthermore, the closely spaced, but long, wire bond busses result in significant cross talk. This method is therefore not very well suited to high-speed circuits (including fast memory). Another limitation is that the technique does not allow for area-array contacts to the die, which is also often combined with high speed die.

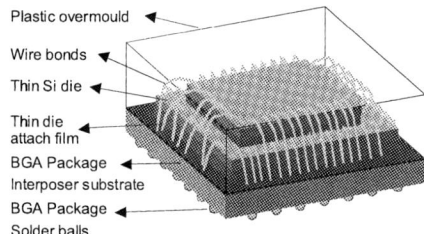

Figure 3: Schematic drawing of a BGA package with multiple stacked die and wire bond interconnects

A final issue with the wire-bond stack is the requirement for "known-good-die", KGD, to avoid compound yield problems. This is why some companies have altered this technique to also include pre-packaged and tested die in such a die and package stack. This will however increase the package cost and lower its volumetric density.

3D DIE STACKING NEEDS

It is clear that the wire-bonded die-stack will not offer the generic 3D packaging solution for advanced systems and scaled semiconductor devices. A number of requirements for 3D packaging and interconnection technologies can however be put forward:

The technology should allow for high-density 3D interconnects. Peripheral interconnects, such as wire-bonds, are inherently limited in density as the pitch has to decrease linearly with the wiring demand, whereas in an area-array situation, the pitch has only to decrease with the square root of the wiring demand.

High speed and low power applications both demand shorter interconnects with low parasitic capacitance. For high speed, also low inductance is of high importance.

Any 3D-technology that crosses through the Si-die area should have minimal impact on the FEOL (Front-end-of-line, the active die area with the transistors) and the BEOL (Back-end-of-line, the on-chip interconnect layers). Large numbers of large 3D-via connections may block large die-areas, where active circuits and interconnects must be excluded. This will cause the die to require a large Si-area on the wafer, increasing the die cost and defeating the purpose of 3D stacking, which is to miniaturize the system and shorten the circuit interconnect lengths. Actually, the loss in die area may be considerably larger than the actual via size, as the routing of the circuit cells becomes more complicated when large areas of the BEOL are blocked.

A 3D stacking technology should allow for different die sizes. In general, when assembling die made in different technologies, using combinations of existing and newly designed die, it is very unlikely that the die sizes will match. Furthermore, when using heterogeneous integration of various technologies, the wafer size of the different die may not match (300mm, 200, 150 mm and even smaller diameter sizes will continue to co-exist for different technologies such as memory, logic, analog, rf, high voltage and compound semiconductors). Wafer-to-wafer bonding for 3D stacks will therefore be limited to 3D IC-stacks where all layers are realized in the same or similar technology. The main applications for this are memory stacks with a single type of memory and high performance logic wafers, where advanced CMOS-wafers are vertically stacked to allow for packing more transistors in a synchronous region of space.

A significant treat to 3D packaging is the so-called "known-good-die" (KGD) problem. When combining n untested die from wafers with a die yield Y_i, the compound yield of the structure will be $Y_m = Y_s \times Y^n_i$ (with Y_s the yield of the interconnect and packaging process). As an example, combining 5 die with a processing yield Y_i of 80% and an assembly yield Y_s of 95%, results in a module yield of only 31%. Lower wafer yields, result in exponentially smaller module yields, e.g. Y_i=70%: Y_m=16%. Such yield loss can not be avoided when using wafer-to-wafer 3D bonding techniques. Technologies that allow for die-to-die or die-to-wafer bonding may introduce a component-screening test to increase the confidence level in the die to a "Good-enough-die" level, e.g. 95 or 97%. This can be done using relatively simple test schemes. For the example above, raising the die yield level to 97% would result in a more acceptable 82% module yield.

3D stacking schemes should also consider the thermal management of the module. The key issue for thermal management in an electronic system is how to transfer the heat generated by a localized heat source (the active silicon die area) to the ambient environment, generally the air surrounding the devices. By stacking the die in a small volume the total heat dissipation of the system may be reduced, due to the shorter interconnect lengths. However the local heat density in the system will dramatically increase. The thermal problem becomes twofold: getting the heat out of the stack to the 'package' boundaries and getting the heat from the small package to the environment. The first problem requires the use of highly thermal conductive materials in the package and the use of thin layers of in the package build-up, in particular for the electrically insulating and 'gluing' layers which generally posses a poor thermal conductivity compared to metals or silicon.

Finally, last-but-not-least, the process for realizing a 3D stacked device should be cost effective. The main implication of this factor is that no single generic 3D-packaging solution will be possible. The 3D technology should be chosen to fit with the requirements set forward by the application. A technology with a very high 3D wiring density may be 'overkill' for an application requiring only a moderate number of interconnects.

In order to achieve the goal of a cost-effective 3D process, a number of technology process requirements may be put forward:

o The technology should maximize collective processing. This favors a wafer-level approach. Although at some stage in the process individual die will need to be handled, because of compound yield issues (which also significantly impact cost).

o The process should maximize the amount of parallel processing:

- Wafers (die) should be prepared separately for 3D stacking
- The process should allow for die screening to obtain "good-enough-die", e.g. using self-test and IDDQ testing methods.
- Preferably a Die-to-Wafer placement is performed (with KGD on KGD), followed by a collective bonding step of the individual die at the wafer level.
- The 3D stack is build by repeating this process with a minimum of sequential processing steps.

Processes that are fully sequential (e.g. wafer-to-wafer bonding, followed by a contact formation process, followed by additional sequential wafer-to-wafer bonding and contact formation processes) suffer from additional yield loss inherent to processes with a large number of sequential steps and also due to the large process strain put on the die that are placed first in the stack.

TECHNOLOGIES FOR 3D

In literature one can find a very large number of different approaches to 3D stacking of die. Generally solutions are sought, starting from an available technology. Apart from the large variety of techniques proposed, also a large variety of achievable package density and 3D interconnectivity are observed.

In an attempt to categories these technologies, one can start from the technology platform ("factory-type") used to create the 3D interconnect structures. We identify 3 major technology platforms for 3D integration, based on the underlying infrastructure:[2]

- 3D-SIP: Packaging infrastructure
- 3D-WLP: Wafer-level packaging infrastructure
- 3D-SIC: IC-foundry infrastructure

The 3D-SIP technology encompasses the packages with wire-bond die-stacks, but also involves package-on-package 3D stacks. It is currently the most mature technology and in high volume production. A relatively low packaging density characterizes 3D-SIP.

The 3D-WLP technology is based on wafer-level packaging infrastructure, as used for flip chip bumping and redistribution metallisations. Using additional technology elements developed for MEMS-technology, such as deep anisotropic Si-etching, 3D electrical connections can be realized at the wafer level. This technology allows for higher integration densities than 3D-SIP. In order to realize cost-effective 3D-WLP stacks, a die-to-wafer and parallel processing route should be explored. Many approaches in this field are proposed and several are being applied in products today.

The interconnects between the die in a 3D-WLP can be at the traditional chip I/O boundaries, or at the global on-chip interconnect layer. In the latter case we can, from a system point of view, consider the 3D-WLP as a heterogeneous "3D-SOC", as discussed in the introduction.

The 3D-SIC approach uses the Si-foundry technology to create very high density vertical interconnects. Many technologies are proposed in this area, but so far they are still in the R&D phase. This type of 3D stacks can be divided in two classed.

o A first class consists of wafer stacks where relatively large circuit blocks "tiles" are interconnected in a 3D fashion. The 3D interconnects mainly correspond to global and possibly intermediate BEOL on-chip interconnects. In this case we can also consider the 3D-SIC as a "3D-SOC". Stacking is in this case also preferably realized using die-to-wafer bonding.

o A second class of high-density wafer stacks aims at connecting small circuits, logic gates and even transistors in a 3D manner. This requires interconnects at the local BEOL wiring hierarchy. Such a device could be considered a true 3D-IC.

As the number of local interconnects exceeds the number of global interconnects on a die with multiple orders of magnitude, an extremely large 3D wiring density must be achieved. This requires extremely small and narrow pitch 3D interconnects. At e.g. a 45nm node, the via pitch should be below 1 μm. Furthermore, the area blocked by these connections will be very large compared to the available die area, significantly reducing the active device density.

The 3D-IC technology requires a wafer-to-wafer bonding approach and is sequential in nature. Wafers with FEOL layers will be stacked first with local 3D interconnects realized after bonding each individual wafer level. Only when all "local" layers are finalized, the BEOL intermediate and global interconnect layers will be added. These layers are common for all layers. Considering the fact that the BEOL layers are one of

the significant bottleneck for the success of future die-shrinks, it can be anticipated that this will also be the case for 3D-IC's, therefore requiring a large numbers of interconnect planes.

For the reasons outlined above, a "3D-SOC" approach to 3D-SIC is more likely to be successful and economically viable than a pure 3D-IC approach. The 3D-IC approach is restricted to special applications, such as the artificial Retina chip from Tohoku university[12], where the 3D-IC approach offers unique possibilities, however at a high cost.

3D TECHNOLOGIES AT IMEC

3D-SIP FOR BUILDING HIGHLY MINIATURIZED SYSTEMS

3D-SIP is of particular interest when it is used as a stacking technology of SIP packages. Consider a system composed of a number of clearly defined sub-systems. Each sub-system could be integrated in a system-in-a-package fashion, using the appropriate packaging technology for that particular subsystem. At the end, the SIP sub-systems can be stacked in the 3rd dimension by a collective process, creating a 3D-SIP system solution.

As the layers of the 3D stack are SIP's by themselves, only a modest 3D interconnect density is required. Also testing of the different SIP layers is greatly simplified.

An example of such 3D-SIP integration scheme, realized at IMEC, is a fully integrated low power rf transceiver shown in figure 4. This device measures only 7x7 mm. It consists of two CSP-type devices (CSP="Chip-Scale-Package"). The top CSP is realized using IMEC's rf-MCM-D technology with integrated passives.[11] The bottom CSP is a double-sided high-density printed circuit board with a high density flip chip die on the bottom side and several discrete passive components mounted on the topside. The connection of this bottom part to the top part is obtained by using solder balls on the topside of the bottom laminate and encapsulation of the topside devices.

Rf front-end CSP   Digital radio base-band CSP

Figure 4: Fully integrated low power rf radio, measuring 7x7x2.5 mm, realized by 3D stacking of CSP packages.3D joining using micro-bumps

A particular interesting application area for 3D-SIP is the realization of distributed, fully autonomous systems for realizing so-called "ambient intelligence" systems. These are sometimes referred to as smart-dust, e-grains or e-cubes. As shown in figure 5, such systems can be divided into clear subsystems: the radio (antenna, rf-front-end base band), the main application (processor, sensors, actuators) and the power management (regulation, storage, generation). Each of these functions can be realized as a SIP-subsystem. These may be very small (a few mm to a few cm), enabling its realization using wafer level processing technologies. These 2D-subsystems can be stacked on top of each other, realizing a dense 3D-SiP system. Figure 6 shows such an "e-cube" module with a volume of only one $1cm^3$ realized at IMEC.

A further evolution of 3D-SIP technology is the embedding of Si-die and SMD components in the 3D-SIP interposer substrate using sequential build-up board technologies.

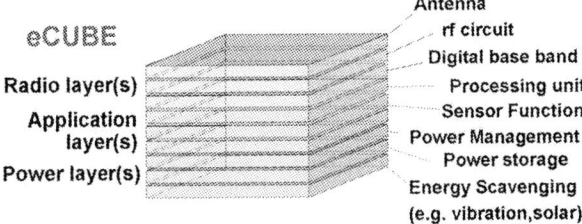

Figure 5: Schematic representation of a 3D-SIP concept "eCube", for the realization of distributed, fully autonomous "ambient intelligent" systems. Each layer in the stack is a fully integrated SIP sub-system.

Figure 6: Photograph (top) and Schematic view of 1cm³ eCube (14x14mm), developed at IMEC for a medical application. The 3D-SIP consists of a rf-SIP, with integrated antenna, a low-power DSP SIP, a 19 channel EEG/ECG sensor die, a power SIP. Small solar cells power scavengers, to provide for long time autonomy, may be added on the module sidewalls.

3D-WLP

Wafer level packaging technologies may be very cost effective to realize 3D stacks. The simplest implementation is the face-to-face bonding using flip-chip or micro-bump connections.

Two different 3D-WLP options can be chosen:
- Realization of 3D-through silicon via connections, followed by stacking by a method similar to flip chip mounting.
- Stacking thin die on-top of other die or substrates and contacting both die using a multilayer thin film technology.

Both technologies are studied at IMEC and described in more detail below.

3D-WLP USING SI-THRU VIAS

The most common approach used for realizing Si-through vias consist of the following steps:

- Etching of a "blind" via hole in the Si-wafer using the Bosch RIE-ICP etching method.
- Dielectric isolation of the Si-hole: Using of CVD oxide or nitride passivation
- Metallization of Si-holes by realizing a solid metal via-"plug". Typically Cu electroplating is used for the via filling, followed by a CMP polishing step to remove the excess Cu plated on the wafer.
- Back grinding of the wafer, exposing the Cu plug, finalizing the 3D via-process.

This process has been shown to be effective for realizing high-density 3D via connections. However a number of important issues remain:

- Only a thin insulation layer is used between the Si substrate and the Cu-plug. This results in a rather high electrical capacitance of the through hole connection, exceeding the capacitance of standard wire-bond pads.
- A rather thick Cu plug is used in the Si-via hole. Due to the large CTE mismatch between Si and Cu this will cause significant thermo-mechanical stresses upon thermal cycling.
- Electroplating of Cu to fully fill the Si-via is a complex process that requires a long process time for each wafer. The use of a CMP polishing step further increases the cost of this technology.

As a solution to these shortcomings, we propose a modified 3D-via build-up [15], shown in figure 7:
- The thin CVD insulating layer is replaced by a 2-5 µm thick polymer isolation layer, deposited by spin or spray coating
- The via is only partially filled with electroplated Cu, similar to a build-up PCB board via, but with smaller dimensions.
- A polymer coating is used to fill the remaining hole in the copper plating.

This method has some significant advantages:
- Lower cost by simplifying the processes, reducing the process time and reducing the required equipment capital investment.
- Strongly reduced capacitance through the use of thicker low-k isolation layers, allowing for high speed and rf 3D-via feed-throughs.

- Strongly reduced thermo-mechanical stresses by using "open" copper metallisations and the use of lower modulus dielectric materials: "compliant" through-hole structure.[9]
- Compatible with common wafer-level packaging redistribution and bumping technologies

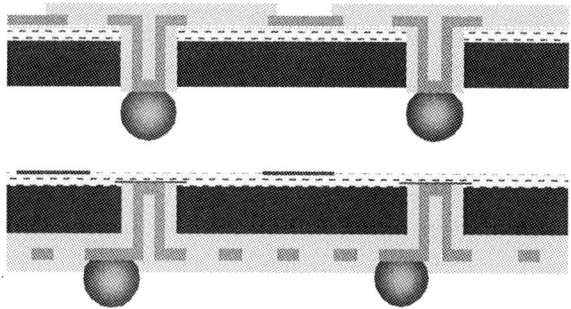

Figure 7: Schematic representation of IMEC's 3D-WLP process with Si through-hole electrical connections. Top: approach with thinning after via formation. Bottom: thinning-first approach: the vias are processed after the wafer thinning process.

3D-WLP BY ULTRA-THIN CHIP STACKING (UTCS)

A different approach to 3D stacking consists in stacking thin die on active device wafers and using multilayer thin film technology[11,16] to interconnect the thin die with the host wafer. Such approach allows for a high level of system level flexibility and integration density. This technique also allows for stacking of die with largely varying dimensions, as well as the integration of thin film passive components in a 3D interconnect stack.

Several approaches of this type of stacking are under investigation.[1]. IMEC's ultra-thin-chip-stacking, UTCS, approach [13] uses very thin (10-20μm) Si die, embedded in a redistribution technology. In order to realize a multilayer die UTCS stack, four basic processes are used.

o Ultra-thin chip-on-die, UTCD, process
o Ultra-thin chip embedding, UTCE, process
o Ultra-thin chip-in-flex, UTCF
o UTCF-stacking process for more than two die layer UTCS structures

The UTCD process is shown schematically in figure 8. A wafer with active, tested die is bonded to a temporary silicon carrier wafer. Using a combination of coarse and fine grinding, the active wafer is thinned to a thickness of 15-20 μm). Plasma etching is used to remove any remaining Si damage and to obtain the desired final thickness. Plasma etching is then used to etch the scribe lanes of the die. The next step consists in dicing the carrier wafer to obtain the UTCD chips for further processing.

For the UTCE process, shown schematically in figure 9, either an active wafer or another dummy carrier wafer is used as host substrate. A polymer glue-layer, such as BCB, is spun on the wafer. A flip chip bonder is used to place with high alignment accuracy KGD-UTCD chips on KGD host wafer die. The actual bonding of these die (polymer glue curing under

pressure) is collectively performed at the wafer level. The next step is the collective removal of the sacrificial layer and the temporary carrier chips of the UTCD die (e.g. thermal or chemical). The next steps consist of depositing thin film dielectric layers for isolation and thin film, electroplated copper patterns for electrical connection. At the end of this process a 2-layer UTCS stack is obtained, as shown in figure 10. An example of a two-die layer UTCE structure is shown in figure 11.

If, for the UTCE process a sacrificial substrate is used, this substrate may be removed, effectively resulting in a very thin flex foil (10 to 30 μm) with embedded active die, the UTCF foil shown schematically in figure 12.

For realizing an n-layer UTCS stack, two different approaches could be used. The UTCE process could be sequentially repeated on another UTCE stack, as shown in figure 13. However, it is also possible to stack UTCF films on UTCE stacks, using micro-bump flip chip connections. This results in a more cost effective parallel process flow.

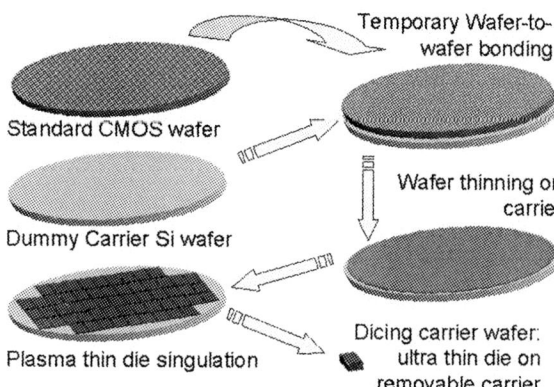

Figure 8 : Schematic representation of the ultra-thin chip-on-die, UTCD, process

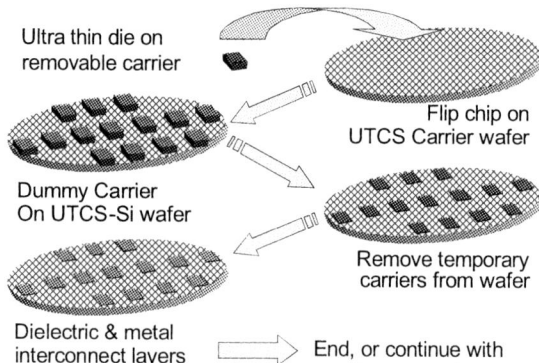

Figure 9: Schematic representation of the ultra-thin chip embedding, UTCE, process

By using thin film lithography, a very high number of interconnects can be realized between the die. This allows for additional interconnects between sub-sections of the die. In the case of the proposed UTCS structure, via connections in the 3rd dimension are realized in the area around the chips. As these thin film vias are realized with pad sizes smaller than 50 μm, a very high interconnect density in the third dimension is obtained.

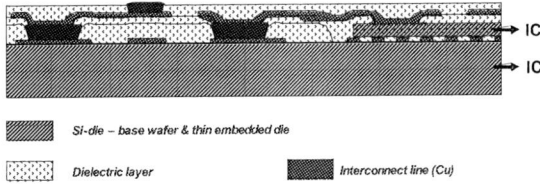

Si-die – base wafer & thin embedded die

Dielectric layer

Interconnect line (Cu)

Figure 10: Schematic cross-section of a UTCE stack.

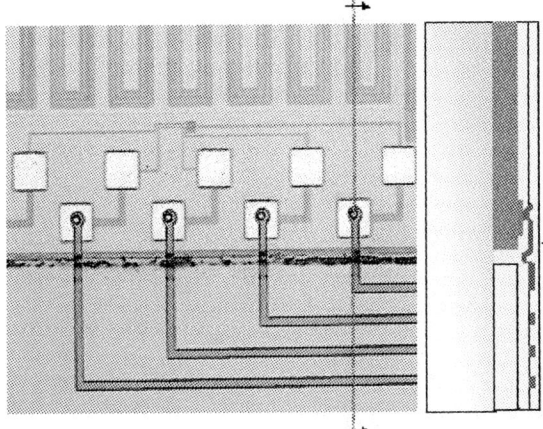

Figure 11: Example of a 15 µm thin Si-die, transferred to a host substrate and electrically connected to that substrate using the UTCE die embedding technique.

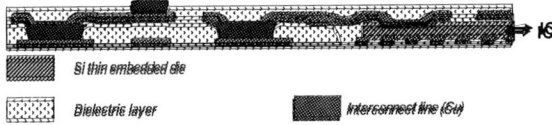

Si thin embedded die

Dielectric layer

Interconnect line (Cu)

Figure 12: Schematic cross-section of the UTCF foil after removal of the sacrificial substrate.

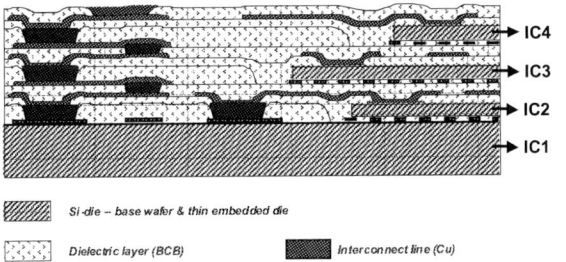

Si-die – base wafer & thin embedded die

Dielectric layer (BCB)

Interconnect line (Cu)

Figure 13: n-layer UTCS stack, realized using a sequence of (n-1) UTCE process steps.

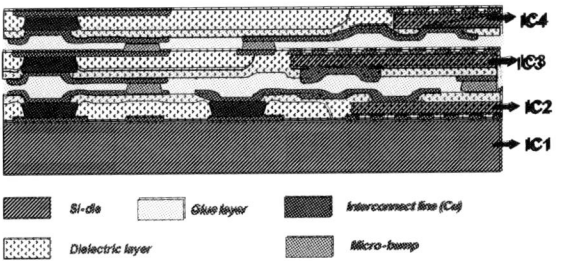

Si-die

Glue layer

Interconnect line (Cu)

Dielectric layer

Micro-bump

Figure 14: n-layer UTCS stack, realized using parallel processing combining one UTCE and n-2 UTCF layers.

In contrast to the traditional die-stacking technologies, the dies that are stacked can have arbitrary sizes at any layer and do not need any particular process or design adaptation.

A further possible evolution of the UTCS technology is to combine it with the 3D-WLP Si via process discussed in 5.2.1, resulting in a very high 3D interconnectivity with large degrees of freedom.

3D-SIC

As explained in section 4, 3D-SIC technology uses IC-foundry infrastructure to realize through-Si via connections. Most approaches in this field realize these through-vias after finalizing the IC process. [6,12]. Our approach differs from this by introducing a small, Si-via and Cu plug, the so-called "Cu-nail". This Cu-nail is processed after the FEOL process (transistors), but before the BEOL process (multilayer damascene interconnect layers), as shown in figure 15. The Cu-nail is realized by plasma etching a ±15µm deep a Si hole with a diameter of 3-5µm. A modified Cu single damascene process is used to fill the hole. A CVD oxide layer is used as thin dielectric insulating layer and a CMP stop layer. A TaN barrier is then deposited. The via hole is then filled with electroplated copper. CMP is used to remove the Cu "overburden". After this process, standard BEOL is used to finalize the Si-die.

After finalizing the wafer process and wafer test, the wafer is mounted on a temporary carrier and thinned down to a Si-thickness of only 10 µm. In the process, the "Cu-nails" are exposed on the wafer backside (see figure 15).

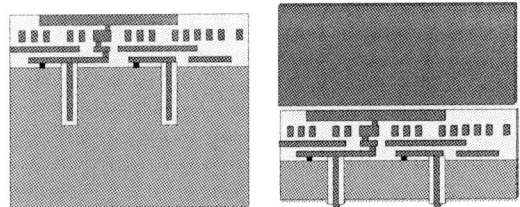

Figure 15: Schematic representation of the "3D-SIC" Cu-nail via process. Left: Standard CMOS wafer with "Cu-nail" before the BEOL process. Right: Thinned CMOS chip on carrier chip with exposed Cu-nails.

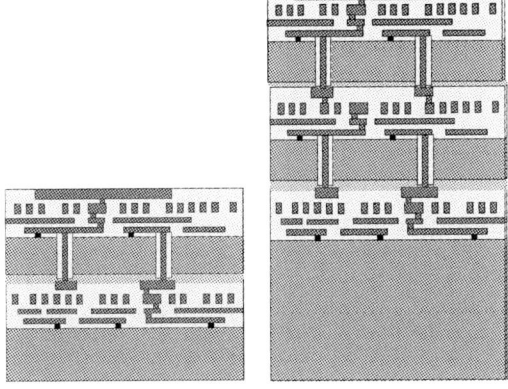

Figure 16: Schematic representation of a 2 and 3-layer 3D-SIC stack using Cu-Cu bonding

25

The 3D stacking is performed as a die-to-wafer bonding process, similar to the 3D-WLP process mentioned above. However, in this case Cu-Cu direct bonding is used, rather than a solder joint or micro bump connection.[14] The stacking process consists of a fast die-to-wafer alignment and placement, followed by a collective wafer-level Cu/Cu bonding process. It can easily be repeated to obtain multi-die stacks. (See fig. 16)

The main advantages of this 3D-SIC approach are:

o Minimal impact on the CMOS wafer design and processing:
  - Only a small exclusion area on the FEOL (small via holes)
  - No impact on the BEOL wiring
  - Small number of additional process steps needed, only one additional litho step, resulting in a low process cost.

o Parallel processing route : wafers are prepared for 3D stacking and only KGD die are stacked involving a minimum number of processing steps. This is required to achieve high 3D-module Yields and low cost processing.

o A very high density 3D-interconnect is possible, as the Cu-nail size is only a few micrometer in diameter. Densities up to, and exceeding, $10^4/mm^2$ are feasible.

One of the possible draw-backs of this 3D-SIC, as well as other 3D stack technologies, is the need for routing 3D-contacts through intermediate die in the stack when going from a connecting two arbitrary die from the stacks. The die, or at least the 3D-contacts, must be arranged in a "wedding-cake" fashion.

A possible solution to this drawback is to combine the UTCS technology, described above in 5.2.2, with the 3D-SIC Cu-nail and stacking technology. This is illustrated schematically in figure 17.

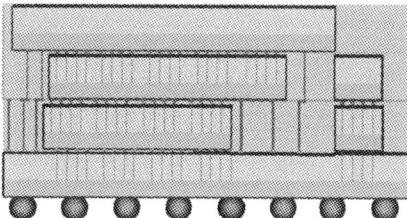

Figure 17: Schematic representation of a combined UTCS and 3D-SIC technology.

3D RESEARCH ROADMAP IMEC

In the previous section, IMEC's approaches to 3D-SIP, 3D-WLP and 3D-SIC technologies were discussed. In Table 1, a comparison of these different approaches is presented. A common aspect of all technologies is the need for die thinning technologies down to 50μm and less. For the 3D-SIC and UTCS technologies, die will be thinned down even further. A schematic representation of IMEC's R&D roadmap is given in figure 18.

CONCLUSION

The further evolution of microelectronic technology and electronics systems will increasingly require 3D interconnect technologies. Different applications will require various 3D-interconnect densities. Therefore, a variety of technologies can be used to reach a cost-effective solution.

In order to be cost-effective, a 3D technology should meet a number of criteria. Parallel processing of the 3D-stack layers should be maximized. Also processing should be performed as much as possible on a wafer or panel level. Also of crucial importance is to provide a solution for the compound yield problems that are inherent to multi-die systems. This favours a die-to-wafer bonding of KGD die to KGD die locations on a wafer.

We propose a classification of the different 3D technologies, based on the underlying manufacturing infrastructure: 3DSIP; 3D-WLP and 3D-SIC.. The approach and roadmap of IMEC to these technologies was presented in detail.

REFERENCES

[1] Proc. of the 1st and 2nd conf. on "3D Architectures for Semiconductor Integration and packaging", RTI international, Burlingame, California, April 13-15, 2004 and Tempe, Arizona, June 13-15, 2005

[2] E.Beyne, "3D Interconnection and packaging: impending reality or still a dream?" proceedings of the IEEE International Solid-State Circuits Conference, ISSCC2004, 15-19 February 2004; San Francisco, CA, USA, IEEE, 2004, pp.138-145.

[3] Phil Garrou, "3D Integration: A status report", proceecings "3D Architectures for Semiconductor Integration and packaging", RTI international, Burlingame, Tempe, Arizona, June 13-15, 2005.

[4] Scott List, "The third dimension: fact or fiction?" idem [3]

[5] Albert Young, "Perspectives on 3D-IC technology" idem [3].

[6] A.Klump et al. "3D Integration of CMOS Transistors with ICV-SLID Technology" idem [3].

[7] M. Karnezos. "3-D Packaging: Where All Technologies Come Together". IEEE/SEMI 29th International Electronics Manufacturing Technology Symposium, July 2004, pp. 64–67.

[8] N.Ranganathan et.al, "High aspect ratio through-wafer interconnect for three dimensional integrated circuits, Proceedings of the 55th ECTC, , Orlando, Florida, May 31-June 3, 2005, pp. 343-348

[9] M.Gonzalez et al, "influence of dielectric materials and via geometry on the thermomechanical behaviour of silicon through interconnects" ". Proc. of 10th Pan Pacific Microelectronics Symposium, SMTA, Hawaii, January 25-27, 2005.

[10] Manuba Bonkohara, "Technologies for 3D assembly and chip level stack" Proceedings of 2nd International Symposium on Microelectronics and Packaging, ISMP2003, IMAPS-Korea, Seoul, Korea, September 24-25, 2003,, pp.85-90.

[11] E.Beyne, "Multilayer thin film technology as an enabling technology for System in Package (SiP) and "above-IC" Processing", idem [7] pp.91-99.

[12] Misma Koyanagi, "3D LSI Technology and Wafer-level Stack", idem [7], pp.101-108.

[13] E.Beyne, "Technologies for very high bandwidth electrical interconnects between next generation VLSI circuits", IEEE-IEDM 2001 Technical Digest, December 2-5, Washington, D.C., S23-p3, 2001.

[14] J.H.McMahon et al., "Wafer bonding of damascene-patterenced metal/adhesive redistribution layers for via-first 3D Interconnect", Proc. of the 55th ECTC, , Orlando, Florida, May 31-June 3, 2005, pp. 332-336.

[15] patent US 10817763

[16] patent EP 0100014, US 6,506,664

Table 1: Classification and comparison of different 3D interconnect technologies at different levels of the interconnect hierarchy.

	3D-SIP	3D-WLP		3D-SIC
Technology	Package interposer	WLP, Post-passivation		Si-foundry, Post FEOL
3D interconnect	Package I/O	UTCS Embedded die	Si-through vias	Si-through "Cu nail" vias
Interconnect density	'package-to-package'	'around' die	'through' die	'through' die
Peripheral	2 - 3 /mm	10 - 50 /mm	10 - 25 /mm	25 -100 /mm
Area-array	4 - 11/mm^2	100 -2.5k/mm^2	16 - 100/mm^2	400-10k/mm^2
3D Si Via pitch	-	-	40 – 100 µm	< 10 µm
3D interconnect pitch	300 – 500 µm	20 – 100 µm	-	-
3D Si Via diameter	-	-	20 - 40 µm	1 - 5 µm
Die thickness	> 50 µm	10 - 50 µm	40 - 100 µm	10 - 50 µm

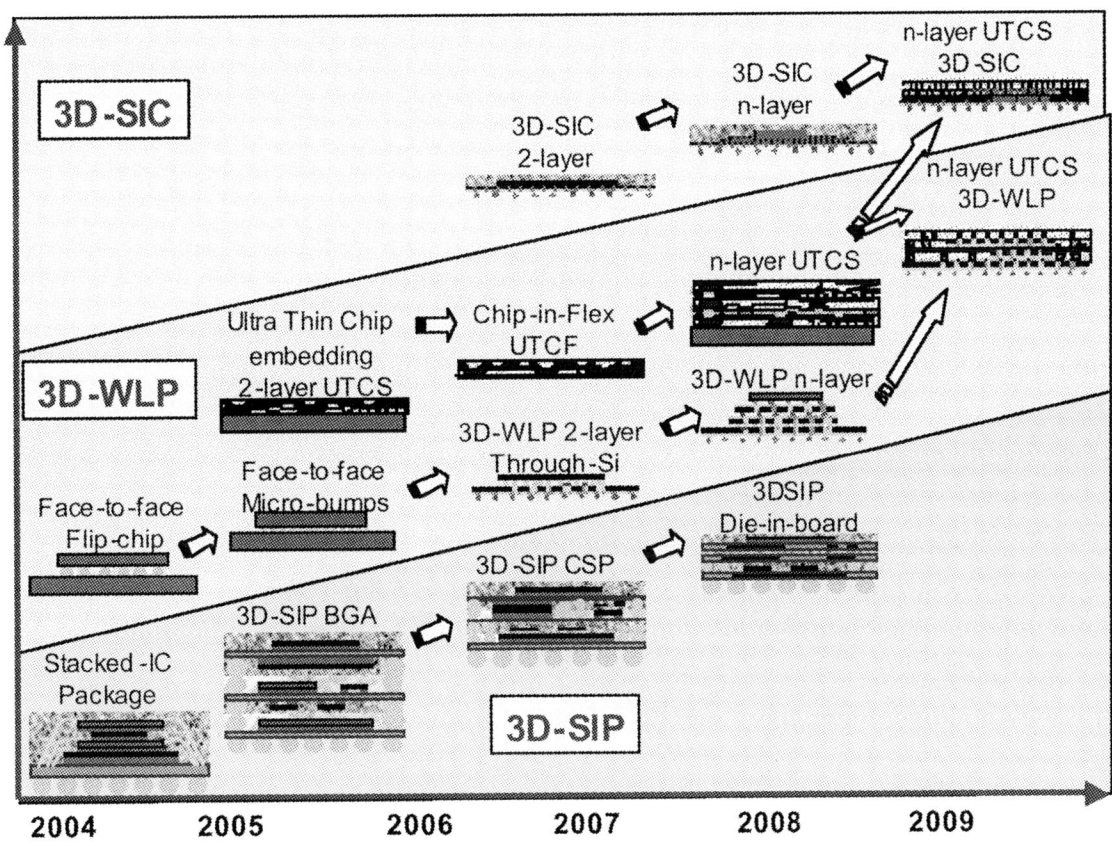

Figure 17: IMEC's 3D packaging and interconnection roadmap for 3D-SIP, 3D-WLP and 3D-SIC technology families.

2006 International Symposium on VLSI Technology, Systems, and Applications

Future Memory Devices - from Stacked memory, Gain memory, Single-electron memory to Molecular memory

Kazuo Nakazato
Department of Electrical Engineering and Computer Science,
Graduate School of Engineering, Nagoya University

ABSTRUCT

Future possibilities for high speed memories will be discussed. In stacked memories, vertical MOSFETs are stacked onto MOSFETs. A 500 MHz 64 Mbit SRAM was demonstrated with reduced cell area, 60% of planar cell. PLEDM was proposed as a scalable DRAM gain cell which leads to single-electron memories. The transfer and detection of one electron was demonstrated using SOI structures at 4.2K. To achieve the room-temperature operations, nodes must be formed in atomic scale, and molecular memories should be investigated..

INTRODUCTION

Memory architectures are now approaching fundamental difficulties. In this paper we discuss the future possibilities for high speed memories, from stacked memory, gain memory, single-electron memory to molecular memory.

STACKED MEMORY

First, stacked memories are described where vertical polycrystalline silicon MOSFETs are stacked onto the substrate MOSFETs (Fig. 1). The vertical structures reduce the cell size by minimizing the footprint of the cell transistor. Furthermore, if the transistor can be formed free from the single-crystal substrate and can be stacked on transistors, the cell size will be reduced drastically. In the stacked vertical MOSFET, amorphous silicon layers were deposited uniformly on silicon dioxide layers, and converted to polycrystalline silicon by low temperature solid-phase growth. These layers were cut by dry etching to form the MOSFET channel. Since the polycrystalline silicon layers have columnar or disc-shaped grains, imposing a vertical MOSFET channel is expected to result in carrier flow mainly without crossing any grain boundaries. Furthermore, since the gate can be formed doubly to quadruply, the gate width can be wider for the same transistor area. These effects increase the ON current, which is essential for high-speed memories. Stacked vertical MOSFETs fabricated using a 130 nm standard silicon process show excellent DC characteristics very similar to single-crystal MOSFETs, although grain boundaries exist inside the channel region ; high ON current of 200 μA/μm, low OFF current of 10 fA/μm, small variation of 0.2 V of threshold voltage, and small subthreshold slope of 70 mV/decade. Using the stacked vertical MOSFETs, 533 MHz high speed 64 Mbit SRAM was demonstrated with reduced cell area size of 1.21μm^2, which is 60% of 130nm node based planar type cell and is almost equivalent to the size of 100nm node. The same reduction effect is surely expected in 100nm node device and beyond.

GAIN MEMORY

Next, we discuss a new DRAM gain cell, PLEDM (Fig. 2), as one of the solutions to overcome the difficulties of present day DRAMs. PLEDM utilizes silicon stacked tunnel transistor PLEDTR, which is a vertical, fully depleted, double-gate SOI-MOSFET with barriers in the channel region. Gate voltage modulates the internal potential in the intrinsic silicon region and the central shutter barrier also moves up and down energetically following the internal potential. The device may be regarded as a three terminal version of the heterostructure hot-electron diode (H2ED) based on the transition from a tunnelling current to a thermionic emission current at a semiconductor heterojunction. PLEDTR enables the construction of a high-density memory because each memory cell occupies the area of just one transistor. A PLEDTR is stacked onto the gate of a conventional MOSFET with a built-in coupling capacitor to realize a memory cell. High-speed write is possible by transferring electrons from the top electrode (bit line) onto the memory node through the ON-state PLEDTR. Since the OFF-state PLEDTR can confine electrons very effectively, the stored information can be kept for a long time. Since the information is read via the current in a MOSFET, this cell has gain and a large S/N ratio. From the measured characteristics it is shown that random access memory operations with 20ns read/ 5ns write times are possible in cells with 5F^2 cell size where F is the minimum feature size. One of the most important points is that the PLEDM has scalability, that is, the stored charge can be reduced according to the reduction of the cell size.

SINGLE-ELECTRON MEMORY

Further into the future, DRAM gain cells can be single-electron memories, where the precise number of electrons is controlled by multiple-tunnel-junctions (MTJ). In the ultimate, single-electron memories can be based on just one electron representing one bit of information. The transfer and detection of one electron on a memory node was demonstrated using silicon-on-insulator MTJs at liquid helium temperatures. To achieve the room-temperature operations, MTJs must be formed in atomic scale, and molecular memories should be investigated (Fig. 3).

REFERENCES

[1] T Kikuchi, S Moriya, H Matsuoka, K Nakazato, A Nishida, H Chakihara, M Matsuoka, and M Moniwa, "A New Vertically Stacked Poly-Si MOSFET for 533MHz High Speed 64Mbit SRAM, " IEDM Dec. 2004

[2] K Nakazato and H Ahmed, "Few Electron Devices and Memory Circuits, " *Silicon Nanoelectronics* (ed. by S. Oda and D. Ferry, Marcel Dekker Inc., New York, 2005), Ch. 10, pp. 239-275

1-4244-0181-X/06/$20.00 ©2006 IEEE

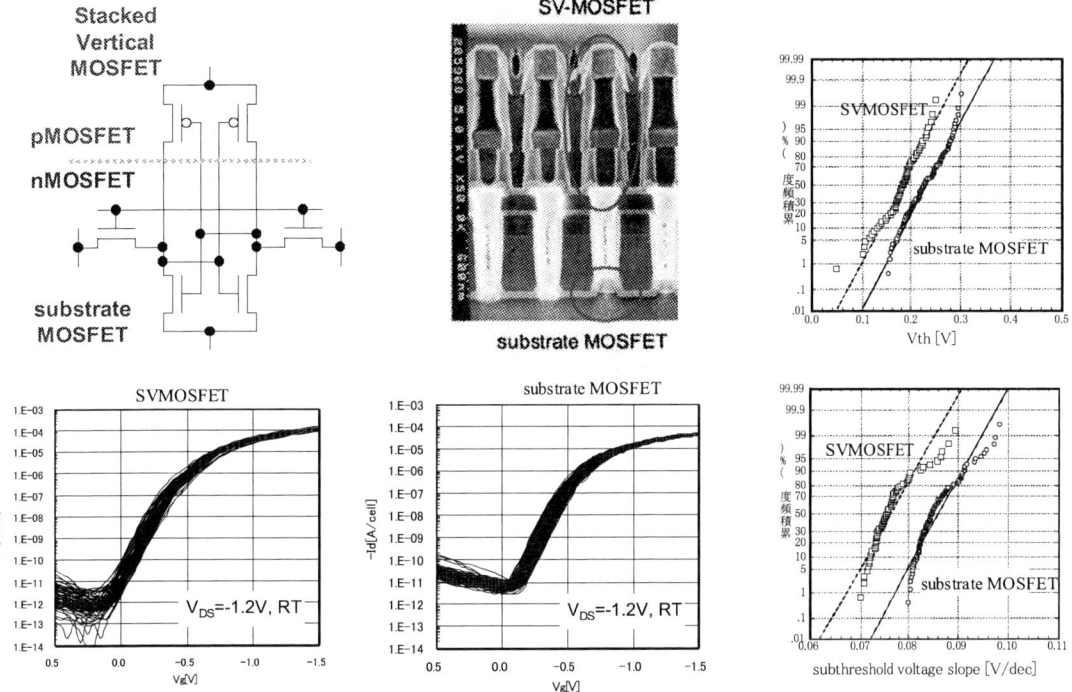

Fig. 1 Stacked SRAM cell using SV(stacked vertical)-MOSFET

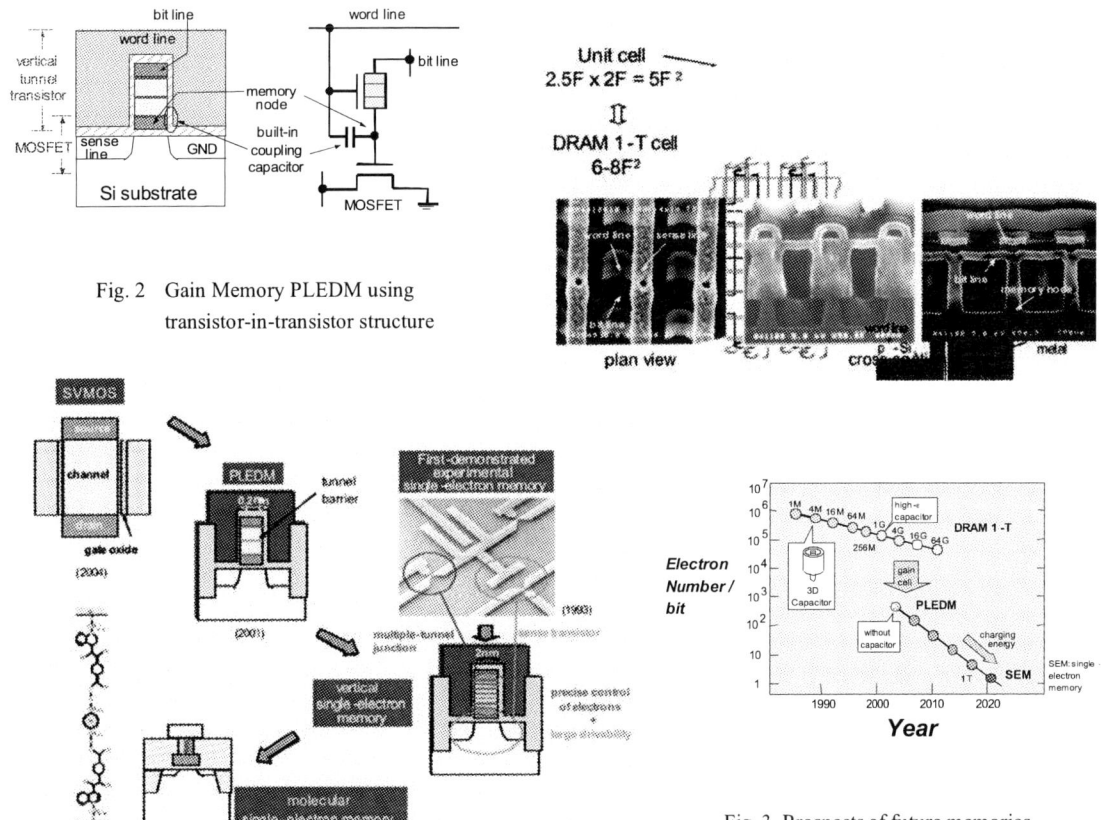

Fig. 2   Gain Memory PLEDM using
transistor-in-transistor structure

Fig. 3 Prospects of future memories

2006 International Symposium on VLSI Technology, Systems, and Applications

Analysis of Electron and Hole Distributions on Scaled NBit Flash Cells

I.C.Yang, K.P. Chen, Y.W. Chang, and T.C. Lu
Macronix International Co., Ltd.
No. 16, Li-Hsin Road, Science Park, Hsinchu, Taiwan 300
E-mail: icyang@mxic.com.tw

ABSTRACT

Charge profiling methodology, re-building the localized charge profile of NBit cell in TCAD environment by comparing I_D-V_G and GIDL curves, is exercised to disclose the distributions of injected electrons and holes after P/E cycle. The mismatch between the profiles of electrons and holes is actually observed. Besides, the difference of injected electrons between long and short channel devices are also examined by the same way. It is observed that as the cell length is scaled down, fewer electrons are needed to achieve the same ΔV_T and the distribution of injected electrons is narrower.

INTRODUCTION

NBit flash EEPROM cell [1][2] that utilizes localized charge trapping in nitride dielectric has attracted much attention due to easy integration for embedded application and two-bit storage capability. The two-bit storage of single cell is achieved by using hot electron programming and band-to-band hot hole erasing with reverse read scheme [1]. For NBit cell's two-bit operations, the location and distribution of the localized electrons and holes play the very important roles. It is helpful if we can extract the profiles of injected electrons and holes after each program and erase operation.

In this paper, we extracted the profiles of injected electrons and holes after one P/E cycle by charge profiling approach [3], which can uniquely re-build the localized charges in TCAD environment by making the simulated I_D-V_G curves and GIDL currents to be coincided with the measured I-V curves of programmed or erased NBit cell. It is the first time to extract the profile of injected holes during erase operation and concretely disclose the mismatch between programmed electrons and erased holes. Such mismatch is suspected to deteriorate after numerous P/E cycles and degrade erase capability or erase speed of NBit flash cell [4]. Besides, we also investigate into the change of programmed charge profiles of NBit cells with decreasing channel length. It is found that the programmed charge profile gets more localized with decreasing channel length. Such change indeed benefits the future scaling of NBit technology.

MISMATCH BETWEEN ELECTRON AND HOLE PROFILES

In this experiment, MXVAND cell [2], our NBit cell, has top-oxide/nitride/bottom-oxide (ONO) thickness of 9/5.4/5.5nm. Cell with gate length 0.32μm is firstly programmed to reach ΔV_T around 2V. And then, the device is erased back by applying $V_G = -4V$, floating V_S and stepping V_D from 4.6 to 5.2V. Four erase shots are done and ΔV_T, which is compared to the initial state, is 0.9/0.34/0.13/-0.01V respectively after each erase shot. The I_D-V_G curves at each program/erase level are shown in Fig.1 (a). Furthermore, GIDL currents at different program/erase levels are also recorded as shown in Fig.1 (b). By fitting both I_D-V_G and GIDL curves, the charge profiles can be uniquely re-built in simulated cell. The extracted electron profile after programmed distributes 200Å over channel and 200Å over junction as shown in Fig.2.

With the existence of the programmed electrons, the profiles of injected holes after each erase shot are also successfully re-built by charge profiling as shown in Fig.3. It is found that, at the first erase shot, the injected holes have already spread wider than the

programmed electron profile to reach 250Å over channel. The profile gets wider in the following three erase shots. When the cell is erased back to initial state, the distribution of injected holes over channel reaches about 300Å. Compared to the profile of programmed electrons, the profile of the injected holes is lower but wider. Fig.4 shows the net charges after neutralized. The mismatch between injected electron and hole profiles make residual holes distribute above the channel region that is about 150Å far from the junction; and residual electrons distribute above the junction. It is the first time to concretely describe the mismatch. Such mismatch may get much worse after numerous P/E cycles and degrade the performance of NBit flash cell.

PROGRAMMED CHARGE PROFILES WITH DECREASING CHANNEL LENGTH

As to the investigation into the programmed charge profiles of NBit cells with decreasing channel lengths from 0.32μm to 0.18μm, programming of all devices are performed at drain side by $V_{GS}=10V$ and stepping V_D from 3.7 to 4.5V with $\Delta V_D =0.2V$ to reach ΔV_T around 2.3V. Charge profiling for each cell is done by well-matched I_D-V_G and GIDL curves, which are not shown here. Fig. 5 shows the extracted charge width and injected electron number. It is observed that the total charge width is reduced and it mainly comes from the narrowing charge width above the channel region. For 0.18μm NBit cell, the distribution of programmed charges above channel is as narrow as 100Å. It is indeed the good news for future scaling of NBit technology.

Fewer electrons to reach the same ΔV_T for the NBit cell with reduced channel length is also observed. It can be clearly interpreted by simulation. By placing the same programmed charge in long and short length device and being read in the reverse direction as shown in Fig.6, one can see that potential difference between program and initial state is larger in shorter one and so the larger program ΔV_T is obtained. Thus, fewer charges are required for shorter length cell to achieve the same program ΔV_T.

Fig.7 and Fig.8 show the growth of the programmed charges in the long (Lg=0.32μm) and short (Lg=0.20μm) length devices under different programming shot numbers. It can be found that the charges are injected into the channel region in the beginning, and then distribute wider gradually with increasing programming shots. The density of the injected electrons in shorter cell is higher than in longer one because of the stronger lateral electric field. Besides, the target ΔV_T value is achieved by fewer programming shots. Consequently, a narrower charge shape with higher peak value is formed in the short channel cell.

CONCLUSION

By charge profiling methodology, it is successful to disclose the distributions of injected electrons and holes after one P/E cycle operation. The mismatch between the profiles of electrons and holes is observed and it is mainly caused by wider erased hole distribution than programmed electron profile. Besides, by extracting the charge profiles of cells with different lengths, it is found that fewer electrons are injected and a more localized electron profile is formed for the cell with scaled channel length to achieve the same ΔV_T.

1-4244-0181-X/06/$20.00 ©2006 IEEE

REFERENCES

[1] B. Eitan, P.Pavan, I. Bloom, E. Aloni, A. Frommer, and D.Finzi, "NROM: A novel loaclized trapping, 2-bit nonvolatile memory cell," *IEEE Electron Device Lett.*, vol.21, pp.543-545, Nov.2000

[2] W. J. Tsai, N. K. Zous, C. J. Liu, C. C. Liu, C. H. Chen, T. H. Wang, S. Pan, C. Y. Lu, and S. H. Gu, "Data retention behavior of a SONOS type two-bit storage Flash memory cell," in *IEDM Tech. Dig.*, pp.32.6.1-32.6.4.

[3] E. Lusky, I. Bloom, and B. Eitan, "Investigation of the spatial distribution of CHE injection utilizing the subthreshold slope and the Gate induced drain leakage (GIDL) characteristics of the NROM device," in *Proc. Non-Volatile Semiconductor Memory Workshop,* 2003, pp. 48-49.

[4] W. J. Tsai, N. K. Zous, M. H. Chou, Smile Huang, H. Y. Chen, Y. H. Yeh, M. Y. Liu, C.C. Yeh, T. Wang, Joseph Ku, and Chih-Yuan Lu, "Cause of erase speed degradation during two-bit per cell operation of a trapping nitride storage flash memory cell," in *IRPS Tech. Dig., 2004.*

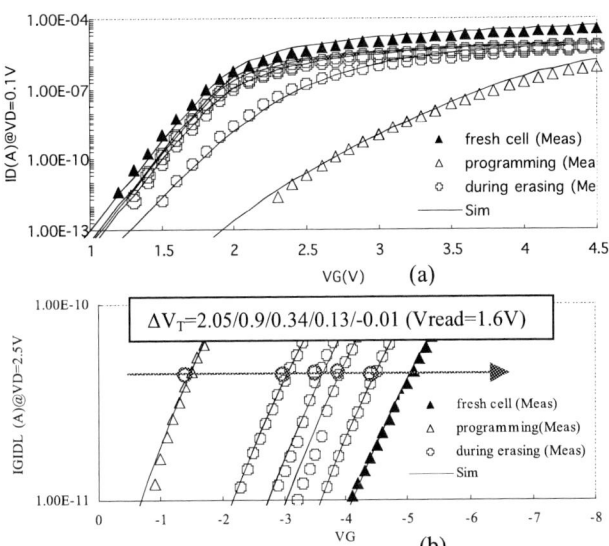

Fig.1 Fitting of (a)I_D-V_G and (b)GIDL curves in the progress of 1 P/E

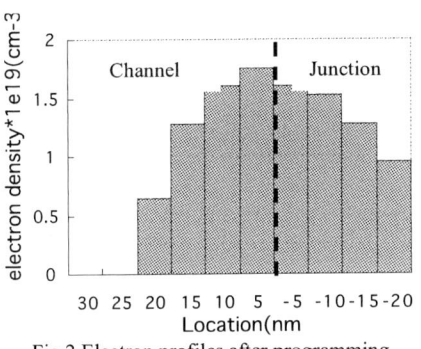

Fig.2 Electron profiles after programming.

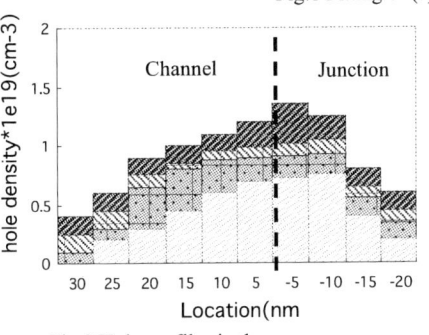

Fig.3 Hole profiles in the erase progress.

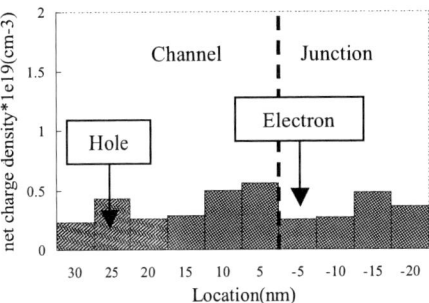

Fig.4 Net charge profile after 1 P/E cycle.

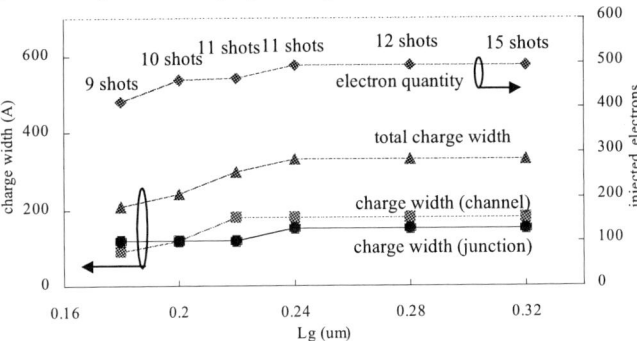

Fig.5 Programmed electron quantity and charge width versus cell length.

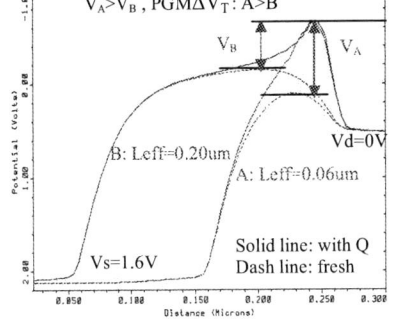

Fig.6 Comparison of surface potential of long and short length cells with the same drain side charge.

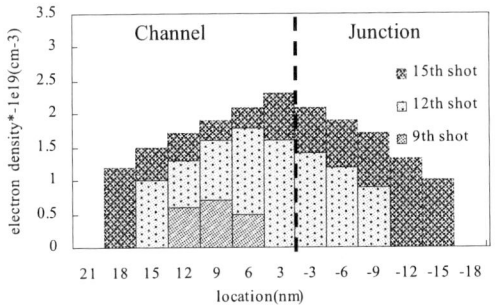

Fig.7 Electron profile in the programming progress of long length cell.(Lg=0.32um)

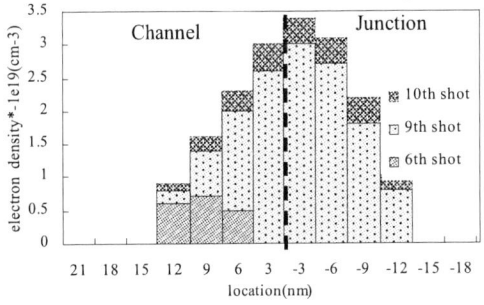

Fig.8 Electron profile in the programming progress of short length cell.(Lg=0.20um)

P-channel SONOS Transient Current Modeling for Program and Erase

Pei-Ying Du, Jyh-Chyurn Guo, H.M. Lee*, H. M. Chen*, Rick Shen*, and C.C.-H. Hsu*
E-mail: jcguo@mail.nctu.edu.tw, Tel : 886-3-5131368, Fax : 886-3-5724361
Institute of Electronics, National Chiao-Tung University, 1001 Ta-Hsueh Rd, Hsinchu, Taiwan, R.O.C.
*eMemory Technology Inc., 3F, No.12, Li-Hsin Rd. 1, Science-based Industrial Park, Hsinchu, Taiwan, R.O.C.

ABSTRACT — Transient current models with time and field dependence are proposed. The time dependence follows asymptotic t^{-1} behavior with slower tunneling rate for erase (ERS) than program (PGM). The field dependence follows FN tunneling in higher field for PGM and intermediate field for ERS with relatively higher corner field for saturation. The models have been justified for P-channel SONOS (P-SONOS) with splits of ONO scheme.

I. INTRODUCTION

NVM devices in SNOS/MNOS family have been investigated for more than 30 years. Recently, SONOS becomes a hot topic and increasing publications report the study on field dependence of PGM and ERS currents [1-3]. However, relatively fewer studies address the transient current behavior and the time dependence of current through the composite dielectric structure. In this work, we derive time dependent fields in bottom oxide (B.O.) for both P-SONOS based on band structures and transient V_T behavior under PGM and ERS. Transient current models of time and field dependence are subsequently developed with simple formulas of physics-based parameters. The analytical model of physical and structure parameters can help on SONOS cell device design for PGM/ERS speed optimization.

II. CELL STRUCTURE & OPERATION SCHEME

P-SONOS cell devices with P⁺ gate were fabricated by 0.18μm CMOS technology with various ONO thicknesses summarized in Table 1.The B.O. and T.O. were formed by thermal oxidation on Si substrate and nitride, respectively. Fowler-Nordheim tunneling (FNT) is utilized for both PGM and ERS with bias conditions shown in Table 2.

III. RESULTS AND DISCUSSION

A. PGM & ERS Speed

Fig.1 shows the ERS characteristics of P-SONOS with B.O. of 32 Å in which regions I, II, and III distinguished by apparently different ERS speeds are interpreted by the band diagram in Fig.2. The electron tunneling from nitride traps to Si substrate contributes the primary source for ERS (regions I, II), while the secondary source from gate through T.O. to nitride acts as a competing factor leading to saturation (region III). Fig.3 shows the T.O. thickness effect on PGM speed for P-SONOS with fixed B.O. and nitride thickness. Obvious degradation of PGM speed is identified corresponding to T.O. thickness ($t_{T.O.}$) increase from 62Å to 70Å.

B. Transient Current Model – Asymptotic t^{-1} Behavior

It is recognized that trap location is one of primary factors affecting the field distribution and current transport. The proposal of trap location at nitride/T.O. interface corresponding to T.O. formation by thermal oxidation on nitride [1] is validated by a universal tunneling characteristics through our study on P-SONOS. Based on the validated trap location, the transient current can be calculated by $J=(dV_t/dt)\times\varepsilon_0\varepsilon_{ox}/t_{T.O.}$. Fig.4 shows the transient currents for P-SONOS under PGM with fixed $t_{B.O.}$ and t_{SiN} but different $t_{T.O.}$ at 62Å and 70Å. A universal curve governed by model of $J(t)=1\times10^{-7}$ A-s/cm² $\times t^{-1}$ is identified for all the transient currents under various bias conditions ($V_{gd}=V_g-V_d$) to follow in long enough time, i.e. an asymptotic t^{-1} behavior. Regarding the ERS condition, all the transient currents shown in Fig.5 corresponding to the mentioned splits of $t_{T.O}$ follow the asymptotic t^{-1} model by $J(t) =3\times10^{-7}$ A-s/cm² $\times t^{-1}$ in time domain of around 10ms~1s. The good match with the transient current by asymptotic t^{-1} model in our study can be

explained by the formulas derived by *H. Bachhofer, et al* [4] in which $J=1/[(\beta\gamma)(t+1/J_0\beta\gamma)]$ can be approximated by $J \doteq (1/\beta\gamma)\times t^{-1}$ at long enough time (t>>1/$J_0\beta\gamma$), β is the tunneling probability through bottom oxide, $\gamma= 1/(k_1C_{ONO})$ and $k_1 \doteq 1.55$. The larger $1/\beta\gamma$ for ERS (3×10^{-7}A-s/cm²) as compared with PGM (1×10^{-7}A-s/cm²) suggests the smaller β for erasing the trapped electrons in nitride than that for programming the free electrons in Si substrate.

C. Transient Current Model – Field Dependence

The band diagram in Fig.6 was proposed for SONOS to define the barrier heights Φ_1 and $\Phi_2+\Phi_t$ responsible for electron injection from substrate (PGM) and ejection from nitride traps (ERS), respectively. $q\Phi_t$ is the mean energy level of the trap centroid from the nitride conduction band edge. Fig.7(a) presents the field dependence of injection current J vs. $E_{B.O.}$ during PGM under various V_{gd}. The field in B.O. is derived by $E_{B.O.}= (|V_G|\pm\Delta V_t)/t_{ONO}$ ("–" for PGM,"+" for ERS) where $\Delta V_t \equiv V_t-V_{t,i}$, $V_{t,i}$ is the initial threshold voltage with neutral traps, t_{ONO} is the effective oxide thickness of the ONO films. There exist two regions for every curve under different V_{gd} in which J vs. $E_{B.O.}$ follows FNT in higher field region given by eq.(1) [4] but dramatic drop in lower field region. Regarding ejection currents during ERS, J vs. $E_{B.O.}$ as shown in Fig.8(a) reveal three regions of current transport featured by initial high ejection, FNT in very narrow region and followed by dramatic current drop at relatively higher $E_{B.O.}$ than condition of PGM. The dramatic current drop is corresponding to region III in Figs.1 and 2 due to gate injection as a competing component. The FNT current given by eq.(2) [4] due to electron ejection from nitride traps can provide a unified curve to cover every narrow FNT region under different V_{gd}. In this study on P-SONOS, $q\Phi_t$ of 1.5eV is extracted to get best match with our experimental results.

$$J = \frac{m_0}{m_{ox}}\frac{q^3}{16\pi^2\hbar}\times E_{ox}^2\frac{1}{q\Phi_1}\exp[-\frac{4\sqrt{2m_{ox}}(q\Phi_1)^{3/2}}{3q\hbar E_{ox}}] \quad (1)$$

$$J = \frac{m_0}{m_{ox}}\frac{q^3}{16\pi^2\hbar}\times E_{ox}^2\frac{1}{q(\Phi_2+\Phi_t)}\exp[-\frac{4\sqrt{2m_{ox}}[q(\Phi_2+\Phi_t)]^{3/2}}{3q\hbar E_{ox}}] \quad (2)$$

The spread of J-E curves under various V_{gd} suggests that barrier heights Φ_1 and $\Phi_2+\Phi_t$ are field dependent. Assuming field-dependent barrier heights $\Phi_1(V_{gd})$ and $\Phi_2+\Phi_t(V_{gd})$ in Fig.7(b) and Fig.8(b) for PGM and ERS, respectively, precise fitting to individual FNT current under different V_{gd} can be achieved. $\Phi_1(V_{gd})$ just follows linear curve with $\Phi_1=3\sim3.15V$ for $t_{T.O.}= 62Å$ and $\Phi_1=2.95\sim3.1V$ for $t_{T.O.}=70Å$. Similar results are demonstrated for ERS shown in Fig.8(b) with good linear $(\Phi_2+\Phi_t)(V_{gd})$ but lower values in the range of 2.49~2.61V for $t_{T.O.}= 62Å$ and 2.44~2.56V for $t_{T.O.}= 70Å$.

IV. CONCLUSION

Transient current models have been extracted with time and field dependence and proven for P- SONOS. The time dependence follows asymptotic t^{-1} behavior with slower tunneling rate (β) for ERS than PGM. The field dependence follows FNT in higher field for PGM and intermediate field for ERS with relatively higher corner field for saturation. The models are useful for SONOS cell device design in terms of PGM and ERS speed optimization.

V. REFERENCES

[1] *Hang-Ting Lue, et al.*, IEEE EDL- 25, p.816, 2004.
[2] *Anirban Roy, et al.*, IEEE ED-37 p.1054, 1990.
[3] *Jyh-Kuang Lin, et al.*, JJAP, 32, p.2748, 1993.
[4] *H. Bachhofer, et al.*, JAP, 89, p.2791, 2001.

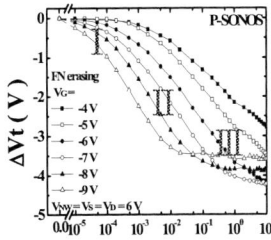

Fig.1 ERS characteristic of P-SONOS device w/ O/N/O (B.O./ Nitride / T.O.) = 32Å/54Å/50Å

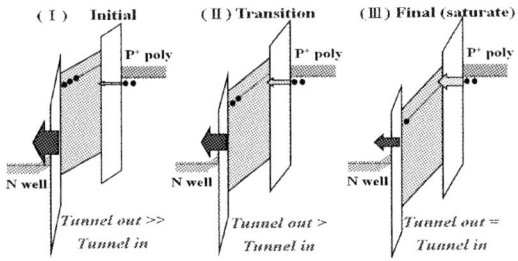

Fig.2 Band diagram of P-SONOS device during ERS process

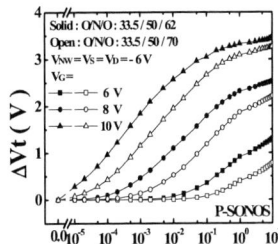

Fig.3 PGM characteristic of P-SONOS device. Thicker top oxide results in slower PGM speed.

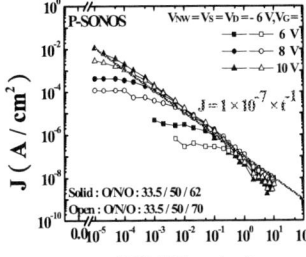

Fig.4 Current density as a function of PGM time for P-SONOS device. The current density at extended long PGM time follows the inverse t model given by $J = 1 \times 10^{-7}$ As/cm$^2 \times t^{-1}$.

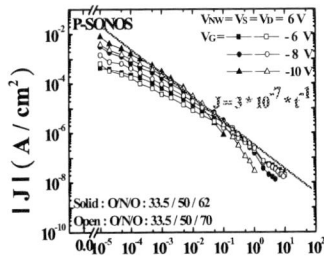

Fig.5 Current density as a function of ERS time of P-SONOS device. The current density in medium ERS time follows the inverse t model given by $J = 3 \times 10^{-7}$ A-s/cm$^2 \times t^{-1}$.

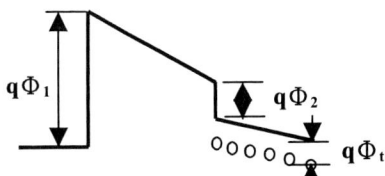

Fig.6 Tunneling band diagram and fitting barrier height parameters

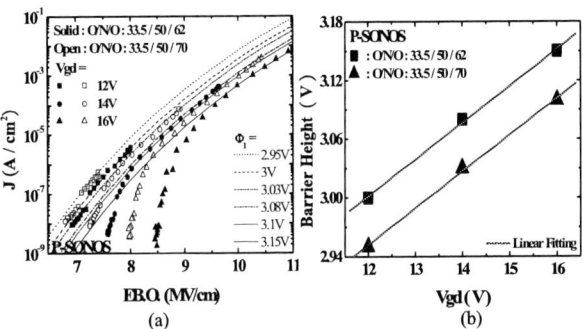

Fig.7 (a) J vs. $E_{B.O.}$ curves and (b) fitting barrier height as a linear function of Vgd of P-SONOS device during PGM.

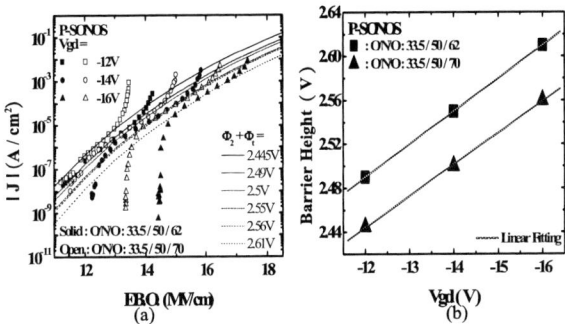

Fig.8 (a) J vs. $E_{B.O.}$ curves and (b) fitting barrier height as a linear function of Vgd of P-SONOS device during ERS.

Device Type Operation	P-SONOS		
Bottom Oxide (Å)	32	33.5	33.5
Nitride (Å)	54	50	50
Top Oxide (Å)	50	62	70

Table 1 ONO thickness of the measured devices

Device Type Operation	P-SONOS
PGM (Program) Electron injection from substrate to SiN	(+) P⁺ Poly (-) (-) S D N-sub (-)
ERS (Erase) Electron ejection from SiN to substrate	(-) P⁺ Poly (+) (+) S D N-sub (+)

Table 2. F-N operation scheme of P-SONOS device

2006 International Symposium on VLSI Technology, Systems, and Applications

Understanding of the Leakage Components and Its Correlation to the Oxide Scaling on the SONOS Cell Endurance and Retention

C. H. Chen[1], P. Y. Chiang[1], Steve S. Chung[1], Terry Chen[2], George C. W. Chou[2], and C. H. Chu[2]

Department of Electronic Engineering, National Chiao Tung University, Hsinchu, Taiwan
[2]Promos Technologies Inc., Science-based Industrial Park, Hsinchu, Taiwan

Abstract- In this paper, the ONO layer scaling and the leakage components in a SONOS cell have been extensively studied. The reliability with focus on both endurance and data retention has been demonstrated. Results have shown that the cell with thinner blocking oxide has better endurance, while it has poorer data retention. However, this can be achieved by a good control of the oxide quality. In terms of the data retention, thermionic and direct tunneling, in relating to the charge loss, are the two dominant leakage components, which can be separated. Moreover, after cycling, we can separate another third leakage component, the trap-to-trap tunneling induced leakage. These results are useful toward an understanding of the leakage mechanisms of SONOS cell as well as the scaling design of ONO layers.

1. Introduction

SONOS cell is simple in structure [1] and offers flexibility for integration with CMOS process by comparing to conventional floating gate cells. Therefore, it has aroused much more interest in the recent years. For the SONOS cell, the endurance and data retention are the two most important reliability issues. Scaled SONOS can be operated with a lower bias; however, data retention has been critical for the scaling of ONO layers. On the other hand, for a thin tunnel oxide SONOS cell, two leakage current components, Frankel-Poole thermionic emission and direct tunneling (DT) [2-5], are believed to be the two dominant leakages for retention. After the cycling, another leakage term, trap-assisted tunneling, occurs. To differentiate these components, it has become very important for an understanding of the ONO layer scaling and cell reliabilities. In this paper, the dominant mechanisms of the above leakages and the scaling effect of ONO layer, in relating to the cell reliabilities, will be extensively studied.

2. Device Preparation

The cells used in this study are shown in Fig. 1(a), with W/L=1.4/0.6(um) and n-poly gate. The split condition of ONO includes thickness of the tunnel oxide 20~25 Å, 54 Å nitride, and 34~65 Å blocking oxide. Tunnel oxide is grown by thermal oxidation with N_2 diluted (1%) at 850°C, followed by LPCVD SiN film with SiH_2Cl_2/NH_3= 1:10 at 790°C and finally the LPCVD grown blocking oxide with SiH_2Cl_2/NH_3= 1:10~11:20 at 725~780°C. In the experiments, we use FN for programming and erase, while source, substrate and drain are grounded.

3. Results and Discussion

A. The Charge Loss Components

Fig. 2 shows the three leakage components for an operation of a SONOS cell. They are thermionic emission, component (1), DT (Direct Tunneling) through either blocking oxide or tunnel oxide, component (2) and (3), and trap-assisted tunneling, component (4). The separation of these leakages has been made by a special technique and with a set of samples with different ONO layer

thickness. Fig. 3 is the data retention with ONO layer of 25/54/65 (tunnel/SiN/blocking-oxide) at a wide range of temperatures. Fig. 4 can be used to separate the DT and thermionic leakages since at sufficient low temperature, the leakage due to DT is described by (2)+(3), while the other component, (1), is regarded as the thermionic component. By assuming that these cells have the same DT through tunnel oxide for a large blocking oxide in Fig. 5, DT loss through the blocking oxide can be predicted. In other words, the DT component through blocking oxide, (3), and through tunnel oxide, (2) can be further differentiated.

B. The Selection of ONO layer Thickness

Fig. 6 is the endurance of two cells with different tunneling oxide thickness, in which thinner tunneling oxide cell shows better endurance characteristic. If the thickness of the tunnel oxide is scaled, most carriers will directly tunnel through the oxide without generating damage. The thicker the tunnel oxide is, the more damage is accumulated in the oxide, as can be verified from the charge pumping measurements in Fig. 7. The endurance of cells with various blocking oxide thickness is shown in Fig. 8. It shows that the degradation depends on blocking oxide thickness and exhibits poorer endurance for thicker blocking oxide cell. It is because the injected electrons is easier to bombard the blocking oxide for a thinner tunneling oxide and the thicker blocking oxide has a larger capacity to accumulate damage. In short, to achieve a better endurance, it is better to choose a thinner tunnel oxide and a thinner blocking oxide thickness. Furthermore, we can plot the ΔQ versus oxide thickness to be a criterion to consider the data retention in Fig. 9, from which we can choose a combination of tunnel oxide and block oxide thickness to achieve our target.

C. Further Examination of the Data Retention

Fig. 10 shows the data retention of the cell for different tunnel and blocking oxide thicknesses. Thinner tunnel oxide has a poorer retention ability. Also, thin blocking oxide exhibits poorer data retention ability. We can further separate the trap-to-trap tunneling after P/E cycles, as shown in Fig. 11. For a better understanding of the excess oxide damage generated after P/E cycles, the data decay rate is plotted in Fig. 12. It shows that thicker block oxide thickness may trap more oxide charge and cause poorer retention ability. The larger the blocking oxide thickness is, the poorer is its data decay rate. Although the thicker blocking oxide thickness provides a better data retention in fresh cells, while after 10^3 P/E cycles, device degrades more seriously. Thus, it is a trade-off between retention and P/E cycles.

In summary, two major reliability monitors, the endurance and data retention, of SONOS cell have been extensively evaluated. The scaling effect of the ONO layers has also been studied. The three dominant leakage current components can be totally separated with a unique separation technique. The design window for choosing appropriate tunnel oxide and blocking oxide thicknesses (Fig. 9) has also been proposed. It can be easily fitted into your design of SONOS technology through the proposed experimental procedures.

1-4244-0181-X/06/$20.00 ©2006 IEEE

Acknowledgments This work was supported in part by the National Science Council, Taiwan, under grant NSC94-2215-E009-008 and in part by the *Promos Tech. Inc.,* Taiwan.

References

[1] M. White et al., *IEEE Circuits and Devices Magazine,* pp. 22-31, 2000.

[2] Y. Yang and M. H. White, *Solid State Electronics,* vol. 44, pp. 948-958, 2000.

[3] L. Lundkvist, C. Svensson and B. Hansson, *Solid-State Electronics*, vol. 10, pp. 221-227, 1976.

[4] S. S. Chung et al., in *Tech. Dig. IEDM,* p. 617, 2003.

[5] A. Goda and M. Noguchi, *IEEE NVSM Workshop,* p. 65, 2003.

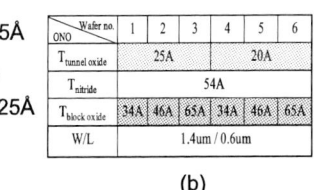

Wafer no. ONO	1	2	3	4	5	6
$T_{tunnel\ oxide}$	25A			20A		
$T_{nitride}$	54A					
$T_{block\ oxide}$	34A	46A	65A	34A	46A	65A
W/L	1.4um / 0.6um					

Fig. 1 (a) The experimental SONOS cell structure with trap rich nitride. (b) The split conditions of samples with different blocking oxide thickness and tunnel oxide thickness.

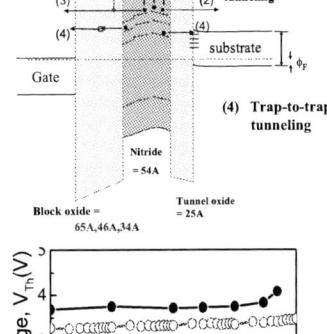

Fig. 2 The charge loss paths and the three dominant leakage components in a SONOS cell,
(1): thermionic emission
(2)+(3) : direct tunneling.
(4): trap-to-trap tunneling

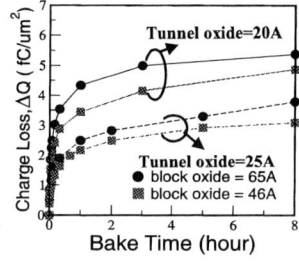

Fig. 9 The relationship between ΔQ and the oxide thickness. It can be used as a criterion of the data retention to find a combination of tunnel and blocking oxide thicknesses.

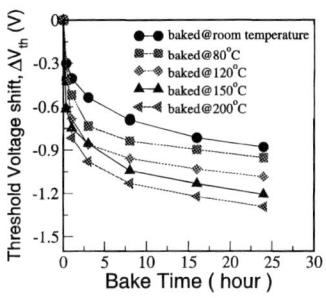

Fig. 3 The data retention of O/N/O = 25/54/65 sample at various temperatures.

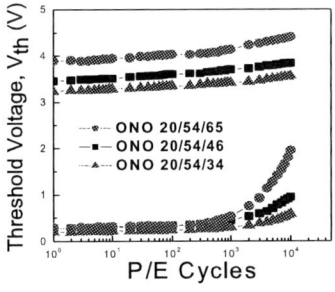

Fig. 6 The endurance of 2 cells with different tunnel oxide thickness, in which thin tunnel oxide shows much better characteristics.

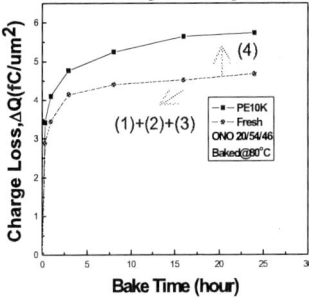

Fig. 10 The calculated charge loss versus bake time for different tunnel oxides. Thin tunnel oxide cell shows much more charge loss as expected.

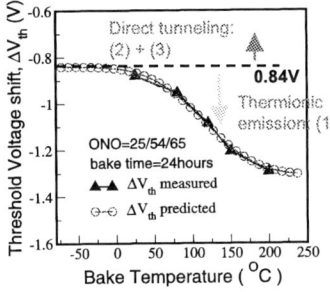

Fig. 4 The measured (triangle) and predicted (circles) ΔV_{th} versus temperature plot. Note that the two major leakage components can be differentiated.

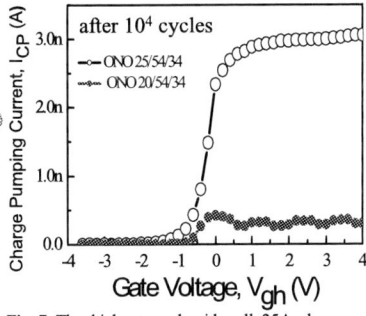

Fig. 7 The thicker tunnel oxide cell, 25A, shows a much larger generated oxide damage after P/E cycles.

Fig. 11 The calculated charge loss versus bake time for different tunnel oxides. Thin tunnel oxide cell shows much larger charge loss as expected.

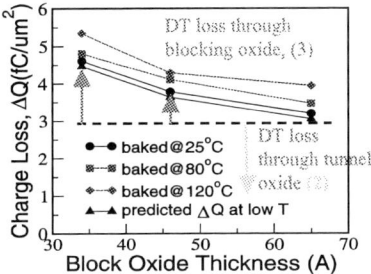

Fig. 5 Further calculations can be done by differentiating the charge loss through tunnel oxide (2) and blocking oxide (3).

Fig. 8 The endurance of three cells with different blocking oxide thicknesses, in which thick blocking oxide shows a very poor endurance.

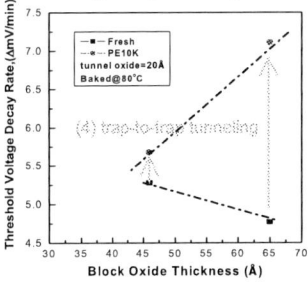

Fig. 12 The calculated decay rate versus block oxide thickness in the first 3 hours for fresh and 10^3 P/E cycles at an elevated temperature 80°C.

2006 International Symposium on VLSI Technology, Systems, and Applications

High-κ Hf-based charge trapping layer with Al_2O_3 blocking oxide for high-density flash memory

S. Maikap[a,*], P. J. Tzeng[a], L. S. Lee[a], H. Y. Lee[a], C. C. Wang[a], P. H. Tsai[b], K. S. Chang-Liao[b], W. -J. Chen[c], K. C. Liu[c], P. R. Jeng[a] and M.-J. Tsai[a]

[a] Electronics Research and Service Organization, Industrial Technology Research Institute, Hsinchu, Taiwan, R. O. C.
[b]Department of Engineering and System Science, National Tsing Hua University, Hsinchu, Taiwan, R.O.C.
[c]Department of Electronics Engineering, Chang Gung University, Tao-Yuan, Taiwan, R.O.C.
[*]Corresponding author: Tel: 886-3-5913913 Fax: 886-3-5917690 E-mail: sidhu@itri.org.tw

ABSTRACT

The high-κ Hf-based charge trapping layer with Al_2O_3 blocking oxide in metal/Al_2O_3/HfO_2/SiO_2/silicon (MAHOS) structure is proposed. The Al_2O_3 as a blocking oxide on high-κ HfO_2 and HfAlO charge trapping layers can improve the program/erase speed and has good retention characteristics, indicating that the MAHOS structure is a promising candidate for future high-speed flash memory. The charge trapping characteristics with different metal gates are also investigated.

INTRODUCTION

The high-κ HfO_2 or Al_2O_3 films are reportedly very interesting for charge storage NVM device applications [1,2]. Due to the retention problem in polysilicon-oxide-silicon-nitride-oxide-silicon (SONOS) flash memory, the high-κ HfO_2 or HfAlO films can be used as trapping layer in the SONOS structure to improve the program/erase speed as well as scaling for future nano-scale flash devices. The high-κ materials with large barrier height such as Al_2O_3 [3] are interesting alternatives as a blocking oxide to improve the vertical scaling, and to increase the electric field across the tunneling oxide resulting a higher program/erase speed. To improve the charge storage and vertical scaling, the Hf-based charge storage layers with Al_2O_3 blocking oxide is proposed for the first time. The charge storage/erase and retention characteristics of the MAHOS structure are investigated. Such a structure with different metal gates is also studied. The pure Al_2O_3 layer in metal/Al_2O_3/SiO_2 (MAOS) structure is also investigated for comparison.

EXPERIMENT

The high-κ HfO_2 or HfAlO charge trapping layers with thickness of 5-10nm were deposited on p-type silicon (100) with resistivity of 15-25 Ω.cm by ALD system using hafnium tetrachloride ($HfCl_4$) and trimethylaluminium ($Al(CH_3)_3$) precursors at substrate temperature of 300°C. The compositional ratio of HfAlO films is (Hf:Al) 1:1. Prior to deposition of Hf-based trapping layer, the p-Si wafers were cleaned by diluted HF dip. Then, 3-4 nm-thick tunneling oxide (SiO_2) was grown by RTO system at the temperature of 1000°C for 20s to 60s. A 10 nm-thick Al_2O_3 as a blocking oxide was deposited in-situ on HfO_2 or HfAlO trapping layer by ALD. The 15 nm-thick Al_2O_3 was deposited on SiO_2 treated p-Si substrate as a control wafer. To increase the charge trapping window, the post deposition annealing (PDA) treatment at 1000°C for 10s in N_2 ambient was used. To study the program/erase speed on memory devices, the Al, Pt and TaN metal gates are used for gate electrodes. The post metal annealing with temperature of 400°C and 5 min was done using N_2 (90%) and H_2 (10%) gases.

MEMORY DEVICE CHARACTERISTICS

Fig. 1 shows the transmission electron microscopy (TEM) image of HfO_2 trapping layer with Al_2O_3 as a blocking layer. After such a high temperature treatment (1000°C, 10s), the Al_2O_3 film can still maintain mostly amorphous. Before deposition of HfO_2 film, the thickness of SiO_2 was 4 nm. After the annealing process, the SiO_2 thickness is reduced (~0.4 nm) due to the reaction at SiO_2/HfO_2 interface. The charge retention characteristics of HfO_2, HfAlO and Al_2O_3 charge

trapping layers are plotted in Fig. 2. The thicker HfO_2 trapping layer has similar retention with pure Al_2O_3 memory devices. The retention is improved with increasing the tunneling oxide thickness. The flat-band voltage (V_{FB}) shift is measured with respect to the V_{FB} of quasi-neutral C-V curve (V_{FBN}). The quasi-neutral V_{FB} (V_{FBN}) is the flat-band voltage from quasi-neutral C-V curve (the sweep gate voltage is V_g~±3V). The V_{FB} shift (Fig. 3) and charge storage density (Fig. 4) of the MAHOS structure are larger as compared with the MAOS memory devices. It can be explained that the conduction band offset of the HfO_2 film with respect to Si substrate is smaller (~1.5 eV) as compared with the one (~2.8 eV) of the Al_2O_3 film. It indicates that more electrons can be injected in the HfO_2 film from Si substrate because the tunneling current is dominated at the same gate voltage. The thick HfO_2 or HfAlO has more trapping sites, resulting the higher V_{FB} shift. It is noted that the higher dielectric constant of Al_2O_3 as a blocking oxide on HfO_2 trapping layer has a great advantage of higher electric field across the tunneling oxide as well as lower electric field across the blocking layer as shown in schematic energy band diagram (Fig. 5). The memory devices with HfO_2 trapping layer shows an excellent charge storage characteristics than that of pure Al_2O_3 layer, indicating that high-κ HfO_2 memory devices with Al_2O_3 as a blocking oxide can be useful for high-speed flash memory.

The metal gate on Al_2O_3 blocking oxide has a great impact to suppress the backward tunneling current. The Pt gate has lower gate current as compared with Al gate (Fig. 6). The backward tunneling current can be reduced in MAHOS memory devices. The Al gate has higher erase voltage (-10V) as compared with Pt gate (-8.2V) due to lower work function of Al gate (Fig. 7), indicating that high work function metal gate is needed to erase the memory devices easily. The programming and erase speeds of HfO_2 charge trapping layer are drastically improved than that of pure Al_2O_3 memory devices with TaN (work function: 4.15eV) metal gate (Fig. 8 & 9). A good endurance of HfO_2 memory devices is also observed (Fig. 10). An excellent retention is observed on 10 years projection for 10 nm-thick HfO_2 charge trapping devices (Fig. 11). The memory window is found to be ~4.2V. The thick Hf-based charge trapping layer has higher retention than that of thin HfO_2 memory devices (Fig. 12). The normalized memory window of 10nm thick HfO_2 layer has been improved (slightly) as compared with 10 nm-thick HfAlO charge trapping layer and it may be due to some unwanted defects in HfAlO films. It is noted that the retention is further improved with thicker (slightly) tunneling oxide.

CONCLUSION

The MAHOS memory devices using Hf-based charge trapping layer with Al_2O_3 as blocking oxide have good program/erase, endurance and retention characteristics. Therefore, SiO_2/HfO_2/Al_2O_3 structure paves a way in future high-speed flash memory device applications.

REFERENCES

[1] T. Sugizaki, et al., Symp. on VLSI Tech., p. 27, 2003.
[2] Y. N. Tan, et al., IEDM Tech. Dig., p. 889, 2004.
[3] C. H. Lee, et. al., App. Phys. Lett. Vol. 86, p.152908,2005.

1-4244-0181-X/06/$20.00 ©2006 IEEE

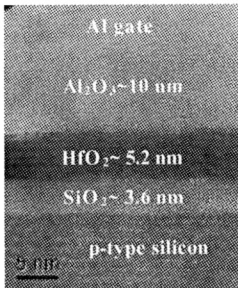

Fig. 1 The cross-sectional TEM image of high-κ HfO$_2$ film with Al$_2$O$_3$ as a blocking oxide, and a 4-nm thick SiO$_2$ is used nominally. The Al$_2$O$_3$ film shows almost no crystallization.

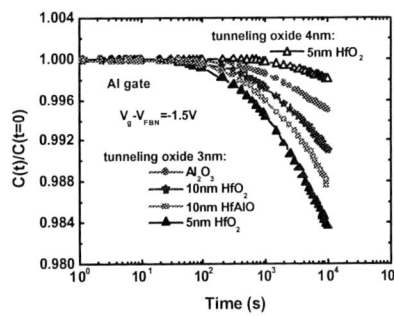

Fig. 2 Normalized discharge characteristics of memory devices with Al$_2$O$_3$ and HfO$_2$ charge trapping layers. A gate bias (V_g-V_{FBN}) of –1.5V is used below the flat-band voltage during the measurement.

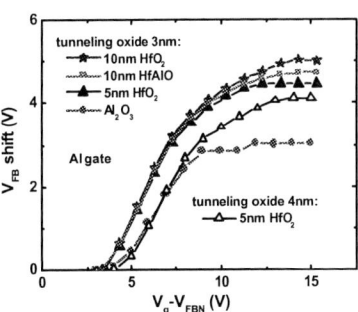

Fig. 3 Flat-band voltage shift is calculated from C-V hysteresis with pure Al$_2$O$_3$ and Hf-based charge trapping memory devices with Al$_2$O$_3$ as a blocking oxide.

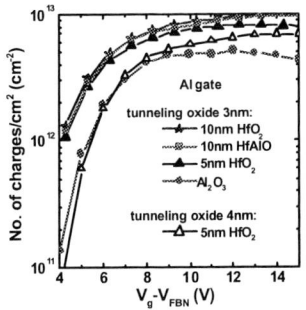

Fig. 4 The densities of charge storage in Al$_2$O$_3$ and high-κ Hf-based charge trapping layers with Al$_2$O$_3$ blocking layer are plotted. The charge storage density is calculated from C-V hysteresis curve. The thick charge trapping layer has high charge storage density.

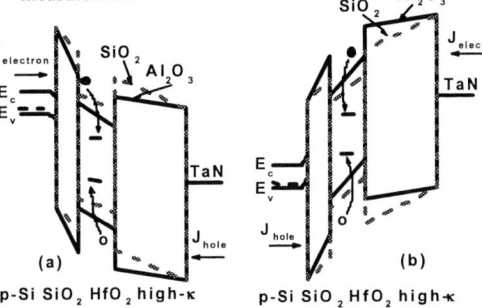

Fig. 5 A schematic energy band diagram of Al$_2$O$_3$ blocking oxide on high-κ HfO$_2$ charge trapping layer with SiO$_2$ as a tunneling oxide. The electric field across the SiO$_2$ tunnel oxide will be increased due to the high-κ Al$_2$O$_3$ blocking layer, and thick Al$_2$O$_3$ blocking layer will also suppress the gate leakage current as a comparison with SiO$_2$ blocking layer with the devices (a) program mode and (b) erase mode.

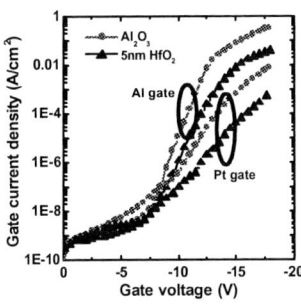

Fig. 6 The HfO$_2$ charge trapping layer with Al$_2$O$_3$ as a blocking layer has lower gate current as compared with pure Al$_2$O$_3$ layer at high field. At low field (>10V), the gate current for all memory devices are same. Using high work function metal (Pt), the gate current is further suppressed.

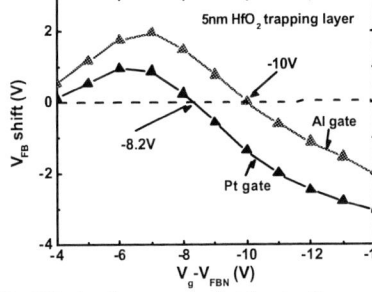

Fig. 7 During the erase, the flat-band voltage shift is higher for high work function metal (Pt). It is corroborating with the suppression of backward tunneling current in Fig. 6.

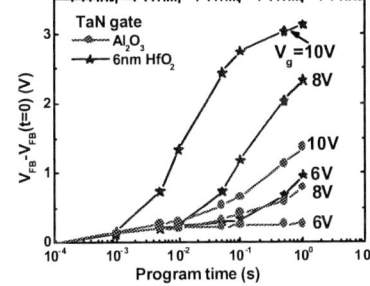

Fig. 8 Programming characteristics of MAHOS and MAOS memory devices with HfO$_2$ and Al$_2$O$_3$ charge trapping layers and TaN metal gate with different gate voltages. The high-κ HfO$_2$ film shows higher programming speed.

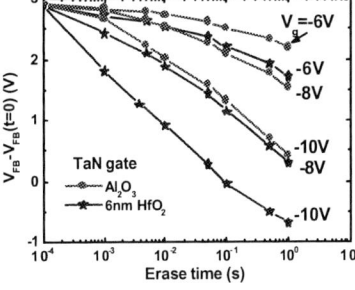

Fig. 9 Erasing characteristics of MAHOS and MAOS memory devices with HfO$_2$ and Al$_2$O$_3$ charge trapping layers with different gate voltages. The high-κ HfO$_2$ film has good erasing speed.

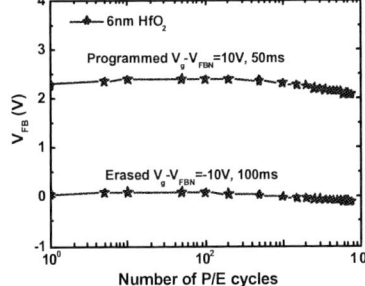

Fig. 10 Endurance characteristics of MAHOS memory devices with HfO$_2$ charge trapping layer and TaN metal gate.

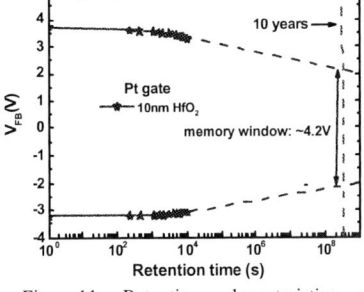

Fig. 11 Retention characteristics of 10nm-thick HfO$_2$ charge trapping layer with Al$_2$O$_3$ as a blocking layer. A promising window is observed after 10 years.

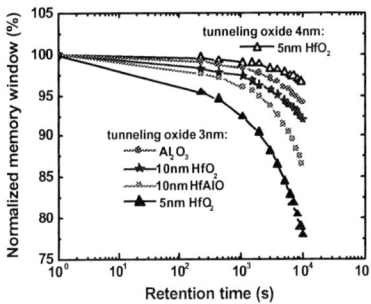

Fig. 12 Retention characteristics of MAHOS devices with HfO$_2$ and HfAlO charge trapping layers. The thick trapping layer or thick tunneling oxide has improved retention.

37

2006 International Symposium on VLSI Technology, Systems, and Applications

V_{DD} Scaling for FinFET Logic and Memory Circuits: the Impact of Process Variations and SRAM Stability

C.-H. Lin[1], K. K. Das[2], L. Chang[2], R. Q. Williams[3], W. E. Haensch[2], and C. Hu[1]

[1] Department of EECS, University of California at Berkeley, CA 94720, USA

[2] IBM Research Division, T. J. Watson Research Center, Yorktown Heights, NY 10598, USA

[3] IBM Microelectronic Division, Essex Junction, VT 05452, USA

E-mail: chl@eecs.berkeley.edu / Tel: +1-510-6432638/ Fax: +1-510-6432636

Introduction

As CMOS technology is fast moving towards the scaling limit, the FinFET is considered as the most promising structure down to 22nm node [1-2]. Both FinFET-based logic and SRAM have been demonstrated recently [3-4]. However, with scaling of the device dimensions, process-induced variations cause an increasing spread in the distribution of circuit delay and power, and affecting the robustness of VLSI designs [5]. SRAM has become the focus of technology scaling since embedded SRAM is estimated to occupy nearly 90% of the chip area in the near future [6]. Due to the area-constrained limit, the device fluctuation in the SRAM cell is significant. In this paper, we explore the performance of FinFET technology in digital circuit applications at 90 nm technology node under various device parameter variations. Comprehensive comparison of FinFET vis-à-vis PD-SOI has been done for logic gates as well as memory structures that are most commonly used in commercial VLSI designs. We also compare the performance of these two technologies at ultra-low voltages for future low-power applications.

Evaluation Methodology

The FinFET compact device models have been modified from Compact Model Council (CMC)-standard planar partially-depleted silicon-on-insulator (PD-SOI) model BSIMPD from University of California, Berkeley [7]. This modified model is similar to the model used in the previous study of FinFET SRAM [8]. Fig. 1 shows the schematic of the modified model. Digital static CMOS, transmission gate circuits, and 6-T FinFET SRAM have been investigated. The basic digital circuit structure used for simulation is similar to [9]. Appropriate fan-out (=FO) has been used for each gate type, based on real designs. Transistor design parameters used in this study is similar to [8]. For FinFET technology, the thicker gate oxide (1.5nm) is allowed for lower gate leakage power. FinFET has lower linear threshold voltage (V_{tlin}) and drain-induced barrier lowering (DIBL). The I_{on} of PFET ends up been identical for FinFET and PD-SOI. The I_{on} of N-FinFET is slightly lower. PD-SOI has higher leakage current than FinFET.

FinFET Logic Circuits

Comparison of FinFET and PD-SOI delay under V_{DD} scaling has been shown for a FO4 inverter delay chain in Fig. 2. Same transistor sizes have been used for both FinFET and PD-SOI for all comparisons in this paper. It is observed that at ultra-low voltages, say 0.6 V, FinFET delay is about 38% less compared to PD-SOI. Similar trends are shown by all other logic gates; plots have been omitted for brevity. Since the transmission gate is very sensitive to (V_{DD}-V_{th}), the increased current drive of FinFET (due to lower V_{tlin}) gets reflected in additional delay improvements at ultra-low voltages. All these simulation data clearly demonstrate that FinFET is an ideal candidate for ultra-low voltage logic applications. At high voltages, the current drive of PD-SOI increases due to worse DIBL, which makes the performance of PD-SOI similar to FinFET. Power number comparison of FinFET and PD-SOI under V_{DD}

scaling is shown in Fig. 3. The average power of the FO4 inverter circuit is plotted versus delay under normal switching conditions. Similar trend is shown by all logic gates. For a given circuit performance, FinFET is more power efficient than PD-SOI for ultra-low voltage application.

Since process variation has significant impact on ultra-low voltage circuits, we study the delay sensitivities of the various logic gates under some of the important device parameter variations like L_{eff} (effective channel length), V_{th} and R_{ds} (source/drain resistance). Our simulation results show that L_{eff} variation has similar impact on the performance of all logic gates for both FinFET and PD-SOI since the device design points are similar for both technologies. For brevity, we have included the plot for just the transmission gate in Fig. 4 at V_{DD} of 0.9V. Transmission gate is very sensitive to V_{th} variation since its resistance is determined by (V_{DD}-V_{th}). In Fig. 5, the transmission gate delay sensitivity has been shown for FinFET and PD-SOI as V_{th} is varied from -80 mV to +80 mV of the nominal value. While the delay varies by 229% for PD-SOI, significantly improved stability is observed for FinFET (121%) at ultra-low voltage (V_{DD}=0.6V) due to the lower V_{tlin} (higher V_g-V_t). Similar reduction in delay sensitivity to V_{th} variation has been observed for other types of logic gates; results of NAND3 have been shown in Fig. 6 as an example. Static gates with stacked NFETs/PFETs like NORs and NANDs are most sensitive to the variation of R_{ds}. In Fig. 7, the NAND4 gate delay is studied as R_{ds} is varied from -100 ohms to +100 ohms of the nominal value; the plots show that FinFET has a slightly reduced R_{ds} sensitivity (18%) as compared to PD-SOI (22%) due to lower gate capacitance of FinFET.

FinFET 6T-SRAM Cell

The fundamental stability problem in the 6T-SRAM is caused by a potential disturb via the pass-gate to the "0" storage node in the read condition (Fig. 8). The static noise margin (SNM) in the read condition is investigated for transistor biasing, sizing, and variations. Fig. 9 shows the butterfly curve with various V_{DD} of FinFET 6T-SRAM. FinFET SRAM is very robust at ultra-low voltages. The beta ratio (current drive ratio between pull-down NFET and access NFET) is a quantized number. A decent SNM can be maintained at 0.6 V and tuned by different beta ratios (Fig. 10). However, there is the trade-off between the read stability and cell area. Since the process variations have significant impact on the minimum-geometry devices such as SRAM. V_{th} variation is studied by considering the worst case of variation in the paired pull-down NFETs (means V_{th} varies in opposite directions in the paired pull-down NFET, Fig. 11). The V_{th} variation due to random dopant and T_{si} fluctuations shows very significant impact on the read stability.

Conclusion

FinFET-based logic and memory circuits for ultra-low voltage applications are analyzed in this paper. The paper clearly shows that FinFET is an ideal candidate for ultra-low voltage applications because of its robustness to various device parameter variations. The paper demonstrates that a decent SNM of FinFET SRAM can be maintained at ultra-low voltages under V_{th} variation, which may be caused by random dopant and T_{si} fluctuations.

Acknowledgment

This work is supported by SRC (2002-NJ-1001) and IBM.

Reference

[1] D. Frank et al., IEDM Tech. Dig., p.553, 1992.
[2] X. Huang et al., IEDM Tech. Dig., p.67, 1999.
[3] B. Rainey et al., DRC Tech. Dig., p. 47, 2002.

[4] E. Nowak et al., IEDM Tech. Dig., p.411, 2002.
[5] D. Burnett et al., VLSI Tech. Dig., p. 15, 1994.
[6] ITRS roadmap 2003 (http://public.itrs.net).
[7] BSIMPD MOSFET Model User's Manual.
[8] R. V. Joshi et al., ESSCIRC, p. 211, 2004.
[9] K. K. Das et al., SOI conf., p. 177, 2004.

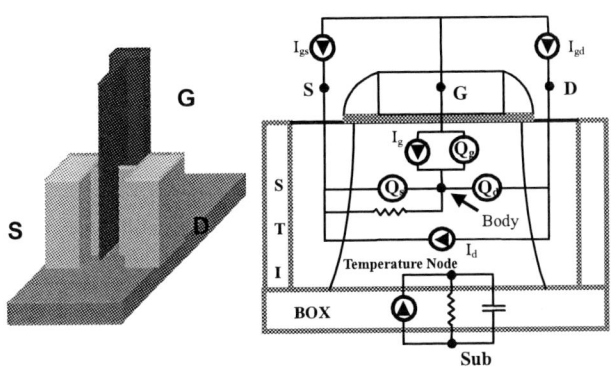

Fig. 1 View of the schematic of FinFET model. The first order FinFET features are included in the model.

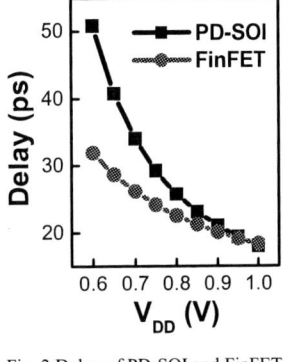

Fig. 2 Delay of PD-SOI and FinFET FO4 inverter chain under V_{DD} scaling. FinFET shows superior performance at ultra-low voltages.

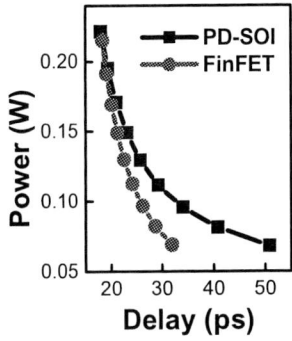

Fig. 3 Power-delay merit of PD-SOI and FinFET FO4 inverter chain. FinFET is more power efficient for given delay at ultra-low voltages.

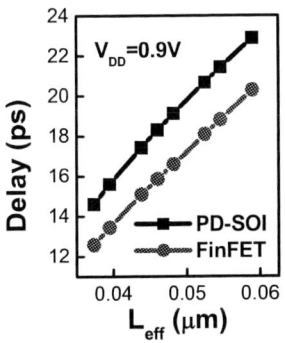

Fig. 4 Transmission gate inverter chain shows similar sensitivity to L_{eff} variation for PD-SOI and FinFET due to fixed device design point.

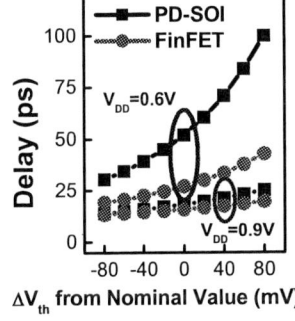

Fig. 5 Transmission gate's sensitivity to V_{th} variation is reduced in FinFET due to lower V_{tlin}

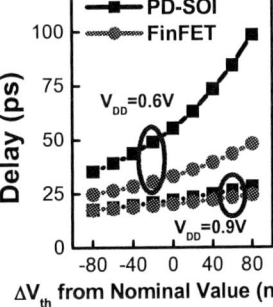

Fig. 6 NAND3 gate's sensitivity to V_{th} variation is reduced in all FinFET logic gates.

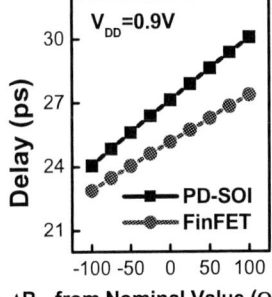

Fig. 7 NAND4 gate's sensitivity to R_{ds} variation is reduced in all FinFET logic gate due to lower gate capacitance.

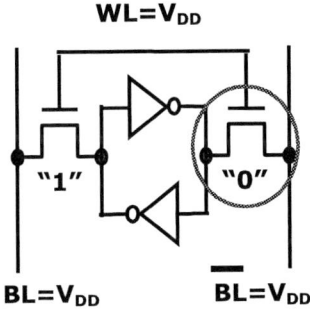

Fig. 8 6T-SRAM under read operation. Access transistor disturbs the internal node (passes a "1").

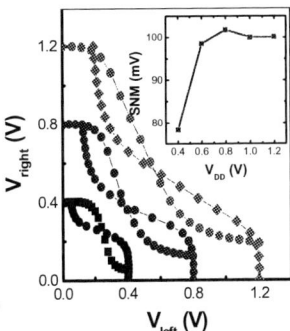

Fig. 9 Butterfly curves of FinFET 6T-SRAM cell under read operation at various V_{DDS}. FinFET SRAM is robust for ultra-low voltage applications.

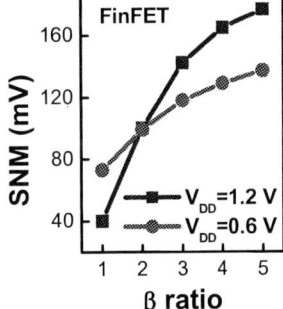

Fig. 10 Wider SNM FinFET SRAM cell can be achieved by increasing beta ratio with the trade-off of cell area. FinFET SRAM is stable for ultra-low voltage applications.

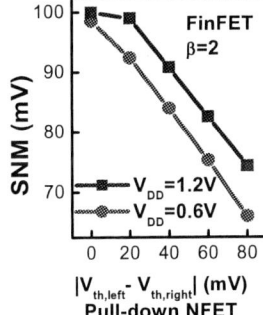

Fig. 11 Random dopant fluctuation- and T_{si}-induced V_{th} variation between paired pull-down NFET has significant impact on SNM.

39

Novel T shape structure PCM and Electrical-Thermal Characteristics

W. H. Wang, D. S. Chao, Y. C. Chen, C. M. Lee, H. H. Hsu, Y. Chuo,
M. H. Tseng, M. H. Lee, W. S. Chen, M. J. Kao, M.-J. Tsai
Electronics Research and Service Organization (ERSO), ITRI, Hsinchu, Taiwan, R.O.C.
E-mail: wenhan@itri.org.tw; Phone: +886-3-5913388; Fax: +886-3-5917690

ABSTRACT

Novel interfacial layer structure was proposed with adjustable resistance ratio and more uniform data distribution capabilities. The reduction of writing power can also be expected with further optimizing the materials and corresponding thicknesses. The performance will be even better with the shrinking of contact area.

INTRODUCTION

Phase-Change Memory (PCM) is one of the most promising candidates for next-generation Non-Volatile Memories (NVM) for having the potential of fast write and read, good read signal margin, very high endurance and an intrinsic scalability [1, 2]. However, PCM suffers from several issues like high reset current and unstable resistance distribution due to the chalcogenide material properties and the high amorphous resistance. Several methods have been demonstrated to solve these problems by either shrinking the contact size or doping other materials into the chalcogenide materials [3, 4], but there is not yet a method that can improve these characteristics at the same time. Aim of this work is to present and to implement a novel PCM cell architecture i) completely compatible with a CMOS technology, either for stand-alone or for embedded applications, ii) with easily controlled resistance ratio, iii) with multi-bit cell feasibility and iv) better resistance uniformity on a wafer.

KEY CONCEPT OF THE NOVEL STRUCTURE

A vertical PCM cell employing the $Ge_2Sb_2Te_5$ (GST) chalcogenide alloy has been integrated into a 0.35μ m CMOS technology and 0.18μ m Cu BEOL. The new key concept introduced in this architecture that keeps the resistance ratio controllable and multi bits feasible and data distribution range improvable is the insert of a interfacial layer with higher resistivity between the plug and the GST material. The conventional and the new structures are schematically depicted in Fig. 1. The simulated electrical field distribution of each structure is shown in Fig. 2. The simulation data and TEM cross-section are shown in Fig. 3 and Fig. 4. Simulation results show that the induced amorphous region is slightly afloat in the chalcogenide material in the conventional structure. On the other hand, the interfacial layer with higher resistivity makes the hot region closer to the plug so that the induced amorphous region completely covers the plug area at a lower temperature than the conventional structure. In this new structure the amorphous region is more or less like a switch to control the current path. When the amorphous region fully covers the plug area the input current would flow horizontally through the interfacial layer bypassing the adjacent high resistance amorphous region so that the overall resistance of stack can be easily controlled by adopting different interfacial layer materials and corresponding film thicknesses.

EXPERIMENT RESULTS

The R-I curve in the conventional structure has a very sharp

transforming characteristic as can be seen in Fig. 5 (a) while in the novel structure the curve has a milder slope, (Fig. 5 (b)) indicating that the high resistance level can be easily differentiated by different input pulses. With this adjustable reset resistance, the threshold voltage can also be properly modified to meet the CMOS capabilities (Fig. 9). For example, a nearly 1V threshold voltage has been reported in papers using 0.1μ m technology [5,6] but it can be easily achieved using 0.24μ m technology with the presented novel T shape structure. The maximum reset resistance in the novel structure (100 K Ω) is an order lower than the conventional structure (1M Ω) under the same operation conditions. A lower reset resistance represents a lower threshold voltage which is required in the scaling CMOS technology. Since the reset resistance in the new structure is mainly determined by the interfacial layer, a more uniform resistance distribution is also expected. Additionally, since the mid point resistance can be easily set, the reference cell design can also be simplified by using a modified single cell as compared with paralleling a high resistance cell to a low resistance cell in the conventional reference cell design. Figs. 6 (a) and 6 (b) show the I-V curves of the conventional structure and the new T structure. It's apparent that the I-V curve in the conventional structure has only one slope in the transition region. The reason is that the strong electrical field near the corner of the plug (Fig. 2 (a)) will induce a sudden large current after the breakdown in the amorphous region. On the other hand, the electrical field distribution in this structure is more uniform (Fig. 2 (b)) so that the I-V curves have a more gradual slope, which is believed to have the effect of lessening the damage by the large current flowing to the interface of PC material and plug material. The cooling rate of the device will be slightly affected (Fig. 7) due to a lower thermal conductivity of the interfacial layer than the plug. But it can be compensated by the lower input current needed to melt the same amorphous volume, which is 30% lower than the conventional structure (Fig. 8). Moreover, a multi-bit storage feasibility has been demonstrated, although only three levels are shown in Fig. 10. It is believed that more operation levels can be achieved with optimized operation strategies. The new structure is to be implemented in the 1Kb chip (Fig. 12).

SUMMARY

In summary, a highly CMOS-compatible PCM structure is presented. By implementing the novel interfacial layer structure adjustable resistance ratio, more uniform data distribution and multi-bit cell are feasible. Moreover, with the optimized interfacial layer material and corresponding thickness, the reduction of writing power can be expected. The improvement will be even more pronounced as the PCM contact size is further downsized.

REFERENCE

[1] Jon Maimon, et al., IEEE, p. 5-2289, 2001.
[2] Stefan Lai et al., IEEE , 2001.
[3] H. Horii, et al., Symposium on VLSI Technology, 2003.

[4] F. Pellizzer, et al., Symposium on VLSI Technology, 2004
[5] Martijn Lankhorst, et al, Nature Material p.347 April 2005

[6] S.J. Ahn, et al., IEEE, 2004.

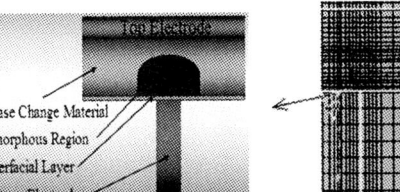

Fig. 1 (a) Conventional T shape structure PCM

Fig. 1 (b) Novel T shape structure PCM

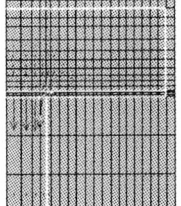

Fig. 2 (a) Electrical field in conventional T structure

Fig. 2 (b) Electrical field in novel T structure

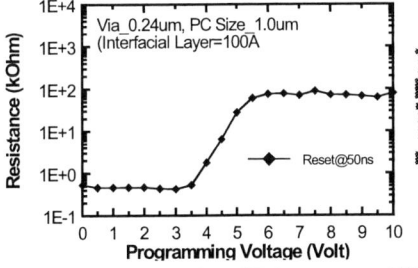

Fig. 3 (a) Simulation result of heat distribution in conventional T shape structure PCM

Fig. 3 (b) Simulation result of thermal distribution in novel T shape structure PCM

Fig. 4 (a) TEM image showing novel T shape structure PCM and corresponding amorphous region

Fig. 4 (b) Schematic demonstration of current flow path in novel T shape PCM

Fig. 5 (a) R-V curve of conventional T shape structure PCM

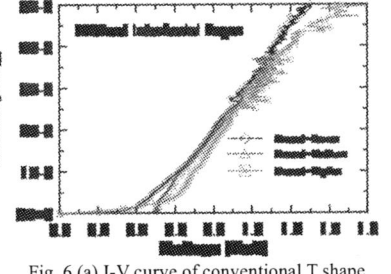

Fig. 5 (b) R-V curve of novel T shape structure PCM

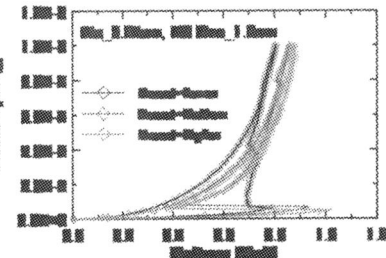

Fig. 6 (a) I-V curve of conventional T shape structure PCM with different reset resistance

Fig. 6 (a) I-V curve of novel T shape structure PCM with different reset resistance

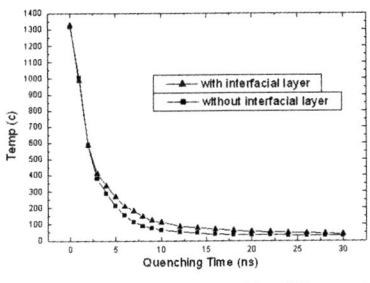

Fig. 7 Simulation result of the difference in cooling speed of two PCM structures

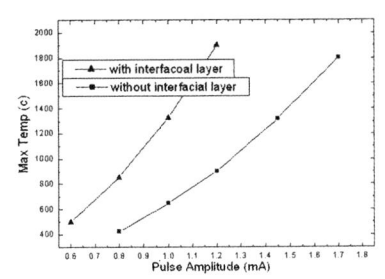

Fig. 8 Simulation result of hot spot temperature in different pulse high of two PCM structures

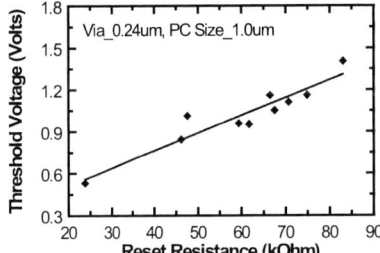

Fig. 9 Threshold voltage as a function of reset resistance

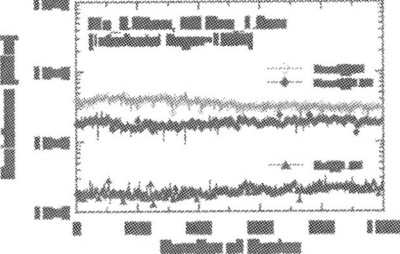

Fig. 10 Cycle-ability test result of the novel showing multi-bit cell possibility

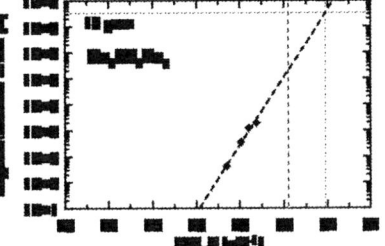

Fig. 11 Data retention capabilities

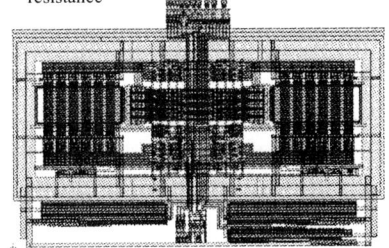

Fig. 12 Photograph of 1 Kb prototype chip

2006 International Symposium on VLSI Technology, Systems, and Applications

A New Read Method by Using DIBL Characteristics in Nitride Storage Device

L.P. Chiang, P. A. Chen, C. H. Hung , C. P. Tsao, H. H. Liao, and C. H. Lin
NVM Division, Winbond Electronics Corp.
Science-Based Industrial Park, HsinChu, Taiwan, R.O.C.
Tel/Fax: 886-3-5678168-7327/886-3-5665516
e-mail: LPChiang@winbond.com.tw

Abstract

A new read operation scheme by using the differential of the threshold voltage (or current) is proposed in nitride storage memory devices. By utilizing the DIBL characteristics between program state and erase state, the principle and application of this cell read operation either in single bit or in two bits is demonstrated. Finally, the application for multi-level operation is proposed.

Introduction

Nitride storage device has recently gained recognition as a low cost, scalable and easy to implementation technology for manufacturing high-density non-volatile memory devices [1]. During cell program/erase, charge trapping in the ONO nitride layer induced threshold voltage instability is the major reliability concern [2]. Furthermore, the threshold voltage instability causes the read operation marginal as the cycling time increases. Especially while device is scaled, read current (either at forward read or reverse read) become smaller, which will cause sensing margin narrower.

In this study, a new read operation scheme by the differential of the threshold voltage (or current) is proposed to increase the read operation margin. By using the DIBL difference between programmed state (electron trapping) and erased state (hole trapping), the cell read operation window is relatively large. This method can be further extended to the application for multi-level operation in nitride storage memory.

Principle

In a conventional nitride storage device, programming is performed by injecting channel hot electrons into the ONO layer and erase is performed by injecting hot holes into the ONO layer by band-to-band tunneling induced hot hole injection. Under such operation condition, the charge trapping is in a localized region near the junction and the band diagrams are shown in the Fig.1. Due to localized injection property, the barrier height near the drain region is higher after programming and the drain induced barrier lowing effect (DIBL) becomes apparent. On the contrary, erase state does not show such behavior.

Fig.2 shows nitride storage device Ids-Vgs characteristics in the program state and erase state respectively with various drain biases. The DIBL effect, (Vth v.s. Vds) for the program state and erase state are then shown in the Fig.3. It is evident that program state Vth varies from 3.6V to 1.8V when drain bias changes from 0.1V to 1.8V. However, the erase state Vth variation is small (<0.4V) under the same measurement condition. Besides, it should also be mentioned that the reverse read (source and drain terminal exchanged between program/erase and read) does not show large DIBL effect because the drain bias can not modulate the barrier height.

Since the current in the sub-threshold region is exponentially dependent on threshold voltage, a threshold voltage difference (DIBL effect) makes the current variation is obvious.

As shown in the Fig.2, the sub-threshold region current ratio for two different drain biases (0.1V and 1.8V) at the programmed state (> 10^3) is significantly larger than erase state (<10). By using such large current differential between program state and erase state, a new read method can be proposed. Besides, Fig. 3 also shows the DIBL effect after cycling. Although the threshold voltage shifts up, the DIBL characteristics between program state and erase state are still existed. This means that the current differential read method can enlarge the device operation margin even under degraded condition.

Application for two-bits and multi-level operation

For nitride storage devices, two-bits storage is a potential advantage for cell size reduction compared with conventional floating gate cells. In such operation, charges are storied in either left or right junction [3]. In our approach, a split-gate type nitride storage device structure [4, 5] is used. Fig. 4 shows forward read characteristics for the program state and erase state, respectively. By applying a proper over ride bias (V_{OL}) at the neighboring bit, the DIBL effect is still apparent at the program state no matter what the neighbor cell states are. Fig.5 shows device program and erase DIBL effect again on split-gate nitride storage device. The DIBL between program state and erases state is ~2V. Because the DIBL effect is determined by the electron trapping magnitude, this method can be also extended to the multi-level application by using various DIBL amplitudes, as shown in the Fig.5, where intermediate states (program state 1, program state 2, ... , etc.) are inserted. By using the exponentially dependence of drain current and threshold voltage at the sub-threshold region, the operation windows for various states are shown in the Fig.6.

Summary

A new read operation to define the logic "0" and "1" of a nitride storage memory cells by using the threshold voltage (or current) differential property (DIBL effect) is proposed. This method can enlarge the cell read operation window and can be used as multi-level operation by properly choosing the programmed levels. Single cell and two-bits cell structures can also applied by such method.

Reference

[1] B. Eitan, P. Pavan, I. Bloom, E. Aloni, A. Frommer, D. Finzi, *Proc. SSDM*, 1999, pp. 522-524

[2] M. Janai, *Proc. IRPS*, 2003, pp. 502-505

[3] B. Eitan, P. Pavan, I. Bloom, E. Aloni, A. Frommer, D. Finzi, *IEEE EDL*, vol.21, pp. 543-545, Nov, 2000

[4] US patent 6707079 B2, "TwinMONOS cell fabrication method and array organization", 200

[5] T. Saito, S. Ogura, T. Ogura, *Proc. NVSMW*, 2003, pp.50-52

1-4244-0181-X/06/$20.00 ©2006 IEEE

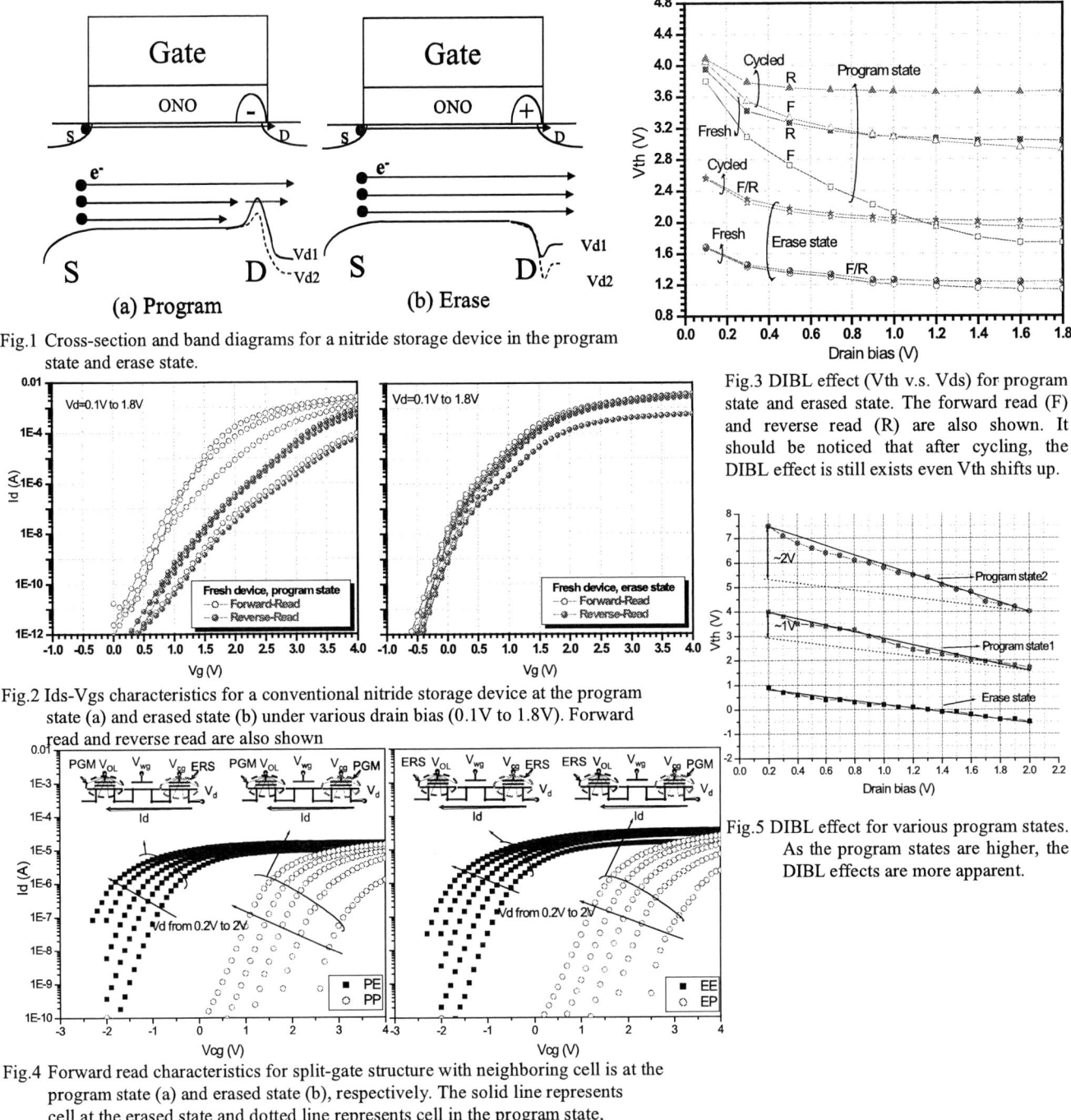

(a) Program **(b) Erase**

Fig.1 Cross-section and band diagrams for a nitride storage device in the program state and erase state.

Fig.2 Ids-Vgs characteristics for a conventional nitride storage device at the program state (a) and erased state (b) under various drain bias (0.1V to 1.8V). Forward read and reverse read are also shown

Fig.3 DIBL effect (Vth v.s. Vds) for program state and erased state. The forward read (F) and reverse read (R) are also shown. It should be noticed that after cycling, the DIBL effect is still exists even Vth shifts up.

Fig.4 Forward read characteristics for split-gate structure with neighboring cell is at the program state (a) and erased state (b), respectively. The solid line represents cell at the erased state and dotted line represents cell in the program state, respectively.

Fig.5 DIBL effect for various program states. As the program states are higher, the DIBL effects are more apparent.

Fig.6 Schematic representation for various states distribution by various flash cell array.

2006 International Symposium on VLSI Technology, Systems, and Applications

Writing Architecture for Magnetic Random Access Memory with Negative Pulse Writing Scheme

C.P. Chang, C.C. Hung, Y.H. Wang, Y.J. Lee, K.L. Su, W.C. Chen, Y.H. Chen, C.S. Lin, M.J. Kao, J.F. Huang, M.-J. Tsai

Electronics Research and Service Organization (ERSO), ITRI, Hsinchu, Taiwan, R.O.C.

E-mail: cpchang@itri.org.tw; Phone: +886-3-5913395; Fax: +886-3-5917677

Abstract

The toggle cell of MRAM needs large current for writing. In this work, the negative pulse scheme for toggle writing is proposed, and the implemented architecture is also described. The scheme can reduce the writing current and improve the toggle yield. A single current source scheme is also designed to provide the symmetrical and bi-directional current.

Introduction

The MRAM has the advantages of non-volatility, high access speed [1] and high density [2], making it suitable for embedded memory applications. Due to the process variations, such as lithography, thin film stress etc., the conventional cross selection asteroid MRAM encounters severe write disturbance problem (Fig. 1). The toggle or "Savtchenko" MRAM switching mode [3] was invented to substantially improve the write selectivity (Fig. 2). Recent progress on the toggle MRAM has demonstrated the feasibility and scalability to the 90nm node [4]. However, the toggle MRAM still suffers from high writing current to rotate the free layers. Besides, toggle yield loss [5] will be one of the major bottlenecks to push MRAM into mass production. The purpose of this work, therefore, is to reduce writing current for the toggle MRAM and improve the write yield.

Negative Pulse Writing Scheme

Applying a magnetic field toward the easy axis direction is an effective way to reduce writing current [6]. The simplest way to generate a bias field is to adjust the thickness of bottom pinned (BP) layer of MTJ stacks and it is verified by micromagnetic simulation (Fig. 3). However, the generated bias field can not be unlimitedly imposed. When the bias field becomes higher, i.e. MRAM can operate at lower writing current, the magnetic moments will become disordered and cross a so-called critical axis (Fig. 4), especially around the end domain region. Due to the repulsion of bias field and thermal excitation, the toggle failure phenomenon thus happen (Fig. 5). A novel toggle waveform with negative pulse (NP) set prior to the normal toggle waveform is proposed in this work (Fig. 6) to modulate random end domains to the required initial state and, therefore, the free layers can be toggled successfully, especially at strong stray field biasing and high temperature excitation. The diagram in Fig. 7a displays that it rotates clockwise accurately for conventional toggle waveform at weak bias field, but counter clockwise at strong bias field. By

implementing the negative pulse writing scheme (Fig 7b), the successful 180 degree rotation can therefore be achieved.

Writing Architecture and Operations

Conventionally, the bi-directional and large current of over 3mA through the bit line in MRAM is generated by two current source circuitry [7]. Due to the routing issue inside the chips, switches may not be properly turned on and off as intended. For example, Fig. 8 shows a conventional writing architecture. If S1 and S4 turn on at the same time, two nodes between bit line may have little difference in potential. Consequently, no current will flow through the bit line and none of the cell will be switched. In addition, it requires more power consumption in the current sources when two dc paths turn on simultaneously. Additionally, a voltage source for MRAM writing is used [8] for low voltage output. However, the output current flowing through the bit line strongly depends on the metal resistance. When the metal resistance is changed due to the process variation, the current through the bit line may be higher or smaller than the expected value. In this work, a writing control circuit based on single current source is proposed, as shown in Fig. 9. A bi-directional current is symmetrically generated. While at stand-by, all switches S1 to S4 are turned off. The MRAM can be operated in the read mode. As for the write mode, S2 and S4 turn on and the current flows through the bit line from bottom toward top in order to generate the negative pulse. Then, S2 and S4 turn off and S1 and S3 turn on to generate a positive current flowing from top toward bottom. A simulation result for the proposed architecture, as shown in Fig. 10, successfully demostrates the symmetrical and bi-directional current waveform.

Conclusions

In this paper, a negative pulse writing scheme was proposed for toggle cell in order to reduce writing current and to enhance the toggle yield. A writing circuit for generating such negative pulse current for low power application was also proposed. It's also resistive to the process variations, such as metal resistance, compared with the conventional voltage source scheme.

References
[1] J. DeBrosse, et al., Symposium on VLSI Circuits, p. 454, 2004.
[2] C. C. Hung, et al., IEDM Tech. Dig., p. 575, 2004.
[3] M. Durlam, et al., IEDM Tech. Dig., p. 995, 2003.
[4] M. Durlam, et al., Symposium on VLSI Technology, p. 186, 2005.
[5] T. Yamamoto, et al., J. Appl. Phys. 97, 10P503, 2005.
[6] B. N. Engel, et al., US Patent 6,633,498 B1, 2003.
[7] M. Durlam, et al., Symposium on VLSI Circuit, p. 158, 2002.
[8] T. Tsuji, et al., Symposium on VLSI Circuits, p.450, 2004.

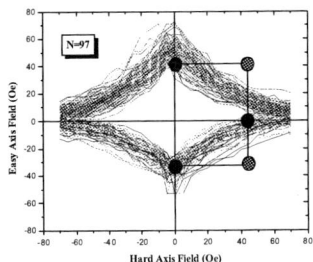

Fig. 1 Asteroid-like write curves superimposed of 97 dice for conventional cell structure and writing scheme.

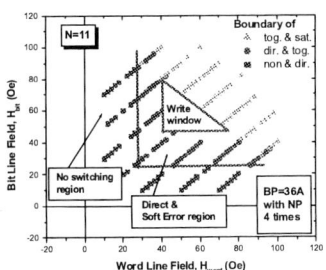

Fig. 2 Superimposed write curves of 11 bits across a wafer from the proposed toggle writing scheme with negative pulse waveform.

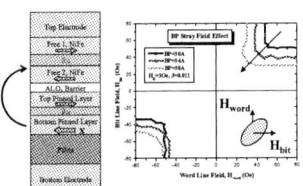

Fig. 3 (a) Schematic of the MTJ stacks for toggle cell with various thicknesses of the bottom pinned layer to generate the bias field and (b) on different thickness of the bottom pinned layer. micromagnetic simulation of switching characteristic

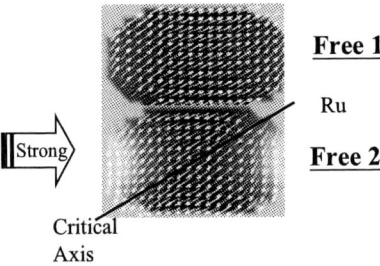

Fig. 4 Micromagnetic simulation of the quiescent state at strong magnetic biasing field generated from BP before toggle writing. Due to the exceeding critical axis of the moments around end domain region, the stuck-at failure mode may happen.

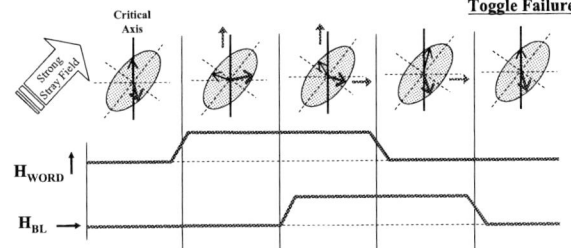

Fig. 5 Write operation of Toggle MRAM[3][4] with conventional pulse waveform at strong bias field. Due to the repulsion of bias field and thermal excitation, the magnetization is disordered and thus lead to the toggle failure.

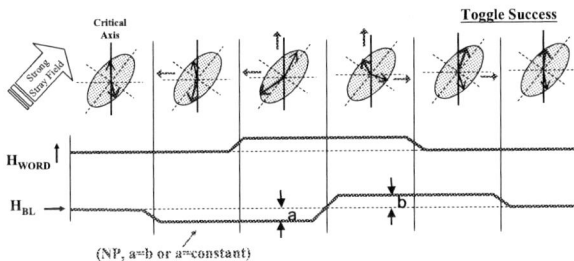

Fig. 6 Write operation of Toggle MRAM with proposed negative pulse writing scheme. The negative pulse procedure modulates end domains to the required initial state for following successfully toggle write.

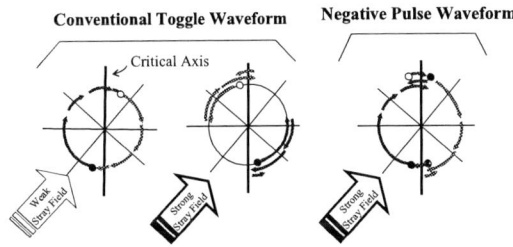

Fig. 7 Rotation diagram of toggle cells (a) in weak bias field, (b) in strong bias field, both for conventional toggle waveform, and (c) for negative pulse waveform which displays the accurate clockwise rotation.

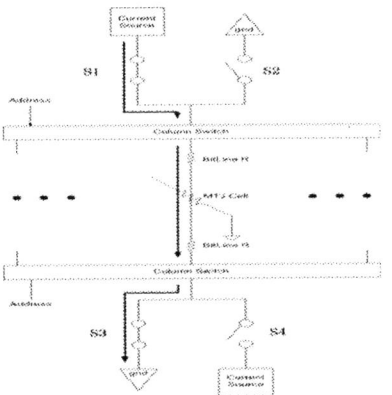

Fig. 8 Writing Architecture with two current sources.

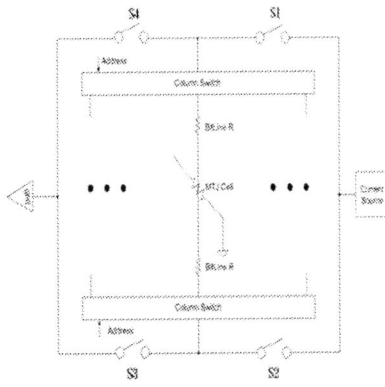

Fig. 9 Writing Architecture with one current source.

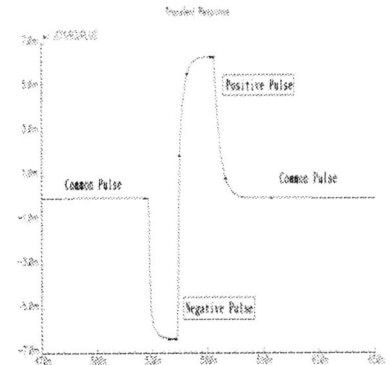

Fig. 10 Simulations result of proposed architecture to generate symmetrical and bi-directional current waveform.

2006 International Symposium on VLSI Technology, Systems, and Applications

Strained Si and the Future Direction of CMOS

Scott E. Thompson and Guangyu Sun
University of Florida, Gainesville FL, 32607
Thompson@ece.ufl.edu

ABSTRACT

Uniaxial process induced strain is being adopted in all 90, 65, and 45 nm high performance logic technologies. Uniaxial strain offers large performance improvement at low cost and minimally increased manufacturing complexity and is scalable to future technology nodes.

INTRODUCTION

The era of "simple" silicon based CMOS device scaling has ended. For at least the next several decades, there is no viable alternative on a cost per function metric to replace silicon CMOS. This presentation will describe how state-of-the-art CMOS is being extended with strained Si at the 90, 65, and 45 nm nodes. In addition, insight will be given into why strained Si or more generically mobility enhancement is the dominant focus to extend CMOS over high k gate dielectrics, multi-gate devices, or fully depleted devices.

STATE-OF-THE-ART CMOS

Strained Si improves MOSFET performance by enhancing the electron and hole mobility. Strain can be implemented into MOSFETs using a global approach where stress is introduced across the entire substrate or locally where stress is engineered into the device by means of epitaxial layers and/or high stress nitride capping layers (see Figs. 1 and 2). For the first and second generation strained Si MOSFETs being adopted in all high performance logic technologies [1-5], the industry is choosing process induced uniaxial stress due to advantages relating to n-channel threshold voltage shift and p-channel mobility enhancement at low strain and high vertical electric fields [6].

At the 90 nm technology node, based on cost and performance considerations, three techniques to introduce strain are being adopted: (1) tensile and/or compressive capping layer post salicide (see Figs. 2 and 4) (2) local epitaxial films grown in the source and drain regions (see Fig. 3) and (3) capping layers on top of the poly-Si gate before source/drain anneal for stress memorization of the poly-Si gate [5]. The combination of these various techniques is additive.

Strained Si is being adopted for two key reasons. First, the processing, engineering, and design costs associated with implementing strained Si is modest. Second, strained Si offers unmatched performance enhancement with 30 - 60% product level improvement in present day production on 300mm wafers [3-4]. The additional processing cost is modest at ~ 1-3 %. Strain does increase the formation of dislocations (see Figure 5 for an example) but can be addressed by modifying the dislocation growth kinetics. Strain increases the variation in performance across all transistor layout styles and requires some design rule modifications.

WHY STRAINED SI.

Based on metrics for cost, process complexity, magnitude of the performance gain, and scalability, strained Si is the clear choice for extending CMOS when simple scaling ended at the 90 nm technology node. At present, other performance enhancement ideas, summarized in Fig. 6, are not ready to compete on the above metrics. A few of the other performance enhancement concepts and key issues that still need to be addressed will next be briefly summarized. Some are additive with strained Si which will likely be a requirement for future enhancement options.

High k gates dielectrics have been worked on for decades and were once though to be the lead material change to extend simple scaling. High k gates may still be adopted slowly after the 45 nm node but a few issues still need to be resolved. Good progress has been made in high k reliability and recovering most of the degraded mobility. However, mobility degradation still exists at low vertical electric fields (see Fig. 7) which reduces the product level performance enhancement to near single digits. The small performance gain versus the high risk and cost makes being first in production with a high k gate dielectric, questionable.

Significantly increased wafer costs also result with high k since dual metal gates are required for integration into the bulk planar CMOS device structure. The integration flow for dual metal work functions is enormously complex and appears somewhat under appreciated. There have not yet been any publications on dual metals with work functions at the conduction and valance bands that can withstand the ~1000-1050C rapid thermal processing midsection anneals. To circumvent the midsection thermal constraints, replacement gate flows are sometimes used (see Fig 8). However dual metal replacement gate flows are complex and further increase wafer cost making it unsuitable for the fast growing consumer market and wireless markets.

The next promising class of device structures that can extend CMOS scaling are fully depleted multi-gate structures. The small area of the bodies in the multi-gate structure requires it to be fully depleted to solve the dopant fluctuation problem. However, once fully depleted, metal gates (and the set of previously mentioned issues) are needed to set the threshold voltage. Other key issues to still solve are (1) fully depleted tri-gates have worse layout efficiency compared to a planar CMOS structure (2) FinFETs suffer high variability due to the difficulty of patterning and fabrication a transistor on a large step and (3) because of the thin body the external resistance tends to be larger which is a significant issue since in state-of-the-art nanoscale MOSFETS the external resistance is comparable in magnitude to the channel resistance.

CONCLUSION

Strained Si is the performance concept extending Moore's Law. High k gates on a strained Si flow are perhaps the next key performance enhancement material change the industry will adopt but this change requires a higher cost metal gate flow and will not become mainstream until after the 45 nm node. The risk / reward ratio is much less favorable for high k gates then strained Si which is why strain should be adopted for nearly all 90 to 45 nm technologies over high k.

1-4244-0181-X/06/$20.00 ©2006 IEEE

REFERENCES

[1] SE Thompson, et al. *A Logic Nanotechnology Featuring Strained Silicon*. IEEE Electron Device Lett. 25:191-193, 2004.

[2] Chidambaram, P. R., 1, et al. *35% Drive Current Improvement from Recessed-SiGe Drain extensions on 37nm Gate Length PMOS*. Symp. VLSI Tech. Dig., 2004.

[3] S. Pidin, et al. *A Novel Strain Enhanced CMOS Architecture Using Selectively Deposited High Tensile And High Compressive Silicon Nitride Films*, Fujitsu, Ltd., Tokyo, Japan, IEDM 2004.

[4] H.S Yang et al, *Dual Stress Liner for High Performance sub-45nm Gate Length SOI CMOS Manufacturing*, IBM Systems and Technology Group, NY 12533 and Advanced Micro Devices, Dresden Germany.

[5] Chien-Hao, C., T. L. Lee, et al. (2004). *Stress memorization technique (SMT) by selectively strained-nitride capping for sub-65nm high-performance strained-Si device application*. Symposium on VLSI Technology, 2004, Digest of Technical Papers (IEEE Cat. No.04CH37526): 56-57.

[6] S E. Thompson, G. Sun, K.Wu, J.Lim and T.Nishida, *Key differences for process-induced uniaxial vs. substrate-induced biaxial stressed Si and Ge channel MOSFETs*. in Tech. Dig. of the IEEE Int. Electron Devices Meeting, Washington, 2004.

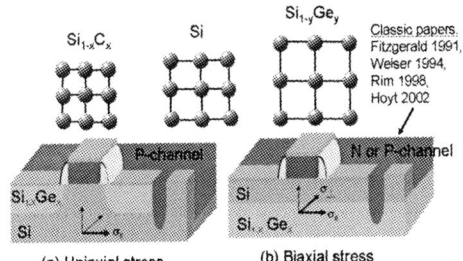

Figure 1: Two techniques to introduce stress (a) uniaxial and (b) biaxial

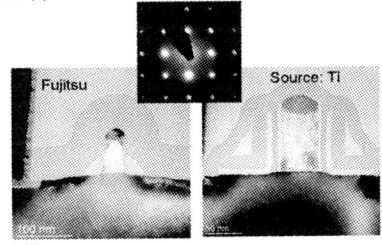

Figure 2: Tensile capping layers for uniaxial tensile stress on <100> channel orientation (source: Chipworks)

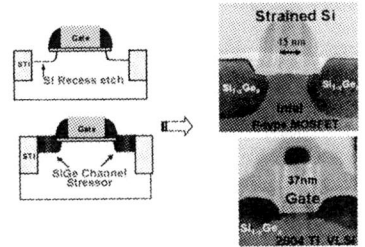

Figure 3: Embedded SiGe in the source and drain to create uniaxial stress

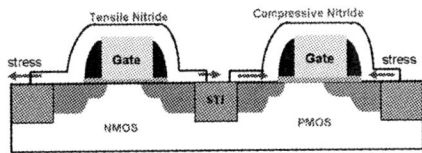

Figure 4: Dual uniaxial stress layers: tensile film for n-channel and compressive film for p-channel

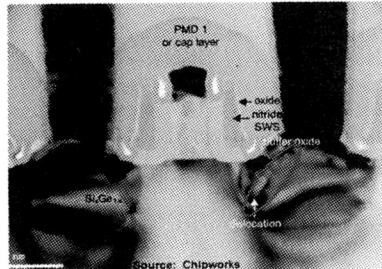

Figure 5: Dislocation in the high stress SiGe layer

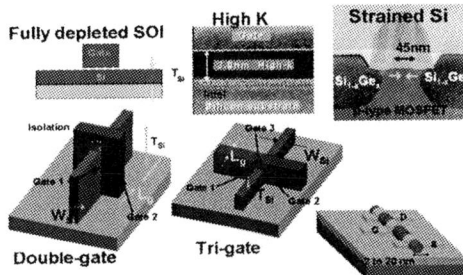

Figure 6: Potential options to continue performance trends

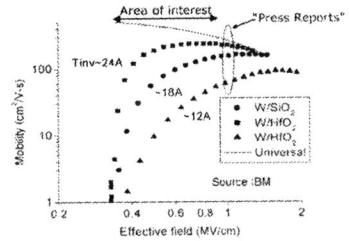

Figure 7: Mobility vs. vertical effective field for SiO_2 and HfO_2

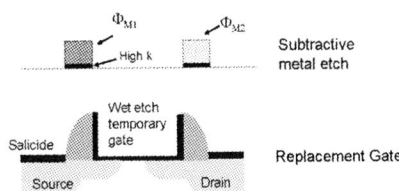

Figure 8: Two integration schemes for high k dielectrics

2006 International Symposium on VLSI Technology, Systems, and Applications

Optimization of Source/Drain Extension for Robust Speed Performance to Process Variation in Undoped Double-Gate CMOS

Ji-Woon Yang, [a]Daniel Pham, Peter Zeitzoff, Howard Huff, and George Brown
SEMATECH, [a]Freescale assignee, 2706 Montopolis Drive, Austin, Texas 78741
Tel: 1-512-356-7460 Fax: 1-512-356-7640; e-mail: Ji-Woon.Yang@sematech.org

Abstract

The speed performance of undoped double-gate CMOS with a gate-source/drain underlap structure is investigated using a 2D device and compact model simulation. The gate-source/drain underlap structure yields optimal characteristics and shows robustness to process variation when the structure is optimized.

Introduction

Double-gate (DG) MOSFETs/FinFETs with ultra-thin bodies (UTBs) will be needed to scale CMOS down to the 32 nm technology node as specified in the *International Technology Roadmap for Semiconductors* (ITRS) [1]. Gate-source/drain (G-S/D) underlap structures have been studied to suppress short-channel effects (SCEs) for undoped DG CMOS without degrading the saturation current (I_{on}) by a bias-dependent effective channel length (L_{eff}) [2]. Several studies have shown optimizations of the G-S/D underlap structure [3-5]; however, either only the underlap resistance was optimized [3,4] or the device structure was inappropriate for employing G-S/D underlap [5].

In this work, we optimized the G-S/D underlap in an undoped DG CMOS to show that it has better speed performance than conventional overlap and greater robustness to process variation when the structure is optimized.

Gate-source/drain underlap

Fig. 1 shows the schematic cross section of the simulation structure with G-S/D underlap, which is defined by the separation from the constant doping edge of the source/drain (S/D) to the physical gate edge (L_{ext}) and S/D lateral doping gradient (σ_L). The effective channel length (L_{eff}) has a strong dependence on bias and decreases with increasing gate voltage (V_{GS}) [4]. This V_{GS}-dependent behavior helps to suppress SCEs with a longer L_{eff} in weak inversion ($L_{eff(weak)}$) and to maintain a high I_{on} with shorter L_{eff} in strong inversion ($L_{eff(strong)}$) for the same physical gate length (L_G) [4].

There is a tradeoff between SCE control and resistance of the G-S/D underlap ($R_{underlap}$), thus I_{on} is maximized with the tuned gate workfunction for the requirement of the off-state leakage current (I_{off}) by optimizing the G-S/D underlap [3,4]. However, the parasitic fringing capacitance (C_{fringe}), which includes an outer fringing (C_{of}) and an inner fringing (C_{if}) capacitance, as shown in Fig. 1, should be considered to achieve optimal speed for the G-S/D underlap because there is also tradeoff between the $R_{underlap}$ and C_{fringe} [4,5].

To predict the speed performance of the DG CMOS with G-S/D underlap, we used a process/quantum mechanics-based DG MOSFET model (UFDG [6]). A 2-D numerical simulator, Medici [7], was used to evaluate the source/drain series resistance ($R_{S/D}$) and C_{fringe} for the G-S/D underlap structure. We used a L_G = 13 nm, equivalent oxide thickness (EOT) = 0.7 nm, gate stack thickness (t_{Gate}) = 13 nm, and operation voltage (V_{DD}) = 0.9V as recommended by the ITRS for high performance 32 nm technology [1]. The silicon channel thickness (t_{Si}) was set to 7 nm. The body doping concentration (N_B) and S/D doping concentration ($N_{S/D}$) were assumed to be $10^{15} cm^{-3}$ and $10^{20} cm^{-3}$, respectively. For the critical parameters of G-S/D underlap, σ_L was fixed at 2 nm/dec. and L_{ext} was varied from 0 nm to 15 nm to achieve the best speed.

Fig. 2 shows the simulation results from Medici for two different G-S/D underlap structures with L_{ext} = 0 nm and 10 nm. Note that SCEs are well controlled for L_{ext} = 10 nm, while they are unacceptable for L_{ext} = 0 nm even when two structures had the same L_G. $L_{eff(weak)}$ for each L_{ext} is extracted by matching the simulation results from UFDG with L_{eff} as a model parameter to the Medici results as shown in Fig. 2. The extracted $L_{eff(weak)}$ vs. L_{ext} is plotted in Fig. 3. It shows

that L_{ext} less than 2 nm yields the conventional overlap structure due to the finite σ_L (= 2 nm/dec.).

Model parameter extraction

To predict the speed performance of a L_G =13 nm DG CMOS with G-S/D underlap using UFDG, we needed to find $R_{underlap}$ because of a deficiency in the model. The total $R_{S/D}$, which includes $R_{underlap}$ and conventional S/D series resistance, was evaluated from the linear extrapolation of the Medici-predicted total resistance (R_{tot}) versus L_G. We evaluated $R_{underlap}$ by subtracting $R_{S/D}$ of L_{ext} = 0 nm (conventional overlap structure) from that of each G-S/D underlap design as shown in Fig. 4. The $R_{S/D}$ as a model parameter in the UFDG was assumed to be the sum of the $R_{underlap}$ and parasitic S/D series resistance (= 107Ω-μm) of the technology node recommended by the ITRS.

Fig. 5 shows the Medici-predicted C_G (V_{GS}) of the L_{ext} = 3 nm and 10 nm with t_{Gate} = 13nm and, for comparison, 0 nm. Since the intrinsic C_G in undoped symmetrical DG devices is negligible at V_{GS} < threshold voltage (V_t) [8], the C_G in weak inversion represents C_{if}. The difference in C_G between the two simulation results of t_{Gate} = 13 nm and 0 nm is C_{of}. Note that C_{if} of L_{ext} = 10 nm is significantly less than the value of L_{ext} = 3 nm due to the increased distance from the gate to the S/D. C_{fringe} versus L_{ext} is shown in Fig. 6. Now, C_{fringe} at low V_{GS} was chosen as the fringing/overlap capacitance for the UFDG model parameter. Since C_{if}, which has a V_{GS} dependence and is vanished by the screening of inversion layer at strong inversion, is not modeled in the UFDG, we consider a best case, with C_{if} neglected, and a worst case, with C_{if} included for all V_{GS}. The actual delay lies between the predictions for these two cases.

Simulation results

I_{on} at V_{GS} = V_{DS} = 0.9 V predicted by UFDG/Spice3 is shown in Fig. 7. For L_{ext} = 0 nm, which yields a conventional G-S/D overlap structure, severe SCEs, as shown in Fig. 2, maximize I_{on} for the midgap workfunction gate. However, the workfunction should be adjusted to meet the I_{off} requirement (= 3×10^{-7} A/μm). The maximum I_{on} is then achieved at L_{ext} = 10 nm, which is an optimal condition for the tradeoff between $R_{S/D}$ and SCE control. The required workfunction for NMOS and PMOS to meet the I_{off} is also plotted in Fig.8.

Now, the intrinsic delay for the technology is predicted by a nine-stage ring oscillator simulation using UFDG/Spice3. The gate width of PMOS is assumed to be two times that of NMOS. The gate workfunctions of NMOS and PMOS are adjusted respectively for each structure as specified in Fig. 8. The extracted $L_{eff(weak)}$, $R_{S/D}$, and C_{fringe} for each L_{ext} are used for the simulation. The minimum average delay is achieved at L_{ext} = 10 nm mainly due to the maximum I_{on} at the structure as shown in Fig. 9. Interestingly, the average delay is not very dependent on L_{ext} from 5 nm to 15 nm because the effect of $R_{S/D}$ and C_{fringe} on the delay is balanced at the region. This insensitivity of speed to L_{ext} clearly reflects that the G-S/D underlap structure is robust to SDE process variation.

Conclusion

The minimum delay was achieved at G-S/D underlap structure mainly due to well controlled SCEs with a longer L_{eff}. The delay was not sensitive to L_{ext} because the effect of $R_{S/D}$ and C_{fringe} on the delay is balanced. This beneficial characteristic of the G-S/D underlap structure is clearly robust to SDE process variation for undoped DG CMOS processes.

References

[1] *International Technology Roadmap for Semiconductors*. Austin, TX: SIA/SEMATECH, 2003
[2] J. G. Fossum, *IEDM Tech. Dig.*, pp.679, 2003
[3] R. S. Shenoy, *IEEE Trans. Nanotechnology*, pp.265, Dec. 2003

1-4244-0181-X/06/$20.00 ©2006 IEEE

[4] V. P. Trivedi, *IEEE Trans. Electron Devices*, pp.56, Jan. 2005

[5] S. Balasubramanian, *Proc. Silicon Nanoelec. Workshop*, pp. 16, 2003

[6] J. G. Fossum, *Solid-State Electron.*, pp.919, June 2004

[7] *Medici-2004.09 Users Manual*, Synopsis, Durham, NC, 2004

[8] J. G. Fossum, *IEEE Trans. Electron Devices*, pp.808, May 2002

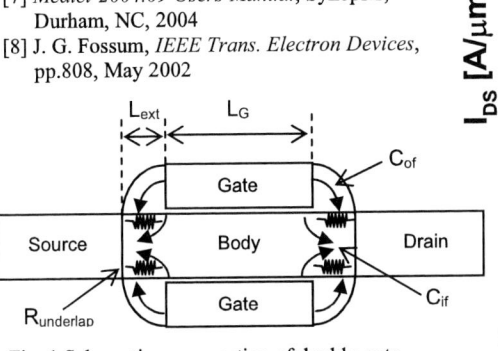

Fig. 1 Schematic cross-section of double-gate MOSFET/FinFET with gate-source/drain underlap showing parasitic fringing capacitances and resistances

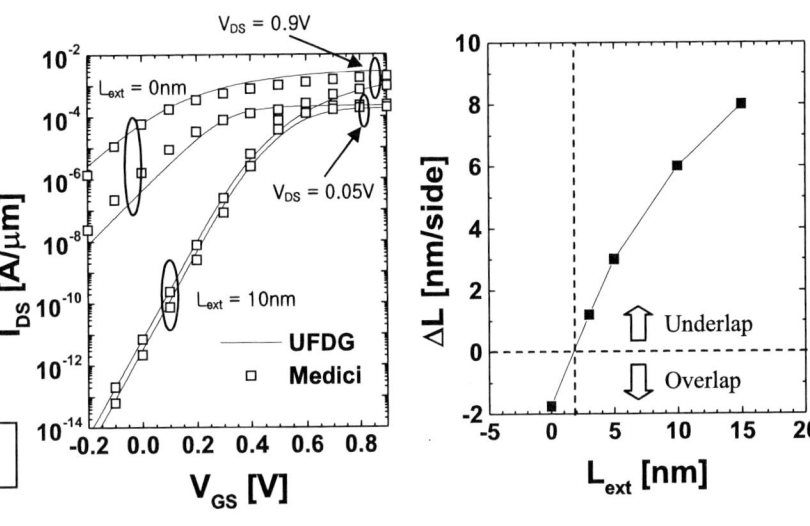

Fig. 2 Medici-predicted current-voltage characteristics of double-gate nMOSFETs with $L_{ext}=0$ nm and 10 nm. UFDG [6] simulation is also conducted to extract L_{eff} in weak inversion.

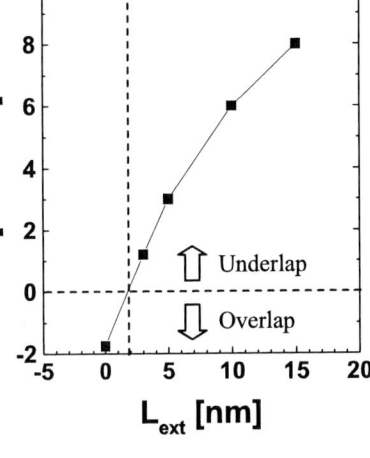

Fig. 3 $L_{eff}(= L_G + 2\Delta L)$ in weak inversion extracted by UFDG [6] vs. L_{ext}.

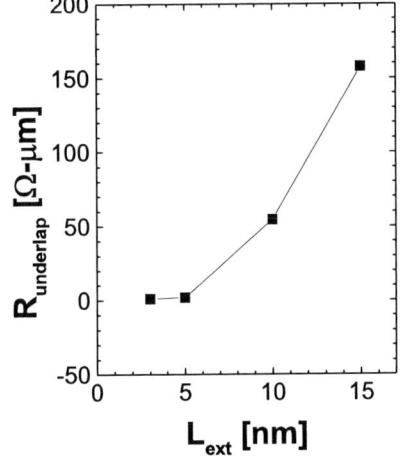

Fig. 4 Dependence of underlap structure resistance on L_{ext}.

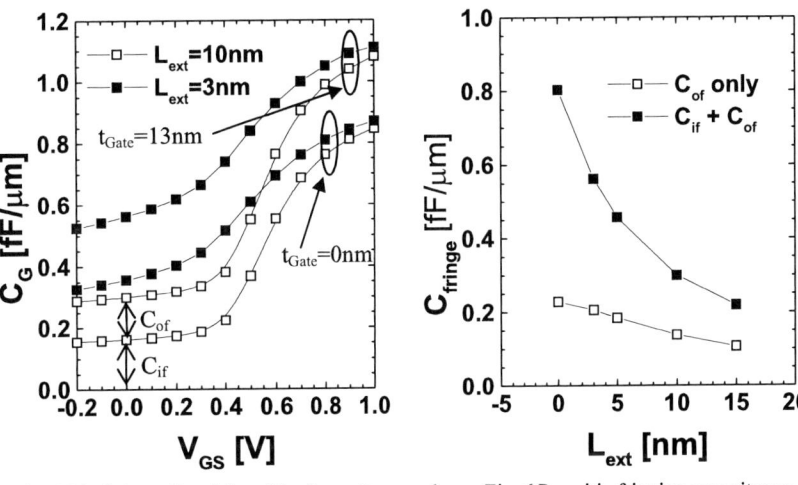

Fig. 5 Medici-predicted C_G of the $L_{ext} = 3$ nm and 10 nm with $t_{Gate} = 13$nm and 0 nm. Parasitic fringing capacitances for $L_{ext} = 10$ nm are indicated.

Fig. 6 Parasitic fringing capacitance extracted from Medici simulation at $V_{GS} = 0$ V as shown in Fig. 5.

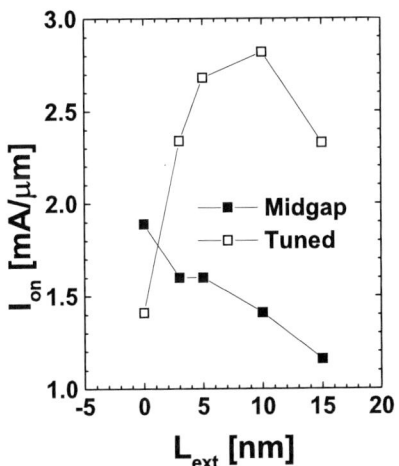

Fig. 7 The saturation current (I_{on}) at $V_{GS} = V_{DS} = 0.9$ V predicted by UFDG/Spice3 [6] for the midgap and tuned (for $I_{off} = 3\times10^{-7}$ A/μm) gate workfunction as shown in Fig.8.

Fig. 8 The required workfunction shift for NMOS and PMOS to meet the $I_{off}(= 3\times10^{-7}$ A/μm) requirement

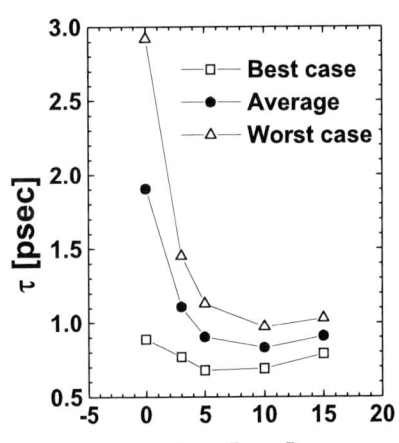

Fig. 9 The intrinsic delay predicted by ring oscillator simulation using UFDG/Spice3 [6].

Mo Gate Deformation Induced by Laser Annealing Process

[1,2]Kentaro Shibahara, [1,2,3]Akira Matsuno, [1,2]Masaki Hino, and [2]Ken-ichi Kurobe

[1]Research Center for Nanodevices and Systems, Hiroshima University
[2]Graduate School of Advanced Sciences of Matter, Hiroshima University
1-4-2, Kagamiyama, Higashihiroshima, 739-8527 Japan
[3]Phoeton Corp.
Atsugi AXT Main Tower 4F IB Room II, 3050, Okada, Atsugi, Kanagawa, 243-0021 Japan
Phone +81-82-424-6267, FAX: +81-82-424-3499, e-mail: ksshiba@hiroshima-u.ac.jp

Abstract

Laser annealing is a promising candidate for ultra-shallow junction formation. However, melt-annealing that utilizes fast recrystallization to achieve non-equilibrium activation tends to accompany undesirable melting at portions other than junction areas. In this paper, deformation of Mo gate is discussed through experimental and simulation work.

1. Introduction

Currently, post-spike annealing methods, such as flash lamp annealing [1-3], solid phase re-growth [4] and laser annealing (LA) [5], are actively investigated for the scaling of source and drain (S/D) junctions and MOSFET performance improvements. We have reported Heat-Assisted Laser Annealing (HALA) [6, 7] that is a combination of substrate heating and laser irradiation showing 20-nm depth ultra-shallow Sb-doped junction formation. The heat-assist at around 450-525°C led to good dopant activation with a relatively wide process window against laser energy density (E_L) and the heat-assist temperature.

In this paper, an issue associated with LA such as a gate deformation due to melting and a countermeasure for it are discussed.

2. Experimental Conditions

KrF excimer laser ($\lambda \sim 248$ nm) was used for LA. Laser light pulse width was about 38 ns. A single laser pulse was irradiated to each point on specimens. In the case of HALA, Si wafers were heated to 450°C or 525°C prior to laser irradiation in a nitrogen atmosphere. Sheet resistance values shown in this paper was obtained with Sb doped n^+/p junctions [6, 7]. Sb^+ was implanted at 10 keV through 5 nm screen oxides for this purpose. Mo was used as a gate material. Because of its very high melting point, 2620°C, robustness against laser irradiation was expected. In addition, it is one of the interesting candidates for the CMOS metal gate, since its workfunction is tunable by nitrogen incorporation [8].

3. Results and Discussions

Laser irradiation for S/D activation sometimes gives rise to melting or deformation of device structures [9]. Especially this problem is very severe for gate pads which are usually located on a field oxide, since the thick filed oxide works as a heat insulator, as explained in Fig. 1. Figure 2 shows an example of the 50-nm thick Mo gate electrode deformation by laser irradiation whose laser energy density (E_L) was moderate one for dopant activation.

To relieve this problem, E_L necessary for sufficient activation should be reduced. HALA can reduce the necessary E_L to about the half of that for the non-heat-assist case, as shown in Fig. 4 [6, 7]. Figure 5 shows optical microphotographs taken after laser irradiation. Assuming the same E_L, HALA deteriorates the gate pad deformation. However, as indicated in this figure, sheet resistance lower than 500 Ω/sq. was obtained by HALA at 350

mJ/cm^2. In the case of non-heat-assist LA, that is R.T. LA, this E_L was too low to activate dopants. As a results, HALA can relieve the gate deformation problem. The deformation was affected by Mo thickness, as shown in Fig. 6. Thicker Mo film showed better robustness against laser irradiation. In Fig. 6, Mo lines located on the gate oxide are also shown. Because of good heat conduction through the thin gate oxide, the Mo gate in active regions did not deform.

To discuss the gate deformation mechanism one-dimensional thermal diffusion simulation [10] was carried out. Figure 7 shows temporal temperatures profiles at the Mo surface and the field oxide bottom. E_L for HALA and the non-heat-assist cases were 300 mJ/cm^2 and 600 mJ/cm^2. In spite that HALA provided better activation as described above, temperature rising at the gate pad was reduced. Therefore, results shown in Fig. 5 was qualitatively explained by the simulation. However, the highest Mo temperature obtained by the simulation was too low to give rise to melt. Table I shows simulated highest temperatures for various Mo and oxide thicknesses. The difference between the highest temperatures for thin 50 nm and thick 100 nm Mo films, was too small to explain clear difference shown in Fig. 6. Two mechanisms which were not included in the simulation were possible candidates to explain these quantitative disagreement between experiments and the simulation. One is the Gibbs-Thomson effect [11], that is, lowering of melting point at structures with nano-meter scale small curvature. Another one is irradiation from backside as a results of diffraction of incident light at gate edges and its reflection at the filed oxide/Si interface (Fig. 8). The latter is considered to result in effective E_L increase at the gate edge.

4. Summary

Gate deformation by laser irradiation for junction formation was evaluated. In the case of R.T. LA, Mo that has very high melting point easily melted. HALA that utilizes low temperature substrate heating was helpful to relieve the problem. The mechanism of the gate deformation was discussed with one-dimensional thermal diffusion simulation.

Acknowledgements

This work was partly supported by NEDO/MIRAI Project. The authors thank to Komatsu Ltd. for their cooperation.

References

[1] K. Yamashita et al., SSDM 2003 Ext. Abst., pp.742-743.
[2] A. Mokhberi et al., IEDM 2002 Tech. Dig., pp. 879-882.
[3] T. Ito et al., 2003 VLSI Tech. Symp. Dig., pp.53-54.
[4] V.I. Kuznetsov et al., IEEE RTP Conf. 2003, pp. 63-74.
[5] A. Shima, IEDM 2003 Tech. Dig., pp.493-496.
[6] K. Kurobe et al., IWJT 2002 Ext. Abst., pp. 35-36.
[7] K. Kurobe et al., Jpn. J. Appl. Phys. **44**, pp. 8391-8395, 2005.
[8] M. Hino et al., Ext. Abst. SSDM Ext. Abst., pp.494-495.
[9] H. Tsukamoto et al., Solid State Electron. **42**, pp.547-556, 1998.
[10] A. Matsuno et al., Nucl. Instr. and Meth. B **237**, pp. 136-141, 2005.
[11] I. Shyjumon et al., The Eur.Phys. J. D, to be published, 2005.

1-4244-0181-X/06/$20.00 ©2006 IEEE

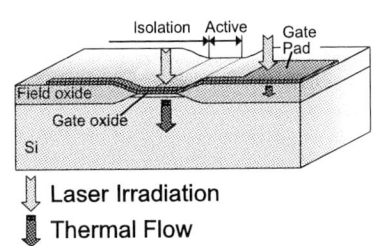

Laser Irradiation
Thermal Flow

Fig. 1 Schematic cross section of MOSFET along gate electrode. Gate pad on a thick field oxide is heated to higher temperature because of low thermal flow through the field oxide.

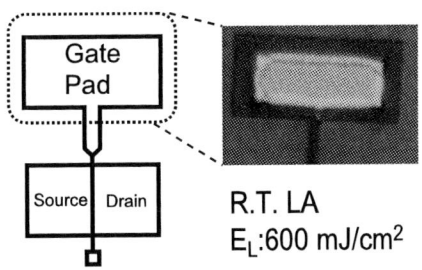

R.T. LA
E_L:600 mJ/cm^2

Fig. 2 Plan-view optical microphotograph of Mo gate pad after laser irradiation to activate dopants and schematic illustration to indicate the position of gate pad.

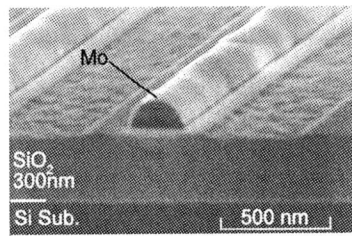

Fig. 3 SEM cross sectional image of Mo L/S formed on thick SiO$_2$. Mo lines deformed because of melting by laser heating. Initial Mo thickness was 50 nm.

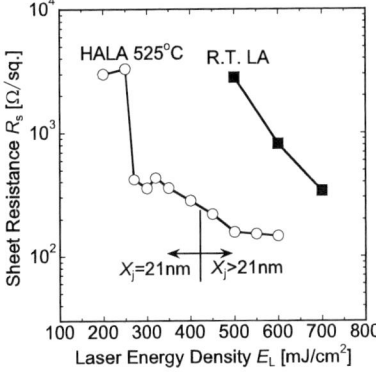

Fig. 4 Relationships between sheet resistance and laser energy density. HALA provides good dopant activation by weaker laser irradiation.

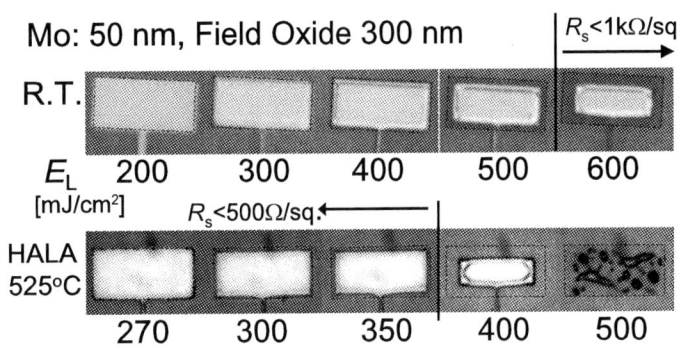

Fig. 5 Plan-view optical microphotographs of Mo gate pad after laser irradiation. Dashed lines indicate original figure of the gate pad.

1μm Line & Space (300mJ/cm^2)

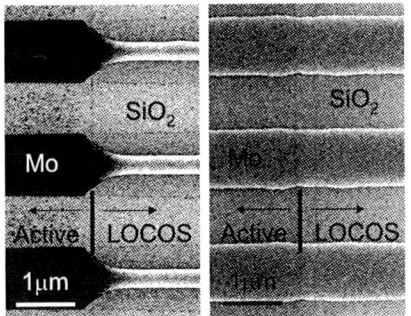

Mo 50nm   Mo100nm

Fig. 6 Plan-view SEM microphotographs of Mo L/S formed on the border of active and isolation regions. Thin 50 nm Mo showed serious deformation. The gate oxide thickness and field oxide thickness were 5 nm and 300 nm, respectively.

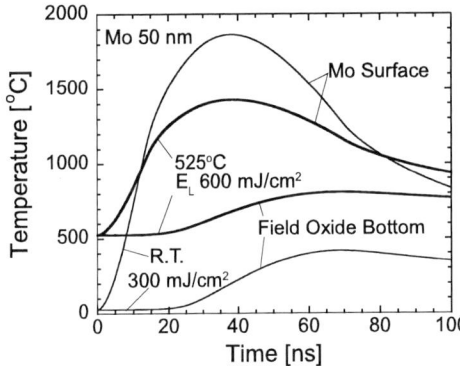

Fig. 7 Simulated temperature at the Mo gate surface and the field oxide bottom.

Table I Simulated maximum temperature for various Mo thicknesses and oxide thicknesses.

	Mo Thickness			
	50 nm	100 nm	50 nm	100 nm
Mo Surface	1430°C	1360°C	1020°C	990°C
Oxide Bottom	810°C	780°C	1000°C	970°C
	300 nm	300 nm	5 nm	5 nm
	Oxide Thickness			

Fig. 8 Model to explain additional heating at the gate edge. Laser light diffracted at the edge and reflected at the field oxide/Si interface heats up the gate edge from its bottom side.

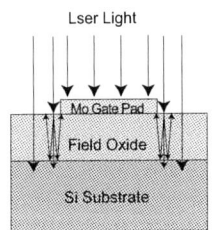

2006 International Symposium on VLSI Technology, Systems, and Applications

Low-Leakage Diode String Design
without Extra Circuits for ESD Applications

Shao-Chang Huang, Yu-Hung Chu, Chen-Chi Kuo, T.Y. Huang, M.H. Song and Mi-Chang Chang

Taiwan Semiconductor Manufacturing Company, Science-Based Industrial Park,
Hsin-Chu, Taiwan, R.O.C. Phone: 886-3-5636688 Ext. 7034706, Fax: 886-3-5670282
E-mail: huangsc@tsmc.com

Abstract – **A novel low-leakage diode string design using separate diode objects in 0.18 m CMOS processes is proposed in this paper. Diode strings can be divided into two (or three) groups with only a large shallow trench isolation (STI) used as spacing. With STI used to separate the groups, the diode string successfully prevents excessive current leakage. The turn-on voltage of the diode string can be derived from the equations concluded from the experiments.**

INTRODUCTION

Diode strings have been used extensively in electrostatic discharge (ESD) protection for integrated circuits (ICs). Under forward bias conditions, a diode can carry a large amount of current with predictable terminal voltages, thus protecting other circuits from permanent damage. To ensure there is no damage to the diodes themselves, they are usually designed at sufficient area sizes. Theoretically, the turn-on voltage of the diode string is equivalent to multiplying the turn-on voltage of a single diode by the number of diodes.

However, the Darlington effect [1] exists between the diodes, and contributes to a lower than expected turn-on voltage for the diode strings. Equation (1) from reference [2] indicates that the turn-on voltage decreases as the bipolar gain increases. In actuality, the value of should be greater than or equal to 1, which leads to the fact that the turn-on voltage of the diode string does not increase in a linear fashion as the number of diodes increases.

$$V_{total} = mV_f - V_o \frac{m(m-1)}{2} \log(\ +1) \qquad (1)$$

where

Vtotal= total voltage across diode chain
Vf= forward turn-on voltage for a particular Io
V_o= nkTln(10)/q
m= number of diodes cascaded
Beta= Parasitic PNP transistor gain
q= electronic charge
k= Boltzman's constant
T= absolute (K) temperature
n= diode ideality factor

In order to prevent the Darlington effect in the diode string, there have been many studies, such as cladded diodes and

boosted diodes, in reference [3], snubber-clamped diode strings, in reference [4], NMOS-controlled lateral SCR (NCLSCR) devices added into the stacked diode string, in reference [5] and Triple-Well diode strings in reference [6]. In reference [3]-[5] additional circuits are implemented in order to overcome the excessive leakage issues, while reference [6] introduces a technique applicable to RF-ESD using only an extra triple-well to avoid the vertical bipolar effect.

While additional circuits or extra process masks were necessary in previous studies, a brand new diode string structure is revealed in this paper. This new design can easily reduce the Darlington effect without the needs for extra circuits or masks.

II. EXPERIMENT

A conventional 5-stage diode string (5D) in p-Substrate CMOS is illustrated in fig. 1. Two new 5stage diode strings with large STI spacing are divided into two groups (4-stage and 1-stage) (4D1D) and three groups (3-stage, 1-stage and 1-stage) (3D1D1D), and are shown in fig. 2 and fig. 3, respectively.

Three kinds of conventional diode strings with two new structures, which were built with large STI spacing, were taken into consideration in our experiment. All the candidates are listed in table . The items shown as 1D, 3D and 5D are conventional diode strings without large STI spacing. The items named mDnD are the first type of the new structure. A large STI spacing was implemented between two groups of diodes. The first group contained m diodes, while the other was composed of n diodes. The mDnDpD items in table are the second type of the new structure. Two large STI spacing were implemented between three groups of diodes, where each group contains m, n and p diodes, respectively.

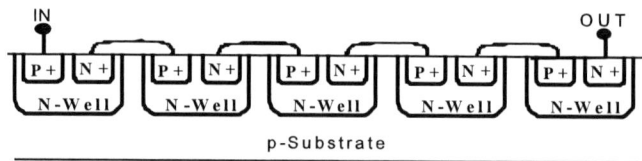

Fig. 1. The cross section of a conventional 5-stage diode string has no large STI spacing (5D).

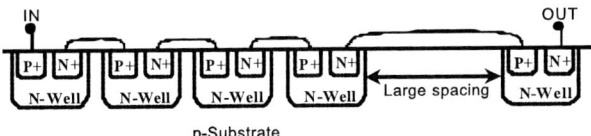

Fig. 2. An example of a 5-stage diode string (4D1D) divided into two groups with a large STI spacing.

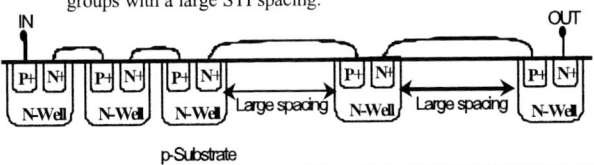

Fig. 3. An example of a 5-stage diode string (3D1D1D) divided into three groups with the large STI spacing.

The I-V curves of the structures mentioned above were measured using an HP4156 with a −1V~8V voltage swing, and were also measured using a TLPG (Transmission Line Pulse Generator) system to determine detailed turn-on characteristics. HBM (Human Body Model) and MM (Machine Model) ESD tests were applied to verify their ESD performances. Only positive ESD zapping was considered, because the diode strings are only designed for forward ESD impulse protection, and are not suitable for ESD current bypass when operating in reverse breakdown mode.

III. TEST RESULTS AND DISCUSSION

Fig. 4 shows a comparison of the I-V curves of the three traditional diode strings and mDnD type diode strings, measured on the HP4156. The I-V curves for the mDnDpD type are shown in fig. 5. It can be seen that under the same operation voltage, the reconfigured diodes will achieve much lower leakage currents than those of traditional diodes. Namely, the reconfigured diode strings have a much larger turn-on voltage than conventional structures.

The detailed data for fig. 4 and fig. 5 listed in table 1. The turn-on voltage of a conventional 5D diode string with a 10μA input current is equal to $2*V_D$, where V_D is the turn-on voltage of a single diode (1D). The vertical bipolar gain leads to an extremely low turn-on voltage. Moreover, the turn-on voltage of a 3D1D structure with a 10μA input current is equal to $3.59*V_D$, while that of a 1D3D structure is merely $1.98*V_D$. The current distribution of a 5D diode string is shown in fig. 6, to illustrate the Darlington effect in conventional structures. The current of a single conventional 5D diode string can be derived from (2).

$$Is= \frac{\beta_5}{\beta_5+1}I_{E5} + \frac{\beta_4}{\beta_4+1}I_{E4} + \frac{\beta_3}{\beta_3+1}I_{E3} + \frac{\beta_2}{\beta_2+1}I_{E2} + \frac{\beta_1}{\beta_1+1}I_{E1}$$

$$= (\ \frac{\beta_5}{\beta_5+1} + \frac{\beta_4}{\beta_4+1} \times \frac{1}{\beta_5+1} + \frac{\beta_3}{\beta_3+1} \times \frac{1}{\beta_4+1} \times \frac{1}{\beta_5+1}$$

$$+ \frac{\beta_2}{\beta_2+1} \times \frac{1}{\beta_3+1} \times \frac{1}{\beta_4+1} \times \frac{1}{\beta_5+1}$$

$$+ \frac{\beta_1}{\beta_1+1} \times \frac{1}{\beta_2+1} \times \frac{1}{\beta_3+1} \times \frac{1}{\beta_4+1} \times \frac{1}{\beta_5+1}\ )\ I_{E5}$$

$$Idd = (\beta_1+1)(\beta_2+1)(\beta_3+1)(\beta_4+1)(\beta_5+1) Iss \qquad (2)$$

The contribution to the turn-on voltage made by the second

group of diodes in an mDnD structure is less than that of one diode, no matter how many diodes are built. According to the test results, the turn-on voltage of an mDnD structure can be summarized as (3).

$$m* V_D < V_{ON-mDnD} < (m+1)* V_D \qquad (3)$$

where

$V_{ON-mDnD}$: the turn-on voltage of an mDnD diode string

V_D: the turn-on voltage of a single diode (1D)

m: the number of diodes in the first group of an mDnD structure

For the mDnDpD type diode string, mDnDpD, we obtain similar equations for the turn-on voltage, (4) and (5) shown as below.

$$m*V_D < V_{ON-mDnDpD} < (m+1)* V_D$$
if m has the greatest value among m, n and p $\qquad (4)$

$$(m+n-1)* V_D < V_{ON-mDnDpD} < (m+n+1)* V_D,$$
$$V_{ON-mDnDpD} < (m+n+p-2)* V_D$$
if m does **not** have the greatest value among m, n and p $\qquad (5)$

where

$V_{ON-mDnDpD}$: the turn-on voltage of a mDnDpD diode string

V_D : the turn-on voltage of a single diode (1D)

m : the number of diodes in mD group

n : the number of diodes in nD group

p : the number of diodes in pD group

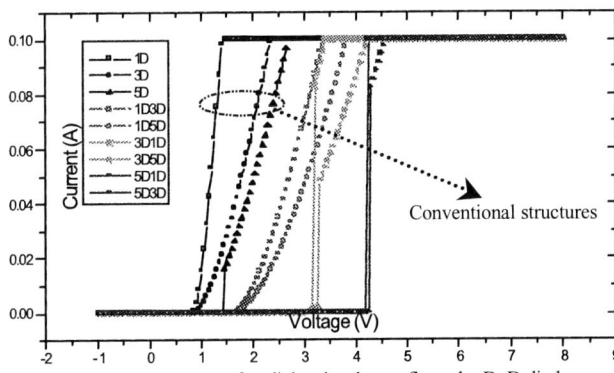

Fig. 4. I-V curves of traditional and reconfigured mDnD diode strings measured using a HP4156.

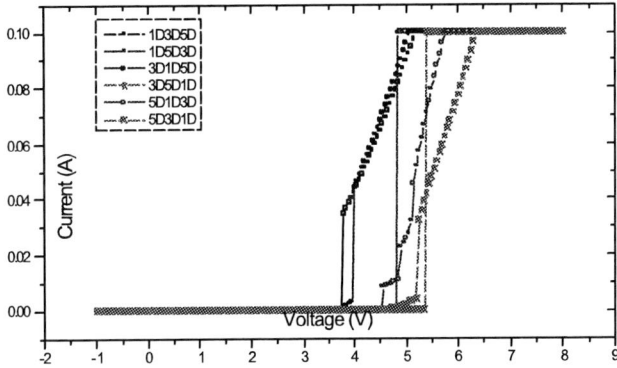

Fig. 5. I-V curves of reconfigured mDnDpD diode strings measured using a HP4156.

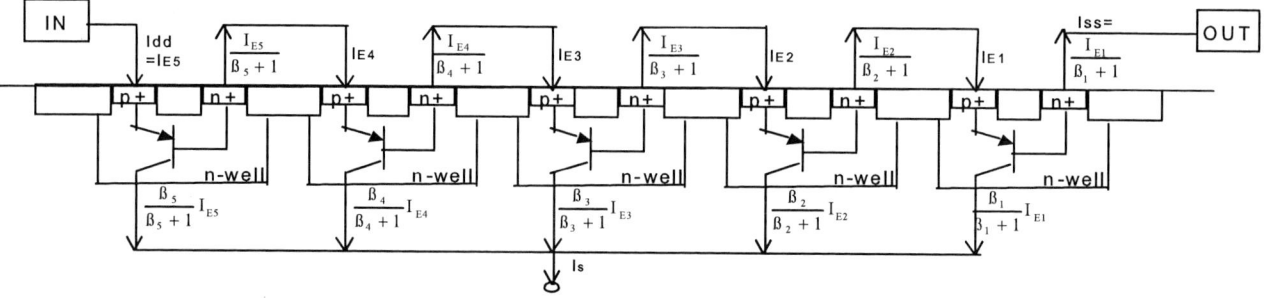

Fig. 6. Diagram showing the large current flow of a conventional 5D diode string caused by the Darlington effect.

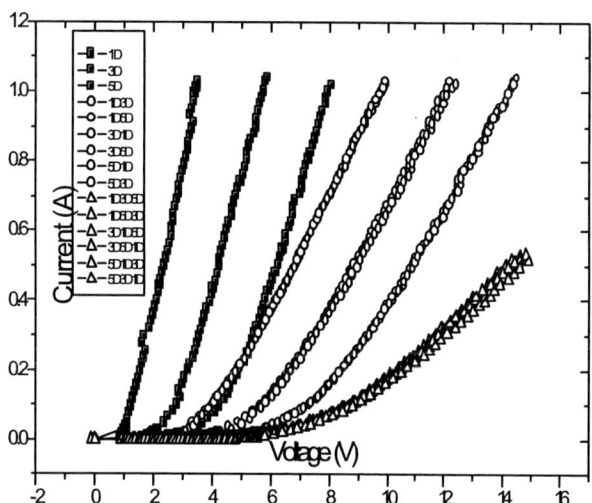

Fig. 7. Turn-on resistance (R_{ON}) measured using a TLPG

TABLE
TURN-ON VOLTAGE (V_{ON}) MEASURED USING THE HP4156.

Structure Name	Equivalent diodes	Von(V) @10μA
1D	1 diode	0.71
3D	1.08 diode	0.77
5D	2 diodes	1.42
1D3D	1.98 diodes	1.41
1D5D	2.01 diodes	1.43
3D1D	3.59 diodes	2.55
3D5D	3.63 diodes	2.58
5D1D	5.18 diodes	3.68
5D3D	5.24 diodes	3.72
1D3D5D	4.44 diodes	3.15
1D5D3D	5.49 diodes	3.9
3D1D5D	4.28 diodes	3.04
3D5D1D	6.61 diodes	4.69
5D1D3D	5.83 diodes	4.14
5D3D1D	5.68 diodes	4.03

TABLE
TURN-ON RESISTANCE (R_{ON}) MEASURED USING A TLPG

Structure Name	Ron(ohm)	String Name	Ron(ohm)
1D	2.42	1D3D5D	12.89
3D	3.10	1D5D3D	12.89
5D	3.62	3D1D5D	12.89
1D3D	5.93	3D5D1D	12.89
1D5D	6.35	5D1D3D	12.89
3D1D	5.93	5D3D1D	12.89
3D5D	6.59		
5D1D	6.35		
5D3D	6.59		

TABLE
ESD PERFORMANCE LIST

Structure Name	HBM (kV)	MM (V)
1D	5.75kV	50V
3D	6.4kV	50V
5D	6.25kV	75V
3D1D	6.05kV	100V
5D1D	6.3kV	100V
5D3D	6.35kV	125V
5D3D1D	6.35kV	375V

All the structures were measured using the TLPG to obtain the turn-on resistance (R_{ON}) value. The test results are shown in fig. 7. The R_{ON} of a conventional diode string is about 3Ω, and the R_{ON} values of the reconfigured mDnD and mDnDpD diode strings are about 6Ω and 12.89Ω, described in table , respectively. Equation (6) can be used to calculate an approximate R_{ON} value.

$$R_{ON} = 3*2^{(x-1)} \qquad (6)$$

where x is the number of diode string groups.

Table lists the ESD test results. For both the HBM and MM tests, only positive zapping was adopted. The lower R_{ON} leads to a poorer HBM immunity value of 5.75kV for 1D. All other devices listed in the table passed the HBM test voltages at higher than 6kV. Since diodes are designed for being used in forward ESD zapping, MM testing is not a major concern because MM has positive and negative waveforms. However, it can be deduced that the greater the R_{ON} value, the better the MM immunity.

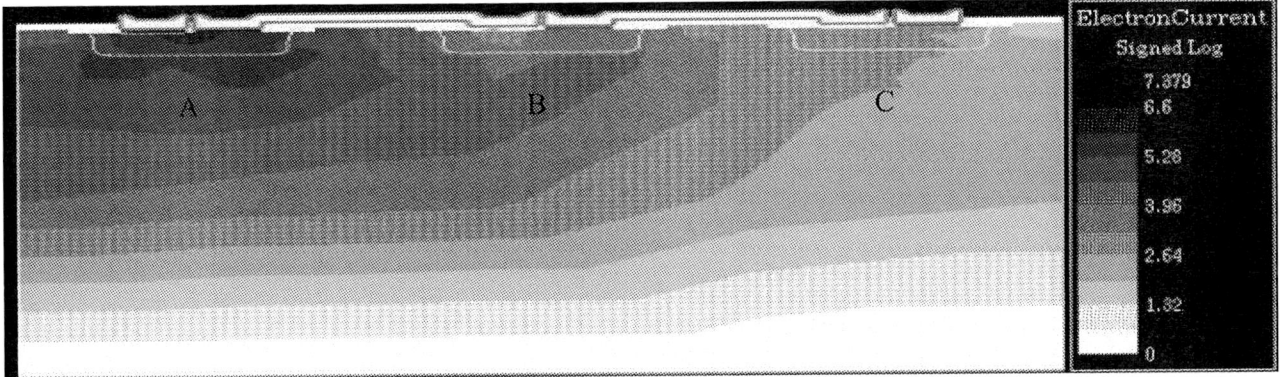

Fig. 8. Current distribution simulat ion of conventional 3-stage diode string (3D).

Fig. 9. Current distribution simulation of a reconfigured (2D1D) diode string.

IV. SIMULATION

T-Suprem-4 and MEDICI were used to perform a current distribution simulation of both a conventional 3D diode string and a reconfigured 2D1D diode string. The simulation results are shown in fig. 8 and fig. 9, respectively. From the figures, it can be seen that diodes A and B, located on the le ft-hand side of the 3D conventional diode string, have an effect on the current distribution of diode C. While diode F, located on the right-hand side of the 2D1D structure, seems to be perfectly isolated from diodes D and E, which is the reason why diode strings divided into several groups with a large STI spacing are capable of reducing the excess carrier effect.

. CONCLUSIONS

There is no doubt that the Darlington effect leads to lower turn-on voltages and large leakage currents for conventional diode strings. In this work, the problem was easily improved by dividing conventional diode strings into two or more groups, separated by a large STI spacing. As a result, both the cost and time to design extra circuits or making process masks can be reduced. This paper also presents several new equations that have been summarized from our experiments to obtain the turn-on voltage and R_{ON} values. This was a major breakthrough and, as a result, designing low leakage diode strings that give better ESD performances will become much easier.

REFERENCES

[1] S. Darlington, "Semiconductor Signal Translating Device," U.S.Patent 2,663,806, Dec. 22, 1953.

[2] S. Dabral, R. Aslett and T. Maloney, "Designing On-Chip Power Supply Coupling Diodes for ESD Protection and Noise Immunity," EOS/ESD Symposium, pp. 239-249, 1993.

[3] Timothy. J. Maloney and Sanjay Dabral "Novel Clamp Circuits for IC Power Supply Protection," EOS/ESD Symposium, 1.1.1~1.1.12, 1995.

[4] S. H. Voldman, G. Gerosa, V.P. Gross, N. Dickson, S. **Furkay and J. Slinkman, "Analysis of Snubber-Clamped Diode-String Mixed Voltage Interface ESD Protection Netweork for Advanced Microprocessors," EOS/ESD Symposium, pp. 43-61, 1995.**

[5] M.D. Ker and W.Y. Lo, "Design on the Low-Leakage Diode String for Using in the Power-Rail ESD Clamp Circuits in a 0.35 m Silicide CMOS Process," IEEE Trans. on Solid-State Circuits, Vol.35, pp. 601-611, 2000.

[6] S.S. Chen, T.Y. Chen, T.H. Tang, J.L. Su, T.M. Shen and J.K. Chen, "Low-Leakage Diode String Designs Using Triple-Well Technologies for RF-ESD Applications," IEEE. Electron Devices Letters, Vol. 24, No. 9, pp. 595-597, 2003.

2006 International Symposium on VLSI Technology, Systems, and Applications

Investigation on RF Characteristics of Stacked P-I-N Polysilicon Diodes for ESD Protection Design in 0.18-μm CMOS Technology

Yu-Da Shiu, Che-Hao Chuang, and Ming-Dou Ker*

SoC Technology Center, Industrial Technology Research Institute, Hsinchu Taiwan.

* Nanoelectronics and Gigascale Systems Laboratory, Institute of Electronics, National Chiao-Tung University, Hsinchu, Taiwan.

ABSTRACT

An ESD protection design by using the stacked P-I-N polysilicon diodes for CMOS RF integrated circuits is proposed to reduce the input capacitance and to avoid the noise coupling from the common substrate. In this paper, the dc I-V characteristics, RF S-parameters, and ESD robustness of the stacked P-I-N polysilicon diodes are investigated in a 0.18-μm salicided CMOS process. This polysilicon diode with small parasitic capacitance and high ESD robustness is fully process compatible to general CMOS process without extra process modification.

INTRODUCTION

For wireless system applications, the ESD protection device in GHz RF ICs is required to have low parasitic capacitance, broadband and constant capacitance, insensitive to substrate coupling noise [1], and high enough ESD robustness. In order to fulfill these requirements, diodes are commonly used for ESD protection in I/O circuit. Because multiple capacitors stacked in series can result in a total capacitance smaller than that of a single capacitor, the diode string is one solution that has been applied for RF ESD protection design. However, the conventional p+/n-well diode string has the p-n junction located within the common substrate of CMOS ICs, which may cause some issues. First, the substrate noise can be coupled into the RF input node through the conventional diodes to seriously degrade circuit performance of RF IC. Second, the leakage current constructed by the parasitic vertical p-n-p bipolar transistors in the diode string is increased exponentially with increasing the voltage difference from anode to cathode [2]. Third, the additional N-well to P-substrate junction capacitance in every p+/n-well diode structure causes some increase in the total parasitic capacitance.

Therefore, a P-I-N polysilicon diode for ESD protection in RF broadband and wireless system applications was proposed [3]. Because the polysilicon diode is isolated far away from the common substrate by the thick field-oxide layer, it is free from the substrate noise coupling problem and can be connected in series to further reduce the total parasitic capacitance.

In this paper, in order to further understanding the impact to RF performance, S-parameters of the stacked P-I-N polysilicon diode under forward-biased condition are investigated in 0.18-μm CMOS process. The spacing between the N+ and P+ regions and the total width of the P-I-N polysilicon diode may affect the device characteristics, which will be investigated in the experimental test chips.

STACKED POLYSILICON DIODE

The structure of the stacked polysilicon diodes is shown in Fig. 1. The polysilicon layer is drawn with separated P+/N+ doping regions to realize the diode structure [3]. Between the N+ and P+, there is a region indicated "I" without doping impurity. To use polysilicon diode as an effective ESD protection device in GHz RF ICs, the RF characteristics of polysilicon diodes with the center region "I" under different layout spacing S and different total junction perimeters W must be investigated. The fabrication of polysilicon diode is fully process compatible to general CMOS process without extra process modification.

MEASUREMENT RESULTS AND DISCUSSION

The ESD stress is performed with a *Paragon* ESD simulator. The measured ESD robustness of the P-I-N polysilicon diodes with different spacing and width are shown in Table I. Each measurement is performed on 3 samples. In Table I, because a polysilicon diode with smaller layout spacing has a lower turn-on resistance, it has a relatively higher ESD level. When the total junction perimeter (width) is increased, the ESD robustness of the polysilicon diodes is increased. The ESD robustness of the stacked P-I-N polysilicon diodes under forward-biased condition with N+/P+ spacing S of 0.6 μm and junction perimeter (width) of 600 μm are shown in Table II. Increasing the number of polysilicon diodes in the stacked configuration does not reduce its ESD robustness.

The S-parameters of these ESD protection schemes have been measured on wafer with two-port ground-signal-ground (G-S-G) probes from 1-to-10 GHz and make a calibration with de-embedding techniques. The RF measurement system (HP-8510B) is used to measure the S-parameters of the polysilicon diodes under different diode dimensions. The parasitic capacitance can be extracted from the measured S-parameters. Fig. 2 shows the parasitic capacitances of the polysilicon diodes under different junction perimeter (width). The parasitic capacitance and ESD robustness are both increased by the increase of junction perimeter (width) under forward-biased condition. Fig. 3 shows the parasitic capacitance of the polysilicon diodes with junction perimeter (width) of 600 μm but different spacing S. When the spacing S of the polysilicon diode is 0.6 μm and junction perimeter (width) is 600 μm, the capacitance approaches 200 fF. A typical request on the maximum loading for GHz RF ESD protection device was specified as ~200 fF [4]. Therefore, the P-I-N polysilicon diode with the spacing of 0.6 μm and junction perimeter (width) of 600 μm is suitable for GHz RF applications.

Fig. 4 shows the measured total parasitic capacitance for stacked polysilicon diodes with different diode numbers. The parasitic capacitance is decreased apparently when the number of stacked polysilicon diodes is increased. Therefore, for higher RF operating frequency applications such as WiMAX which required much lower input parasitic capacitance, the stacked P-I-N polysilicon diode is a good choice for RF ESD protection. The 1-to-10-GHz S21 and S11 of polysilicon diodes with different numbers of stacked diodes are shown in Fig. 5. The insertion loss S21 of these polysilicon diodes with the same device dimension is decreased when the operation frequency is increased. However, the S21 of the stack-3 polysilicon diodes is with a little decrease during the 1-to-10 GHz frequency band. With less degradation on RF performance and high enough ESD robustness, the proposed stacked polysilicon diodes will be a useful solution to GHz RF ESD protection.

CONCLUSIONS

ESD protection design with stacked P-I-N polysilicon diodes for GHz RF ICs has been experimentally investigated with ESD robustness and RF characteristics in a 0.18-μm salicided CMOS process. The experimental results have shown that the stacked polysilicon diodes with a low enough turn-on resistance in forward-biased condition are good enough to be the ESD clamp device for RF ICs. These polysilicon diodes can be further stacked to significantly

1-4244-0181-X/06/$20.00 ©2006 IEEE

reduce the total input capacitance and to achieve high enough ESD robustness for GHz RF circuits.

REFERENCES

[1] A. Wang, H. Feng, R. Zhan, G. Chen, and Q. Wu, "ESD protection design for RF integrated circuits: new challenges," in *Proc. of IEEE Custom Integrated Circuits Conf.*, 2002, pp. 411–418.

[2] M.-D. Ker and W.-Y. Lo, "Design on the low-leakage diode string for using in the power-rail ESD clamp circuits in a 0.35-μm silicide CMOS process," *IEEE Journal of Solid-State Circuits*, vol. 35, pp. 601–611, 2000.

[3] M.-D. Ker and C.-Y. Chang, "ESD protection design for CMOS RF integrated circuits using polysilicon diodes," *Microelectronics Reliability*, vol. 42, pp. 863-872, 2002.

[4] C. Richier, P. Salome, G. Mabboux, I. Zaza, A. Juge, and P. Mortini, "Investigation on different ESD protection strategies devoted to 3.3V RF applications (2 GHz) in a 0.18μm CMOS process," in *Proc. of EOS/ESD Symp.*, 2000, pp. 251-259.

Table I ESD robustness of the P-I-N polysilicon diodes under different layout parameters.

HBM ESD level		Junction Perimeter (Width) (μm)			
		600	300	240	100
Spacing (μm)	0.4	**6 kV**	**3 kV**	**2.5 kV**	**1 kV**
	0.6	**5 kV**	**2.5 kV**	**2 kV**	**500 V**
	0.8	**4 kV**	**2 kV**	**1.5 kV**	N/A
	1.0	**3 kV**	N/A	N/A	N/A

Table II ESD robustness of the stacked P-I-N polysilicon diodes under different numbers of stacked diodes.

Junction Perimeter (Width)=600 μm , Spacing=0.6 μm			
Stack numbers	Stack 1	Stack 2	Stack 3
HBM ESD level	**5 kV**	**5 kV**	**5 kV**

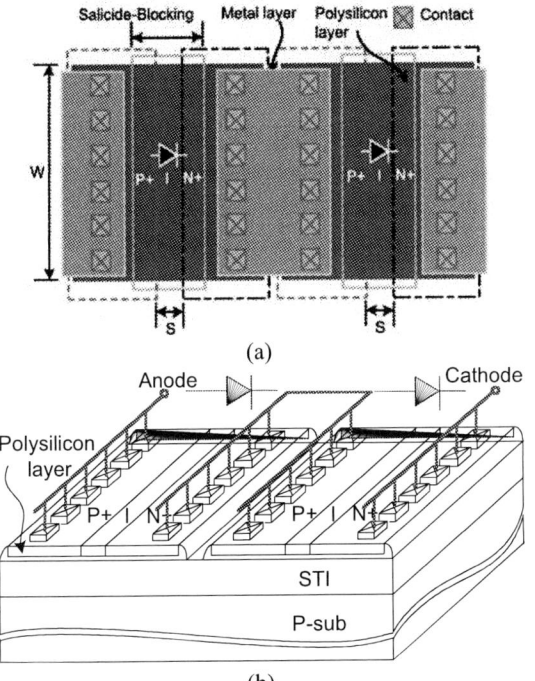

Fig. 1 (a) Layout top view, and (b) device structure, of the stacked polysilicon diodes realized in bulk CMOS technology.

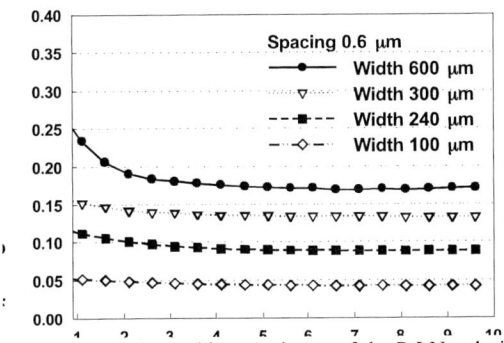

Fig. 2 The extracted parasitic capacitance of the P-I-N polysilicon diodes under different junction perimeter (width).

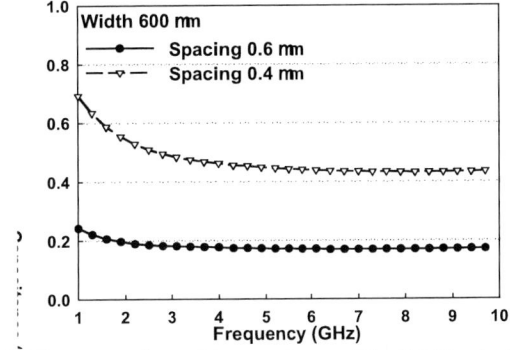

Fig. 3 The extracted parasitic capacitance of the P-I-N polysilicon diodes with different spacing of the center region "I:".

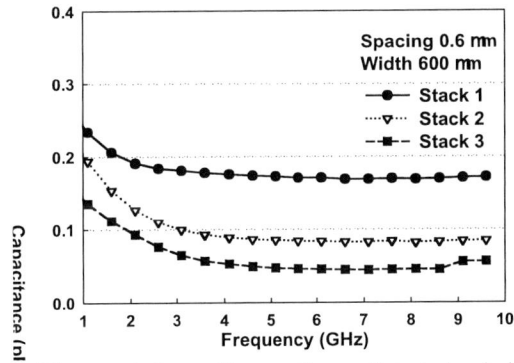

Fig. 4 The extracted parasitic capacitance of stacked polysilicon diodes measured with different numbers of stacked diodes.

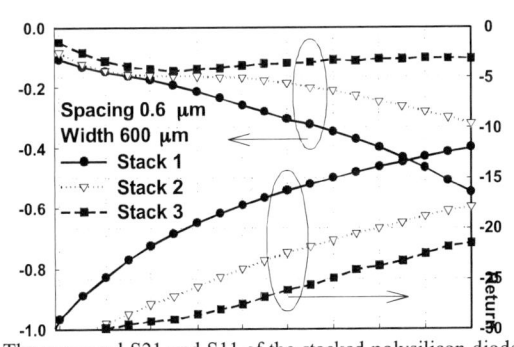

Fig. 5 The measured S21 and S11 of the stacked polysilicon diodes with different numbers of stacked diodes.

2006 International Symposium on VLSI Technology, Systems, and Applications

65nm SOI CMOS Technology for High Performance Microprocessor Application

Samuel K.H.Fung, P.A.Grudowski[1], C.H.Wu, V. Kolagunta[1], N. Cave[1], C.T.Yang, S.J.Lian, V. Adams[1], O. Zia[1], B. Min[1], N. Grove[1], K.H.Chen, W.J.Liang, D.H.Lee, H.T.Huang, J. Cheek[1] and H.C.Tuan

R&D, Taiwan Semiconductor Manufacturing Company, Science-Based Industrial Park, Hsinchu, Taiwan, R.O.C.

[1]Freescale Semiconductor, Austin, TX, USA

Abstract

This paper presents a state-of-the-art 65nm SOI CMOS transistor technology target for high performance microprocessor application. N/PFET shows short channel control meeting manufacturing margin at 32/35nm respectively. By using Dual Contact Etch Stop Layer (CESL) process, PFET Ion is 875uA/um at Ioff=100nA/um (V_{DD}=1.2V), which is 65% increase over 90nm node. The 65nm technology also offers SRAM cell sizes ranging from $0.499um^2$ to $0.64um^2$ for various speed and density requirement.

Key Technology Feature

In order to achieve >30% speed enhancement over 90nm node, four key enabling features are employed in 65nm node – (1) Dual-CESL process [2] to enhance PFET mobility without degrading NFET, (2) Nickel silicide process for low gate resistance at gate length <32nm., (3) careful thermal cycle and implant tuning for abrupt extension junction while keeping very low reverse diode leakage and junction capacitance, (4) Minimize STI stress to avoid Ion degradation in narrow transistor.

In a dual-CESL process, a first film say compressive SiN is deposited and patterned on PFET region and then a second film say tensile SiN is deposited and patterned on NFET region (Fig. 1). Alternatively, tensile SiN can be deposited first while producing similar result. Fig. 2 shows a top down and cross-section SEM showing the boundary between the tensile and compressive film. During the second SiN patterning, the first SiN film is potentially attacked. Carefully process control is necessary to avoid weakening the first SiN stress. Another challenge for this technology is the contact formation at the overlapping region where the SiN film is twice as thick as other region.

Dual-CESL stressor is chosen over embedded SiGe stressor [1] mainly because the process complexity is less. Though there is always doubt whether compressive CESL can give similar hole mobility enhancement compared to SiGe stressor. As demonstrated in our 65nm silicon as well as other literatures [2], highly compressive CESL can give comparable hole mobility enhancement as in SiGe stressor. At V_{DD}=1.2V, we demonstrate PFET DC Ion 875uA/um at I_{OFF}=100nA/um, which is 65% increase over

90nm node (Fig. 3). This result is comparable to the best SiGe PFET published. For NFET, Dual-CESL gives little benefit over single CESL process. As a result, NFET Ion-Ioff is comparable to 90nm node. We expect >10% NFET Ion improvement at later stage of development by applying Stress Memorization Technique (SMT) and advanced anneal technique like LSA [3].

Minimizing reverse diode leakage (I_{REV}) and junction capacitance (Cj) are very important for both AC performance and history effect for PD-SOI technology. As we reduce thermal cycle for more abrupt junction, Cj and I_{REV} will increase. As a result, 1st switch body potential becomes higher and then worsen history effect. Higher Cj will reduce gate-to-body coupling and hence reduce AC switching current. By careful LDD/HALO tuning, we have demonstrated NFET with floating body DIBL=153mV/V at Lg=32nm which is the best ever reported for PD-SOI technology. Even though diode leakage is so low, short channel control at 32/35nm for N/PFET is meeting manufacturing requirement (Fig. 4).

It is well known that compressive STI stress degrades NFET Ion especially for small width device. By using less compressive STI fill material, we have minimized Ion degradation for narrow NFET (Fig. 5). There is ~10% Ion increase from wide to narrow device which can be explained by smaller thermal resistance at narrow width [4].

Logic Performance and SRAM

The combination of 65% PFET Ion increase and aggressive gate length scaling gives ~50% boost to intrinsic inverter speed over 90nm node (Table 1). However, design effort is required to accommodate the small N/P drive current ratio (1.3) compared to 90nm (2.2). SRAM cell sizes ranging from $0.499um^2$ (for highest density) to $0.64um^2$ (for highest cell current) are offered in this technology for various speed and density requirement. We have demonstrated functional SRAM cell size at $0.499um^2$ (Fig. 6) and also yield for 1Mb $0.64um^2$ SRAM cell.

Acknowledgement

The work done in this paper is part of the join development program between TSMC and Freescale Semiconductor.

1-4244-0181-X/06/$20.00 ©2006 IEEE

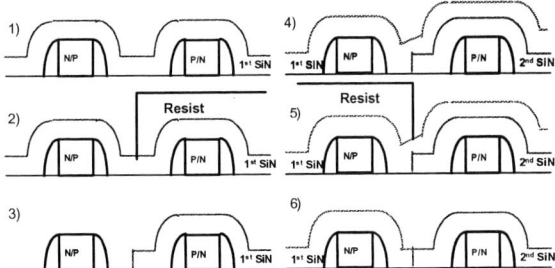

Fig. 1 Dual CESL process flow has two options : tensile first or compressive first.

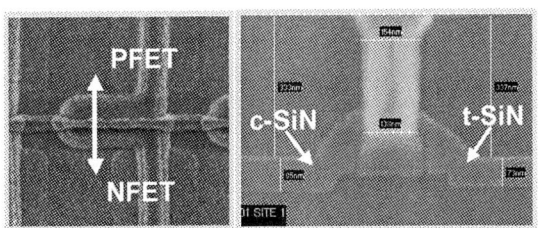

Fig. 2 Top down and cross section SEM for dual CESL process.

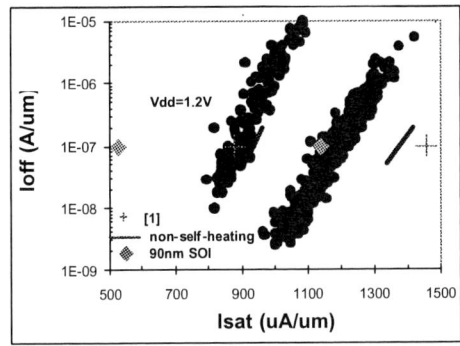

Fig. 3 N/PFET Ion-Ioff at Vdd=1.2V

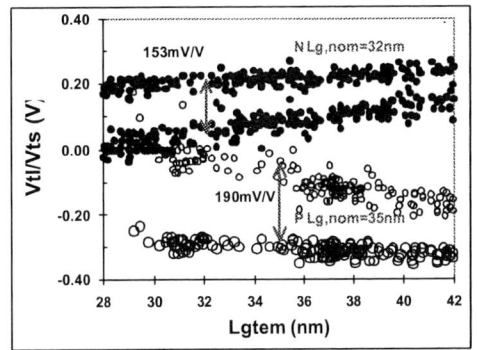

Fig. 4 N/PFET Vt roll-off at Vds=0.05/1.2V. Very low DIBL is achieved by tuning diode leakage.

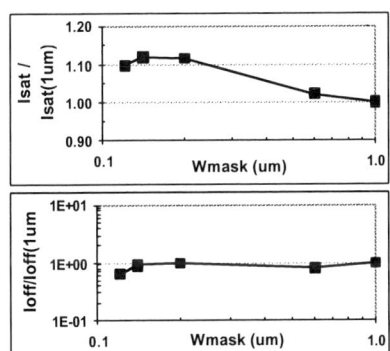

Fig. 5 NFET Ion/Ioff width trend.

	90nm	This work	
N Lg [nm]	40	32	-20%
N DC Isat [uA/um]			+1.8%
P Lg [nm]			-12.5%
P DC Isat [uA/um]			+65%
N/P Ratio			
Project Relative Instrinic Ring Speed	1	1.49	+49%

Table. 1 Summary of performance Enhancement from 90nm to 65nm node.

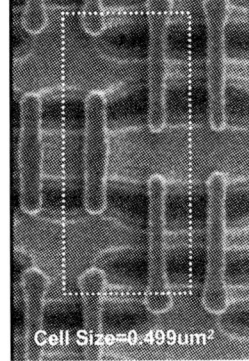

Fig. 6 SRAM with cell size=0.499um² top down SEM

Reference

[1] P. Bai et al, IEDM 2004, pp.657-660.

[2] E. Leobandung et al., VLSI 2005, p.126

[3] S.K.H.Fung et al., VLSI 2004.

[4] W. Jin et al., IEDM 1999.

2006 International Symposium on VLSI Technology, Systems, and Applications

A Promising Planar Transistor with in-situ Doped Selective Si Epitaxy Technology (**GORES MOSFET**) for 32nm node and beyond

Y. Kikuchi, Y. Tateshita, T. Kataoka, J. Wang, Y. Miyanami, N. Yamagishi, T. Ikuta, Y. Yamamoto, S. Hiyama, H. Ugajin, H. Ikeda, S. Fujita
R.Yamamoto, S. Kanda, T. Imoto, S. Kashiwadate, Y. Tagawa, H. Iwamoto, T. Ohno, T. Kobayashi, M. Saito, S. Kadomura and N. Nagashima
Semiconductor Technology Development Group, Semiconductor Solutions Network Company, Sony Corporation
Atsugi Tec. 4-14-1 Asahi-cho, Atsugi-shi, Kanagawa, 243-0014 Japan
Phone: +81-46-230-6744, Fax: +81-46-230-6556, E-mail: Yoshiaki.Kikuchi@jp.sony.com

Abstract

We demonstrated 40nm gate length "**G**ate **O**verlapped **R**aised **E**xtension **S**tructure: **GORES MOSFET**" without halo implantation and proofed that the ultra shallow junction (USJ) could coexist with the reducing parasitic resistance in GORES MOSFET. It is the new concept planar transistor with the gate overlapping the in-situ doped epitaxial extension to break through the trade off relation between reducing parasitic resistance and suppression of the short channel effect (SCE) [1]. As a matter of course, it has the USJ and the low parasitic resistance (LPR) for 32nm node with in-situ doped selective Si Epitaxy (ISDSE).

Introduction

Recently it has been difficult to improve the transistor performance due to scale down along ITRS. In this situation, three technologies are mainly noticed as the items of transistor performance improvement. The technologies are "The mobility improvement", "USJ and LPR" and "High-k dielectrics and metal gate". This time we focused on "USJ and LPR". As to both USJ and LPR, several kinds of studies were done, for example, Laser Spike Anneal (LSA), Flash Lamp Anneal (FLA) and Solid Phase Epitaxy (SPE) [2-5]. But LSA, FLA and SPE can not form the USJ which is required by 32 nm node of ITRS yet. We propose a new concept planar transistor GORES MOSFET. It has the raised extension with a facet (shown in Fig.1 (a)) on which the gate dielectric and the gate electrode are formed, so USJ and reducing parasitic resistance are realized; furthermore the two are not trade off. As above, GORES MOSFET can control the SCE more than a conventional planar transistor and an Elevated Source/Drain Extension transistor, we insist. Fig.2 shows the simulated GORES junction image which has USJ and thick extension. GORES MOSFET is similar to the conventional planar transistor and fabricated with a damascene flow, so it is easy to apply both "The mobility improvement" technology and "High-k dielectrics and metal gate" technology to GORES MOSFET. In this paper, we report the USJ formed by ISDSE and the properties of GORES MOSFET.

Device Fabrication

Fig.1 shows GORES process flow. After a dummy gate etching, a SiN spacer was formed to prevent the undesirable epitaxial growth on the side of the gate. Next, the raised extension with a facet configuration was formed by ISDSE as shown Fig.1 (a). The extension impurity concentration (N_{ext}) of nMOSFET and pMOSFET were about $2.5E19cm^{-3}$ and below. After that, the SiN spacer was removed and the SiO_2 spacer, which could control the gate overlap length (dL) on the extension, was formed. The side wall spacer is constituted by SiN and SiO_2 (Fig.1 (b)). After that, the deep source/drain was formed by an implantation and a Spike RTA at 1050°C. By means of a damascene flow, a dummy gate and a spacer SiO_2 are removed (Fig.1 (C)). As a result, a part of the extension was exposed. On the extension, a gate dielectric and a gate electrode were formed (Fig.1 (d)). The final structure is Fig.3 and we could fabricate a desirable gate overlapped formation and a uniform thickness gate oxide.

Results and Discussion

A, Junction characteristic
This time, with ISDSE and Spike RTA, we could materialize the USJ (Xj=5.1nm@5E18) and the LPR (Fig.4). Little diffusion by the Spike RTA could be observed in the both MOSFETs (Fig.5). The ISDSE and the Spike RTA enable to form the USJ and to reduce the parasitic resistance for 32nm

node and beyond.

B, GORES MOSFET characteristic
Id-Vg and Id-Vd characteristics of GORES are shown in Fig.6. Fig.7 shows Roll off characteristics without halo implantation and about 40nm gate length GORES MOSFET is demonstrated. In addition, we measured gate dielectric capacitance because of the GORES' unique gate dielectric shape, and the Tinv were 2.11/2.16nm (nMOSFET/pMOSFET). We analyzed the advantage of GORES MOSFET with the attention to both SCE and parasitic resistance by comparison with a conventional extension transistor. We discussed the effect of the ultra shallow junction and the scalability of parasitic resistance of GORES MOSFETs.Drain Induce Barrier Lowering (DIBL) is one of the most sensitive quantities and Lmin (=Lg@Ioff=100nA/um) was defined as the quantity that characterizes a short channel effects. In Fig.8 DIBL@Ioff=100nA/um,Vd=1V of GORES is compared with that of conventional S/D extension transistor at same SCE for nMOS and pMOS, respectively. We found that DIBL of GORES for both channel is improved by reducing the S/D extension junction depth. In particular, the improvement of pMOSFET is better than that of nMOSFET in GORES MOSFET, because the improvement of the pMOSFET's junction depth is better than that of nMOSFET's one. Next we paid attention to the parasitic resistance extracted by the shift-and-ratio method [6], which was expressed as Rsd. As shown in Fig.9, the overlap length dependence of Rsd dose not appear over 3nm in the experience, but from the simulation it is expected that Rsd has the strong dependence under 2nm. Because the less the overlap length is shortened, the more the parasitic resistance is increased, we insist that GORES is very effective to lower the parasitic resistance. And more, without worsening SCE, by increasing N_{ext} we can optimize Rsd and the overlap length. Fig.10 shows Rsd dependence of Ion and Lmin for nMOSFET and pMOSFET respectively. The drive current of both channel MOSFETs increase as the parasitic resistance decrease by changing impurities density or Epi thickness. On the one hand, Lmin only becomes barely long, even if the parasitic resistance decreases for nMOSFET and N_{ext} can be increased without worsening SCE, and furthermore, with halo implantation, the SCE will be improved (Fig.11 and 13), on the other hand, Lmin becomes large as the parasitic resistance increases for pMOSFET and the GORES SCE is better than the conventional one by about 10nm (Fig.12 and 14).

Conclusion

In this work, we fabricated GORES transistor which had USJ and LPR by using in-situ doped selective Si epitaxy technology. In nMOSFET, we proofed that the USJ could coexist with the reducing parasitic resistance and indicated that the transistor characteristic could be improved with more concentration of extension impurity and halo implantation. In pMOSFET, as compared with the SCE of conventional transistor, that of the GORES MOSFET could be improved by about 10nm.

References

[1] Y. Tateshita *et al.*, SSDM, p.904 (2005)
[2] N. Yasutake *et al.*, VLSI Tech. Dig., p.84 (2004)
[3] A. Shima *et al.*, VLSI Tech. Dig., p.174 (2004)
[4] K. Suguro *et al.*, IWJT proceedings, p.18 (2004)
[5] K. Adachi *et al.*, VLSI Tech. Dig., p.142 (2005)
[6] Y. Taur *et al.*, IEEE Electron Device Lett. Vol.13, No.5, p.267, 1992.

1-4244-0181-X/06/$20.00 ©2006 IEEE

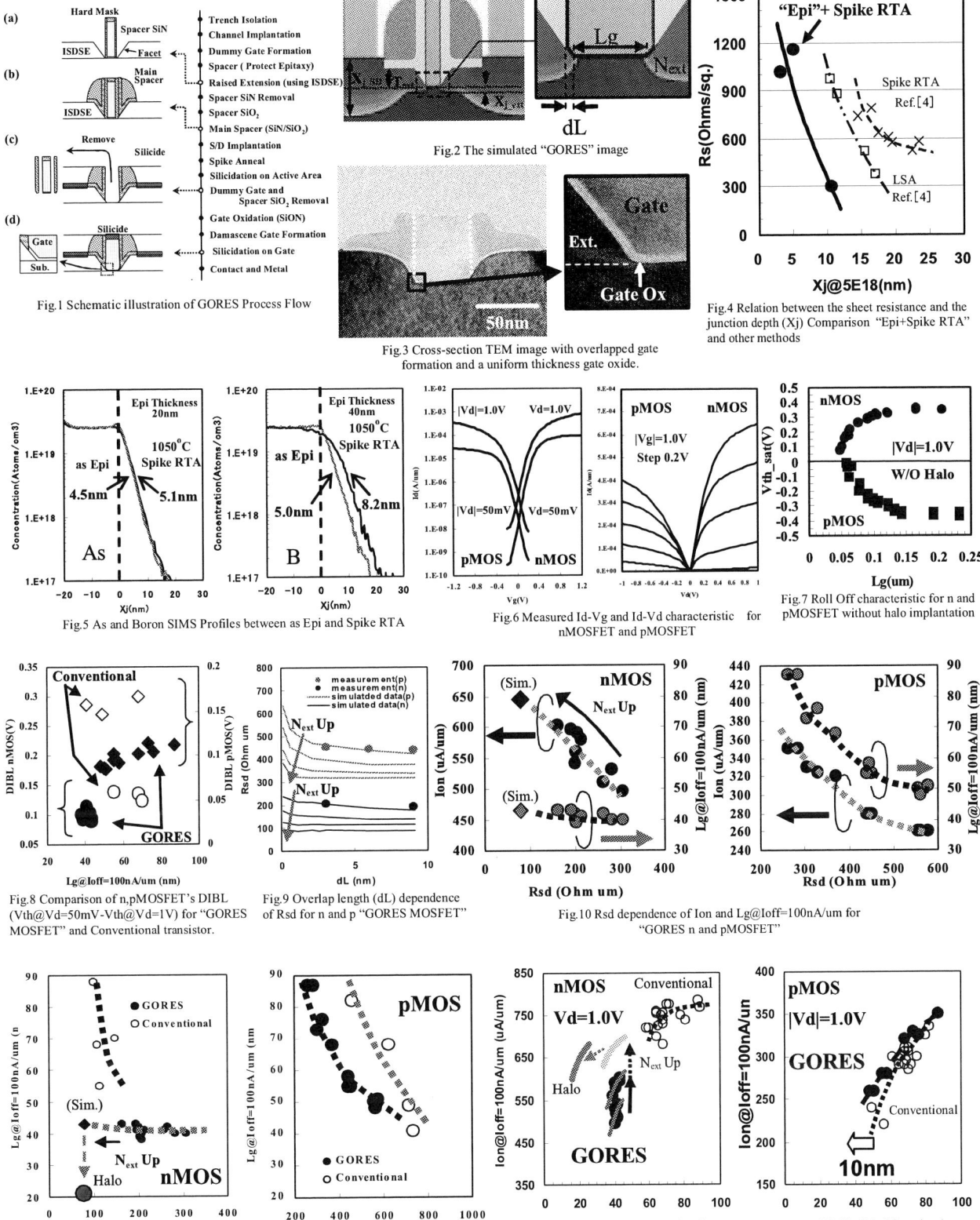

Fig.1 Schematic illustration of GORES Process Flow

Fig.2 The simulated "GORES" image

Fig.3 Cross-section TEM image with overlapped gate formation and a uniform thickness gate oxide.

Fig.4 Relation between the sheet resistance and the junction depth (Xj) Comparison "Epi+Spike RTA" and other methods

Fig.5 As and Boron SIMS Profiles between as Epi and Spike RTA

Fig.6 Measured Id-Vg and Id-Vd characteristic for nMOSFET and pMOSFET

Fig.7 Roll Off characteristic for n and pMOSFET without halo implantation

Fig.8 Comparison of n,pMOSFET's DIBL (Vth@Vd=50mV-Vth@Vd=1V) for "GORES MOSFET" and Conventional transistor.

Fig.9 Overlap length (dL) dependence of Rsd for n and p "GORES MOSFET"

Fig.10 Rsd dependence of Ion and Lg@Ioff=100nA/um for "GORES n and pMOSFET"

Fig.11 Rsd dependence of Lg@Ioff=100nA/um for "GORES" and Conventional nMOSFET

Fig.12 Rsd dependence of Lg@Ioff=100nA/um for "GORES" and Conventional pMOSFET

Fig.13 Comparison of nMOSFET's Roll off characteristic for "GORES" and Conventional MOSFET

Fig.14 Comparison of pMOSFET's Roll off characteristic for "GORES" and Conventional MOSFET

61

2006 International Symposium on VLSI Technology, Systems, and Applications

Strain-Induced Channel Backscattering Modulation in Nanoscale CMOSFETs

Hung-Wei Chen, Hong-Nien Lin*, Chih-Hsin Ko, Chung-Hu Ge, Horng-Chih Lin*, Tiao-Yuan Huang*, and Wen-Chin Lee

Taiwan Semiconductor Manufacturing Company Ltd., Hsinchu, Taiwan, R.O.C.
*Institute of Electronics, National Chiao-Tung University, Hsinchu, Taiwan, R.O.C.
TEL: 886-3-6665174 Fax: 886-3-5637525 E-mail: hwchen@tsmc.com

ABSTRACT

The channel backscattering ratios as well as the ballistic efficiency of strained CMOSFETs were studied for both nondegenerate and degenerate-limited cases. We found that the simple nondegenerate assumption can predict strain-induced change of ballistic efficiency with fair accuracy. The mechanism of drain current dependence on strain-induced mobility change was also investigated based on channel backscattering theory.

INTRODUCTION

Uniaxial strain engineering has been widely adopted in 90 nm node and beyond to boost device performance drastically with little cost penalty [1,2]. Recently, it has been reported that channel backscattering of strained CMOSFETs is strain-polarity-dependent [3,4]. The concept of channel backscattering theory is illustrated in Fig. 1. The saturation drain current in nanoscale devices is ultimately determined by carrier injection velocity (υ_{inj}) and channel backscattering ratio (r_{sat}) [5] as summarized in Table I. Fig. 2 shows r_{sat} as a function of carrier mean-free-path (MFP) for backscattering (λ) and k_BT layer thickness (l). The ballisitic efficiency (B_{sat}) can then be derived from r_{sat}, which represents the ratio of the drain current to its ballistic limit.

M. Lundstrom reported that the fractional change in drain current (ΔI_{dsat}) can be expressed in terms of the fractional change in mobility ($\Delta\mu$) and the ballistic efficiency [5]:

$$\frac{\Delta I_{dsat}}{I_{dsat}} = (1 - B_{sat})\frac{\Delta\mu}{\mu} \qquad (1)$$

For nanoscale CMOSFETs, $\Delta\mu/\mu$ does not translate to $\Delta I_{dsat}/I_{dsat}$ due to finite ballistic efficiency [6]. Therefore, B_{sat} is a key index when assessing nanoscale device performance.

In this paper, we discuss the channel backscattering behaviors of strained CMOSFETs using the nondegenerate and degenerate-limited assumptions. We also investigate that B_{sat} is affected by different uniaxial strain polarities. Correlation between ΔI_{dsat} and $\Delta\mu$ under uniaxial strain is also examined.

EXPERIMENTAL

Devices were fabricated using state-of-the-art process-strained Si (PSS) techniques [2], as shown in Fig. 1. PSS_HS CMOSFETs and PSS_LS CMOSFETs were studied, where PSS_HS and PSS_LS represent devices with high stress level and low stress level, respectively. The flow chart of backscattering parameters extraction is shown in Fig. 2. The backscattering ratio (r_{sat}) in the saturation region is extracted from a temperature-dependent analytical model [3,7]. In Table I, the full-range expression of saturation drain current for nondegenerate and degenerate cases are summarized. The related items used in this temperature-dependent analytical model are also deduced in both cases. One subband occupation is assumed in this work to obtain a simplified full-range expression of drain current.

RESULTS AND DISCUSSION

In this study, PSS_HS CMOSFETs with subthreshold swings, DIBLs, threshold voltages (Fig. 3) and inversion C-V characteristics (not shown) similar to those of control were investigated. In the PSS_HS split, the drain current improvement of NMOSFET and PMOSFET are 19% and 36%, respectively, as compared to control wafers (Fig. 4). While for the PSS_LS split, the drain current improvement of NMOSFET and PMOSFET are 7% and 23%, respectively (not shown).

For process-strained Si, the assumption of one-subband occupation is adequate due to more carrier repopulation in the first subband by means of strain-induced band splitting. As shown in Fig. 5, electrons have a preference to repopulate the 2-fold valleys under uniaxial tensile strain because of the large difference between the lowest subband energies of the 2-fold valleys ($\Delta 2$) and the 4-fold valleys ($\Delta 4$) [10]. Likewise, holes tend to repopulate the first subband with lowest energy under uniaxial compressive strain [11].

Fig. 6 exhibits the extracted ballistic efficiency (B_{sat}) of control and PSS_HS CMOSFETs under both nondegenerate and degenerate-limited cases. In the degenerate-limited case, the MFP for backscattering, λ'_0, can be expressed as [9]:

$$\lambda'_0 = \lambda_0 \frac{3\sqrt{\pi}}{4}\eta_F^{0.5} \qquad (2)$$

where λ_0 is the MFP for backscattering in nondegenerate case, $\eta_F = (E_F - \varepsilon_1)/k_BT$, and ε_1 is the first subband energy. Thus, higher MFP for backscattering can be obtained in the degenerate-limited case, which leads to lower r_{sat} (higher B_{sat}) value. As the gate length becomes shorter than 100 nm, B_{sat} of NMOSFETs is improved under tensile strain while that of PMOSFETs is degraded under compressive strain. In spite of the B_{sat} offset between the nondegenerate and degenerate-limited cases, the trend of B_{sat} modulation under different uniaxial strain polarities is consistent in both cases.

When the gate voltage is above the threshold voltage, the inversion charge density normally exceeds 10^{13} /cm^2 [3], which invalidates the nondegenerate assumption. However, as shown in Fig.7, the nondegenerate and degenerate-limited cases show very similar ΔB_{sat} vs ΔI_{dsat} dependence. This justifies the use of the nondegenerate assumption in evaluating the influence of uniaxial strain on ΔB_{sat} and ΔI_{dsat}.

Fig. 8 depicts ΔB_{sat} and $\Delta\upsilon_{inj}$ of PSS_HS and PSS_LS CMOSFETs. B_{sat} of PSS_LS CMOSFETs shows negligible difference from that of control CMOSFETs likely due to insignificant change of λ_0 and l_0 under low stress level (not shown). In general, PSS NMOSFETs show improved ΔB_{sat} and $\Delta\upsilon_{inj}$, which are both beneficial to drain current. However, PSS PMOSFETs exhibit improved $\Delta\upsilon_{inj}$ but degraded ΔB_{sat}, which reveals a trade-off in overall drain current enhancement.

Next, the correlation among $\Delta\mu$, ΔI_{dlin}, and ΔI_{dsat} under strain are investigated. In Fig. 9a, ΔI_{dsat} of PSS_HS NMOSFETs and PMOSFETs are about half of $\Delta\mu$. However, ΔI_{dlin} of PSS_HS NMOSFETs is almost equal to ΔI_{dsat}, while ΔI_{dlin} of PSS_HS PMOSFETs is almost 2X ΔI_{dsat} (Fig. 9b). The above-mentioned observation is consistent with previous finding [6]. The strong dependence of ΔI_{dlin} on $\Delta\mu$ for PSS_HS PMOSFETs is beneficial to transient circuit operation, but the mechanism is still under investigation. It is worthy of noting that B_{sat} deduced from the nondegenerate case can fit the correlation between ΔI_{dsat} and $\Delta\mu$ well, and it supports the nondegenerate assumption in evaluating channel backscattering characteristics of strained CMOSFETs.

CONCLUSIONS

We have shown that no significant ΔB_{sat} difference exists between the nondegenerate and degenerate-limited cases. It indicates that the impact of uniaxial strain on ΔB_{sat} can be fairly projected by the nondegenerate assumption. Furthermore, experimental ΔI_{dsat} and $\Delta\mu$ can be well correlated by B_{sat} deduced from the nondegenerate case. It is therefore essential to consider B_{sat} as a performance index in nanoscale CMOSFETs.

REFERENCES

[1] T. Ghani et al., IEDM Tech. Dig., 978, 2003. [2] C. H. Ge et al., IEDM Tech. Dig., 73, 2003. [3] H.N. Lin et al., Symp. VLSI Tech., 174, 2005. [4] H.N. Lin et al., IEDM Tech. Dig., s6p4, 2005. [5] M. Lundstrom, IEEE EDL-22, 293, 2001. [6] P. Bai et al., IEDM Tech. Dig., 657, 2004. [7] M. J. Chen et al., IEDM Tech. Dig., 39, 2002 [8] F. Assad et al., IEEE TED-47, 232, 2000. [9] A. Rahman et al., IEEE TED-49, 481, 2002. [10] H. Irie et al., IEDM Tech. Dig., 225, 2004. [11] E. Wang et al., IEDM Tech. Dig., 147, 2004.

1-4244-0181-X/06/$20.00 ©2006 IEEE

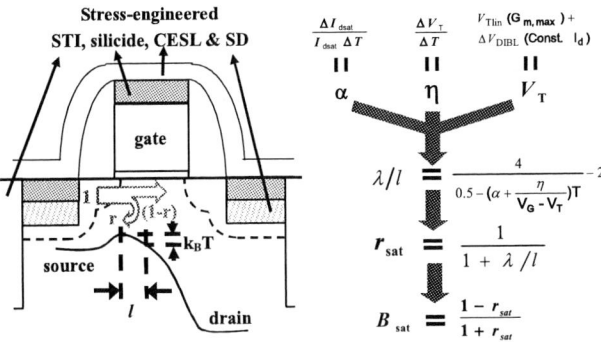

Stress-engineered STI, silicide, CESL & SD

Fig. 1 Schematic view of PSS MOSFET. The inset illustrates that injected carriers from the source into the channel are scattered back by a backscattering ratio r in k_BT layer region where the thickness of k_BT layer is l.

Fig. 2 Flow chart of the methodology used to extract backscattering parameters, where λ, l, r_{sat}, and B_{sat} represent carrier mean-free-path for backscattering, k_BT layer thickness, backscattering ratio, and ballistic efficiency, respectively.

Table I Comparison of full-range expression of saturation drain current, injection velocity, and the ratio of mean free path (MFP) for backscattering to k_BT layer thickness in (a) nondegenerate case and (b) degenerate case.

Full-range Expression of drain current [5,8,9] :	
$I_d = W Q_i(0)\left(\frac{1-r}{1+r}\right)\left[\upsilon_{inj}\frac{\Im_{1/2}(\eta_F)}{\Im_0(\eta_F)}\right]\left[\frac{1-\frac{1-r}{1+r}\frac{\Im_{1/2}(\eta_F-U_D)}{\Im_0(\eta_F)}}{1+\frac{1-r}{1+r}\frac{\Im_0(\eta_F-U_D)}{\Im_0(\eta_F)}}\right]$	
(a) Nondegenerate (In saturation region: $V_{ds}\gg k_BT/q$)	**(b) Degenerate (In saturation region: $V_{ds}\gg k_BT/q$)**
$I_{dsat}=W\left(\frac{1-r_{sat}}{1+r_{sat}}\right)\left[\upsilon_{inj}C_{eff}\right](V_G-V_T)$	$I_{dsat}=W\left(\frac{1-r_{sat}}{1+r_{sat}}\right)\left[\upsilon_{inj}\frac{\Im_{1/2}[(E_F-\varepsilon_1)/k_BT]}{\ln(1+e^{(E_F-\varepsilon_1)/k_BT})}C_{eff}\right](V_G-V_T)$
$\upsilon_{inj}=\sqrt{\frac{2k_BT}{\pi m^*}}=A\sqrt{T}$	$\upsilon_{inj}^{'}=\sqrt{\frac{2k_BT}{\pi m^*}}\frac{\Im_{1/2}[(E_F-\varepsilon_1)/k_BT]}{\ln(1+e^{(E_F-\varepsilon_1)/k_BT})}$
	As $\eta_F(=(E_F-\varepsilon_1)/k_BT)\rightarrow\infty$ (degenerate limit) :
	$\upsilon_{inj}^{'}=\frac{4\eta_F^{0.5}}{3\sqrt{\pi}}\sqrt{\frac{2k_BT}{\pi m^*}}=A'\sqrt{V_G-V_T}$
$\frac{\lambda_0}{l_0}=\frac{4}{0.5-[\eta/(V_G-V_T)+\alpha]T}-2$	As $\eta_F(=(E_F-\varepsilon_1)/k_BT)\rightarrow\infty$ (degenerate limit) :
	$\frac{\lambda_0'}{l_0'}=-\left[\left(\frac{5}{T}+\frac{\eta}{V_G-V_T}\right)\Big/\left(\alpha+\frac{1.5\eta}{V_G-V_T}\right)+2\right]$

$Q_i(0)$: inversion charge density r : channel backscattering ratio
υ_{inj} : nondegenerate thermal velocity of a hemi-Maxwellian distribution
$\Im_{1/2}(\eta_F)$: Fermi-Dirac integral, where $\eta_F=(E_F-\varepsilon_1)/k_BT$, $U_D=qV_d/k_BT$, and ε_1 is first subband energy
λ_0 : MFP for backscattering in the nondegenerate case λ_0' : MFP for backscattering in the degenerate case
l_0 : k_BT layer thickness in the nondegenerate case l_0' : k_BT layer thickness in the degenerate case
m^* : effective mass

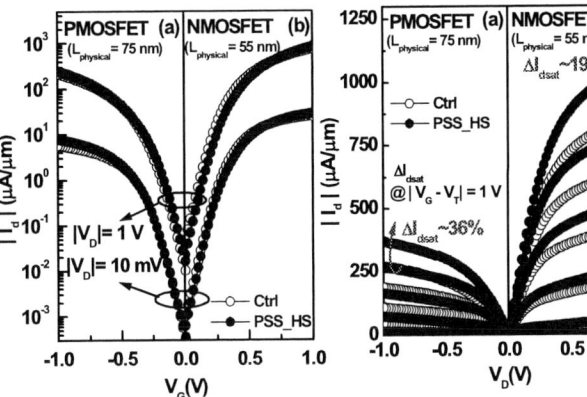

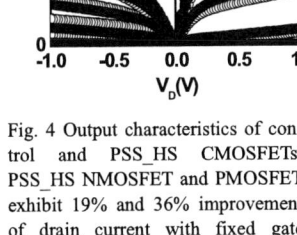

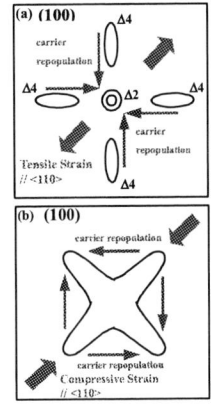

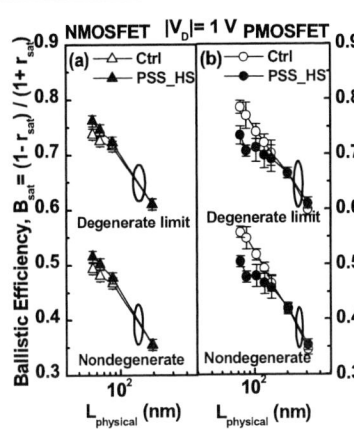

Fig. 3 Transfer characteristics of control and PSS_HS CMOSFETs. The swings of NMOSFETs are 88 mV/dec (ctrl) and 87 mV/dec (PSS_HS), while those of PMOSFETs are 94 mV/dec (ctrl) and 95 mV/dec (PSS_HS), respectively.

Fig. 4 Output characteristics of control and PSS_HS CMOSFETs. PSS_HS NMOSFET and PMOSFET exhibit 19% and 36% improvement of drain current with fixed gate overdrive of 1V.

Fig. 5 Schematic diagram of 2D equienergy lines for (a) tensile strain // <110>, more carriers repopulate first subband of the 2-fold valleys (Δ2) of conduction band, and (b) compressive strain // <110>, more carriers repopulate the first subband with lowest energy of valence band.

Fig. 6 Extracted ballistic efficiency (B_{sat}) of control and PSS_HS CMOSFETs. A significant B_{sat} offset is observed between the nondegenerate and degenerate-limited cases

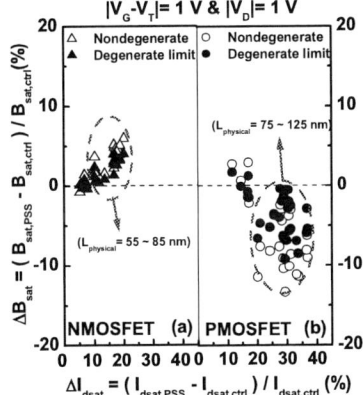

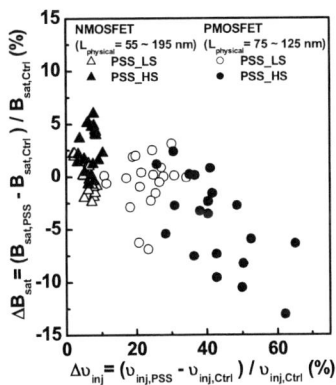

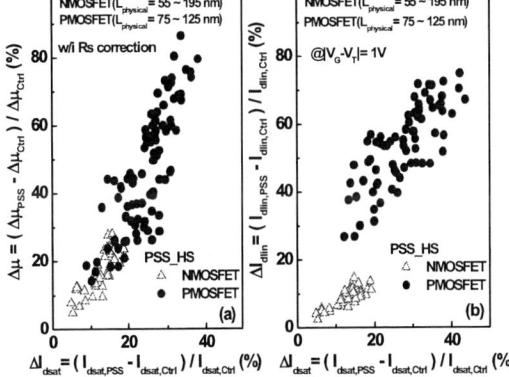

Fig. 7 Dependence of ΔI_{dsat} on ΔB_{sat} of control and PSS_HS CMOSFETs. ΔB_{sat} vs ΔI_{dsat} is quite similar between nondegenerate and degenerate-limited cases.

Fig. 8 ΔB_{sat} vs $\Delta\upsilon_{inj}$ for PSS_HS and PSS_LS CMOSFETs, where the nondegenerate assumption is adopted. PSS_HS NMOSFETs have both enhanced ΔB_{sat} and $\Delta\upsilon_{inj}$ as compared to PSS_LS. PSS_HS PMOSFETs have enhanced $\Delta\upsilon_{inj}$ but degraded ΔB_{sat} as compared to PSS_LS.

Fig. 9 (a) Dependence of ΔI_{dsat} on $\Delta\mu$ for PSS_HS CMOSFETs with series resistance correction. (b) Dependence of ΔI_{dsat} on ΔI_{dlin} for PSS_HS CMOSFETs.

63

2006 International Symposium on VLSI Technology, Systems, and Applications

PSDG MOSFET

Deyuan Xiao, Gary Chen, Roger Lee, Daniel Lu, Leong Tan, Yung Liu, C.C. Shen, Jong Woo Kim
Memory Technology Development Center, Semiconductor Manufacturing
International (Shanghai) Corp. 18 Zhangjiang Road
Shanghai 201203, PRC

ABSTRACT

A new planar split dual gate MOSFET device (PSDG MOSFET) is reported for the first time. The theoretical calculation, 3D device simulation as well as the experiment data show that with the two independent split dual gates in PSDG MOSFET, it can provide dynamical control of the device characteristics, such as threshold voltage (Vt) and sub-threshold swing (SS) as well as the Idsat of the device.

INTRODUCTION

One of the fundamental issues facing scaling of CMOS transistors is the ability to control the transistor leakage current (Ioff), while at the same time maintaining high drive current (Ion). Multiple Vt and Vdd and aggressive power management through circuit means will be the general practice. Reducing Vdd is essentially the only option available to device technologists for lowering the active power. Doing so unfortunately reduces the currents of the transistors. The currents can be lifted by reducing the Vt and thinning the gate dielectric, but those actions raise the sub-threshold leakage and gate leakage. This is today's speed/power dilemma. [1] One solution to this is to go to a fully depleted design [2]. Double-gate FinFET, [3] Omega MOSFET [4] and Tri-Gate Fully-Depleted transistors [5] have been offered for future transistor design.

In this paper, we, for the first time, introduce a novel planar split dual gate (PSDG) MOSFET device which enables independent biasing and provides the ability to dynamically adjust the threshold voltage and sub-threshold swing (SS).

PSDG MOSFET DEVICE STRUCTURE

Figure1 shows the angled top view on a PSDG NMOS device structure. The gate is spited by etching the poly to the gate oxide and separated both sides physically and electrically.

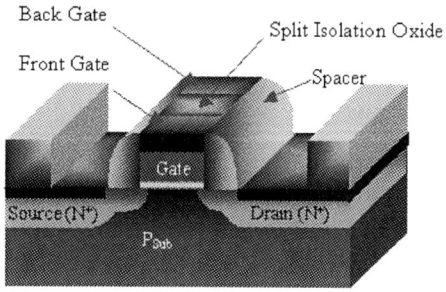

FIGURE 1. ANGLED TOP VIEW ON A **PSDG NMOSFET**

The PSDG MOSFET layout is shown in Figure 2

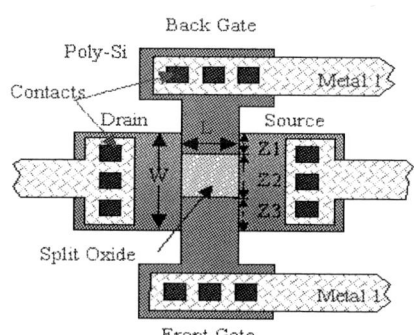

FIGURE 2. **PSDG MOSFET** DEVICE LAYOUT

PSDG MOSFET DEVICE CHARACTERISTICS

To derive the basic characteristics of the Split Dual Gate Metal-Oxide-Semiconductor Field Effect Transistor (PSDG MOSFET), we make the assumption that by applying different biasing on the two gate terminals, the field perpendicular to the current flow along the gate direction is not uniform and take all the assumptions described in reference [6], then we derived the drain current as:

$$
\begin{aligned}
I_D &= \frac{1}{L}\mu_n \int_0^Z \int_0^{V_D} |Q_n(y,z)| dz\, dV \\
&= \frac{Z1}{L}\mu_n C_o \left\{
\begin{array}{l}
\left(V_{GF} - 2\psi_B - \dfrac{V_D}{2}\right)V_D \\
- \dfrac{2}{3}\dfrac{\sqrt{2\varepsilon_s eN_A}}{C_o}\left[(V_D + 2\psi_B)^{3/2} - (2\psi_B)^{3/2}\right]
\end{array}
\right\} \\
&+ \frac{Z2}{L}\mu_n C_o \left\{
\begin{array}{l}
\dfrac{1}{2}V_D(V_{GB} + V_{GF}) - (2\psi_B + \dfrac{V_D}{2})V_D \\
- \dfrac{2}{3}\dfrac{\sqrt{2\varepsilon_s eN_A}}{C_o}\left[(V_D + 2\psi_B)^{3/2} - (2\psi_B)^{3/2}\right]
\end{array}
\right\} \\
&+ \frac{Z3}{L}\mu_n C_o \left\{
\begin{array}{l}
\left(V_{GB} - 2\psi_B - \dfrac{V_D}{2}\right)V_D \\
- \dfrac{2}{3}\dfrac{\sqrt{2\varepsilon_s eN_A}}{C_o}\left[(V_D + 2\psi_B)^{3/2} - (2\psi_B)^{3/2}\right]
\end{array}
\right\}
\end{aligned}
$$

Considering the symmetric split dual gate structure (Z1=Z3), we can get the drain current as below:

When both channels are in weak inversion, drifting is assumed to

$$
I_D = \frac{W}{L}\mu_n C_o\left\{\frac{1}{2}V_D(V_{GB}+V_{GF}) - (2\psi_B + \frac{V_D}{2})V_D - \frac{2}{3}\frac{\sqrt{2\varepsilon_s eN_A}}{C_o}\left[(V_D+2\psi_B)^{3/2} - (2\psi_B)^{3/2}\right]\right\}
$$

be the dominant transport mechanism for carriers in the charge sheet. Again for the symmetric split dual gate structure with Z1=Z2=Z3, we get the drain current density as below:

$$J_D = \frac{qWD_n n_i \exp(\frac{-q\psi_B}{kT})}{3L}(1 - e^{-\beta v_D}) \times$$

$$[(e^{\beta(V_{GF}-V_T)} + e^{\beta(V_{GB}-V_T)}) + \frac{e^{\beta(V_{GB}-V_T)} - e^{\beta(V_{GF}-V_T)}}{\beta(V_{GB} - V_{GF})}]$$

PSDG MOSFET 3D DEVICE SIMULATIONS

The Integrated Systems Engineering (ISE) 3D Devise and Dessis TCAD tools were used in our simulations. Figure 3 shows the simulated plot of I_D versus front gate voltage V_{GF} (for small V_D) with two different back gate voltage V_{GB}. Its noted that Ioff=7.5×10^{-11}A/μm with V_{GB} =0V, while Ioff=3.4×10^{-11}A/μm with V_{GB} =-1.5V; I_{dsat}=1.344 mA/μm. L_g=100nm, t_{ox}=16Å

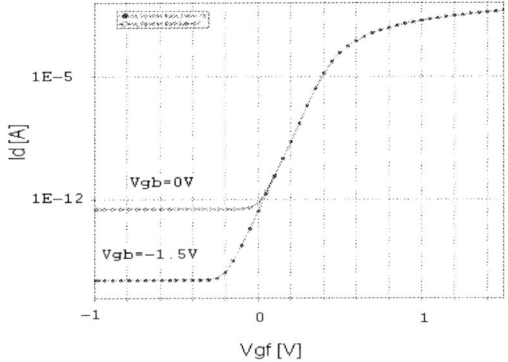

FIGURE 3 SIMULATED PLOT OF I_D VERSUS FRONT GATE VOLTAGE V_{GF} WITH TWO DIFFERENT BACK GATE VOLTAGE V_{GB}

PSDG MOSFET FABRICATION AND RESULTS

The PSDG MOSFET is fabricated with conventional planar CMOS compatible process. Figure 4 shows the measured curves (exponential) of drain current I_D versus front gate voltage V_{GF} with the back gate at different bias voltages V_{GB}.

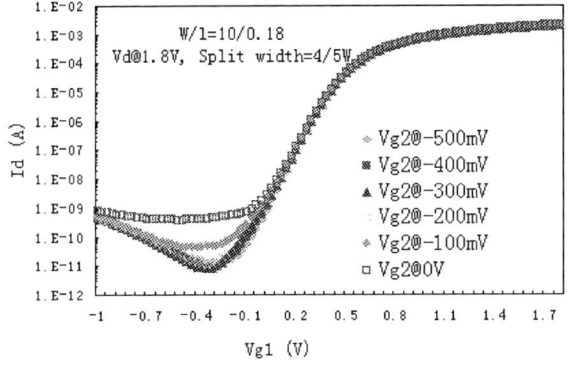

FIGURE 4 MEASURED PLOT OF I_D VERSUS FRONT GATE VOLTAGE V_{GF} WITH TWO DIFFERENT BACK GATE VOLTAGE V_{GB}

Figure 5 shows the measured curve of a n-channel PSDG MOSFET sub-threshold swing versus one of the gate bias (V_{G2}). From the curve, it can be seen that with the two independent gates

bias, the threshold voltage (Vt) can be modulated. Thus, it can provide dynamical control on device characteristics such as threshold voltage (Vt) as well as the sub-threshold swing.

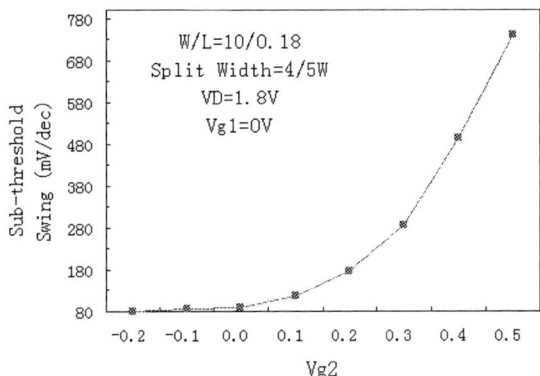

FIGURE 5 MEASURED CURVE OF A N-CHANNEL PSDG MOSFET SUB-THRESHOLD SWING VERSUS ONE OF THE GATE BIAS

CONCLUSIONS

The planar split dual gate MOSFET device structure, characteristics as well as the 3D device simulations have been described for the first time. The two independent gates provide dynamic control on device characteristics such as threshold voltage (Vt) and sub-threshold swing (SS) and may used for circuit power management, such as sleeping mode design. The fabrication process is simple compare to other split gate MOSFETs: Current planar technology requires fabricating multiple transistors, each with a different gate oxide thickness and different doping to get different Vt (threshold voltages).

From the energy band point of view, all published MOSFET devices including the FinFET like transistor, Omega MOSFET as well as the Tri-Gate transistors all is two-dimensional devices. Only our PSDG MOSFET is the first true three-dimensional device.

ACKNOWLEDGMENTS

Authors thank Scott Lee, David Yuan, Yongsheng Yang and Sean Xing for help in 3D device simulation. Also thank Daisy Yang, James Hong and Willy Yan for help in device fabrication

REFERENCES

[1] Chenming Hu, 2004 Symposium on VLSI Technology Digest of Technical Papers, pp.4-5
[2] R.Chau et al, IEDM 2000, pp. 45-48.
[3] Peter Singer, Semiconductor International, Vol.26(12), pp.28 (2003)
[4] T. Park et al, 2003 Symposium on VLSI Technology Digest of Technical Papers
[5] B.Doyle et al, 2003 Symposium on VLSI Technology Digest of Technical Papers
[6] Semiconductor Devices, 2/E by S. M. Sze, Copyright © 2002 John Wiley & Sons. Inc.

Vapor and Solution Deposited Organic Thin Film Transistors

Tom Jackson

Center for Thin Film Devices and Materials Research Institute, Department of Electrical Engineering,
Penn State University, University Park, PA 16802, tnj1@psu.edu

Introduction

OTFT device performance rivals or exceeds that of a-Si:H devices, and low OTFT processing temperatures allow fabrication on a range of surfaces including cloth, paper, or polymeric substrates. To find significant commercial application, OTFTs must demonstrate characteristics that differentiate them from other device technologies, especially a-Si:H TFTs. Potential advantages for OTFTs include the possibility of device and system fabrication on substrates not readily accessible to a-Si:H devices and, likely more importantly, reduced cost manufacturing.

For practical device and system use, OTFTs must demonstrate the uniformity, reproducibility, reliability, and integration with other devices, needed for real applications. As a candidate application we have considered the integration of vapor-deposited OTFTs with organic light emitting diodes (OLEDs) and have fabricated small test displays that allow us to investigate device characteristics and passivation and isolation requirements for integrating these organic devices.

All-Organic Display

Displays are an obvious area of interest for organic electronics. Flat panel displays and especially active-matrix liquid crystal displays (AMLCDs) are among the largest and fastest growing areas of microelectronics. Active matrix liquid crystal displays with diagonals as large as 82" have been demonstrated [1] and the largest manufacturing facilities now produce several million m^2 of finished displays per year (~0.1 m^2/s of finished work). The well-developed technology and large manufacturing capability of current AMLCDs may make displays a difficult area for organic electronics impact.

Although organic thin film transistor (TFT) replacements for the amorphous silicon (a-Si:H) devices currently used in AMLCDs may be possible, applications where organic devices provide new function may be more viable. One possibility is flexible displays. a-Si:H devices and even displays have been demonstrated on flexible substrates [2,3], however, organic thin film transistors (OTFTs), particularly when combined with organic light emitting diodes (OLEDs), offer advantages, especially for fabrication on low-cost, low-temperature substrates.

Figure 1 shows a simple all-organic flexible substrate display. This display has 48 × 48 pixels, is fabricated on a polyester substrate, and uses vapor deposited pentacene organic TFTs in a simple two transistor per pixel design to drive small-molecule OLEDs.

Solution-Processed Organic Electronics

To date, most high mobility (≥ 1 cm^2/V·s) OTFTs have used vapor-deposited organic semiconductors as the active material. The OTFT fabrication process for these materials is similar to conventional inorganic TFT fabrication and is likely to have similar costs with some savings related to reduced processing temperature and low substrate cost. Solution processed OTFT active layers are of interest because they may allow simplified device fabrication and further reduced cost. Most solution processed OTFTs have used polymeric organic semiconductors; however, such materials typically have relatively low field-effect mobility, usually < 0.1 cm^2/V·s. Many small molecule organic semiconductors, including pentacene, are not conveniently soluble; however, soluble precursors have been used to fabricate solution-processed pentacene OTFTs. By using a relatively high-temperature step (200 °C) to convert the precursor to pentacene, field-effect mobility as large as about 0.9 cm^2/V·s has been

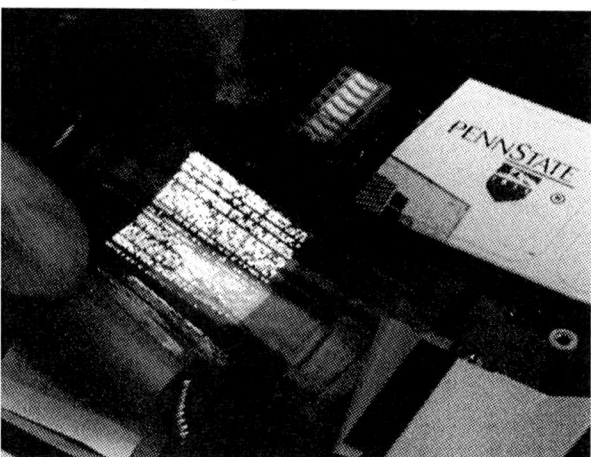

Figure 1: All-organic flexible substrate AMOLED display. The 48 x 48 pixel display uses pentacene organic TFTs, two OTFTs per pixel, and small-molecule OLEDs on a polyester substrate.

demonstrated [4]. However, this temperature is too high for low-cost polymeric substrates and field-effect mobility falls to near 0.1 cm^2/V·s for precursor conversion at 130 °C.

In collaboration with John Anthony (University of Kentucky) we have been investigating pentacene modified to change its molecular crystal ordering and other characteristics. By adding bulky groups at the 6,13 positions of pentacene it is possible to synthesize soluble functionalized pentacene derivatives [5]. Using 6,13-bis(triisopropylsilylethynyl) pentacene (TIPS pentacene) we have fabricated OTFTs with field-effect mobility > 1 cm^2/V·s (figure 2).

Surprisingly, solution-deposited organic semiconductors can have good molecular ordering even for rapid deposition techniques. For drop-cast TIPS pentacene films x-ray diffraction shows strong molecular ordering and atomic force microscopy shows molecular steps. The molecular ordering is likely responsible for the high mobility OTFTs obtained from this material. These results demonstrate that high-mobility devices are possible from low-temperature solution deposition and may allow device fabrication at lower cost than vapor deposited materials.

Challenges

Many device and system challenges remain for OTFT-based applications. Device reliability, uniformity, and reproducibility all need improvement and device details remain problematic. For example, most organic semiconductors resemble insulators in many respects (useful for obtaining low switch off currents) and injecting

and/or extracting charge is often not straightforward. In contrast with inorganic devices, doping is typically not used to enhance injection and metal-semiconductor junctions are often not barrier-free. Contact related injection barriers can limit OTFT performance and the large electric field at such barriers and/or large local power dissipation can limit device stability and reliability. Contact performance also often depends on the details of device fabrication. For example, solution processed devices may require surface energy modifications for both conductor and dielectric surfaces.

Conclusion

The technological landscape appears ripe for an explosion of flexible and electronics anywhere applications. Both evolutionary and revolutionary approaches are of interest and will likely find application dependent use. Details of device structure, performance, and electrical and environmental stability are critically related to the success of various application possibilities. The wide range of device and application possibilities as well as physical phenomena makes this one of the most interesting and exciting areas of device physics and engineering.

Acknowledgements

Support from the Army Research Laboratory, NSF (PSU MRSEC), and Kodak for this work is gratefully acknowledged.

References

1. S. S. Kim, "The World's Largest (82-in.) TFT-LCD", *2005 SID Symp. Digest of Tech. Papers* 1842-7, 2005.
2. A. Nathan, A. Kumar, K. Sakariya, P. Servati, S. Sambandan, and D. Striakhilev, "Amorphous silicon thin film transistor circuit integration for organic LED displays on glass and plastic," *IEEE J. Solid-State Circuits,* **39**, 1477-86, 2004.
3. J. A. Nichols, M. H. Lu, M. Hack, and T. N. Jackson, "a-Si:H Phosphorescent OLED Active Matrix Pixels on Polymeric Substrates," 62nd *Device Research Conf.*, 59-60, 2004.
4. A. Afzali, C. D. Dimitrakopoulos, and T. L. Breen, *J. Am. Chem. Soc.*, **124**, 8812-13, 2002.
5. C. D. Sheraw, D. L. Eaton, J. E. Anthony, and T. N. Jackson, *Advanced Materials*, **15**, 2009-11, 2003.
6. J. A. Nichols, D. J. Gundlach, and T. N. Jackson, *Appl. Phys. Lett*, **83**, pp. 2366-8, 2003.

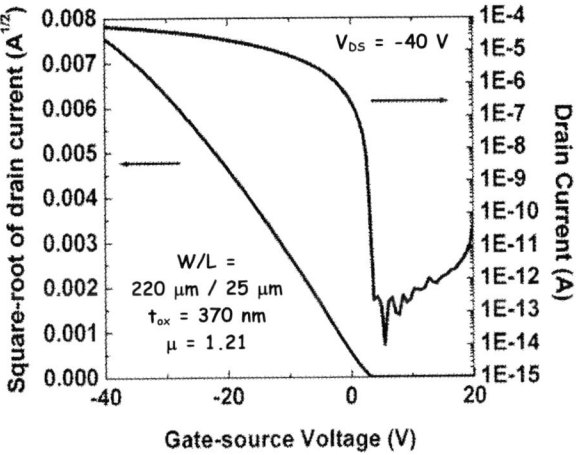

Figure 2: √I$_D$ – V$_{GS}$ and log(I$_D$) – V$_{GS}$ **characteristics for a TIPS-pentacene solution deposited organic thin film transistor with mobility > 1 cm^2/V-s.**

2006 International Symposium on VLSI Technology, Systems, and Applications

Printed transistors and passive components for low-cost electronics applications

Vivek Subramanian, Josephine B. Chang, Steven E. Molesa, Steven K. Volkman, David R. Redinger
Department of Electrical Engineering and Computer Sciences
University of California, Berkeley
Berkeley, CA 94720-1770
USA

INTRODUCTION

In recent years, there has been great interest in the use of printing as a means of realizing ultra-low-cost electronic systems. Since printing potentially offers a method for fabricating electronic systems using fully-additive processing techniques, it is expected to result in a substantial reduction in overall process complexity. This in turn will lead to a large reduction in cost per unit area for circuit fabricated using printing technology when compared with circuits fabricated by conventional means.

Given the low cost per unit area of printed electronics, various applications have garnered substantial interest as potential deployment points for printed electronics. In particular, three applications that have received tremendous attention are displays, sensors, and RFID tags. In this work, we review the application-specific material and device requirements of printed electronics as they relate to these applications, and discuss the state of the art of printed electronics technology being developed by us for the same.

ANALYSIS OF APPLICATION-SPECIFIC REQUIREMENTS

As mentioned above, the main consequence of a fully-printed process flow is a reduction in the overall process complexity. A typical fully-printed process flow reduces the total number of process steps in a typical fabrication process by up to 60%. This in turn results in a dramatic reduction in cost per unit area. Depending on the assumptions made with regards to process throughput, capital expenditure, and raw materials costs, the overall cost reduction in printed electronics relative to conventional subtractive processing is 2 to 3 orders of magnitude. Interestingly, the cost per transistor in printed electronics is actually not cheaper than the cost per transistor in silicon, since silicon transistor densities are orders of magnitude higher than that achievable using any existing commercially viable printing techniques. This cost structure has two consequences. First, ideal applications for printed electronics are applications where form factor is relatively independent of transistor count (e.g., displays, some sensors, and RFID tags operating at lower frequencies). Second, the real benefits of printed electronics derive from a reduction in process complexity; therefore, there is tremendous need to drive towards a fully-additive, all-printed solution, which in turn places constraints on material needs and device structures.

Displays

Displays are a tremendous driver for printed electronics. Since display form factors are determined by the viewing requirements and transistor counts typically do not determine display size, this application fits well within the cost paradigms of printed electronics. Additionally, since the incumbent technology is amorphous silicon, the performance requirements are certainly within the realm of those achievable using printed materials. Devices required for display applications include transparent conductors, low-resistance conductors, transistors, and capacitors, in addition to the display element itself. The display element (LCD, OLED, etc.) will not be discussed here. Based on the other needs, it is clear that material needs include transparent conductors and low-resistance metallic conductors, printable semiconductors, and printable dielectrics. For display applications, there is also generally a benefit resulting from moving towards all-transparent electronics, since this allows the use of larger electronics within sacrificing display brightness or aperture ratio; this in turn allows the use of lower-performance parts.

Sensors

The drive towards printed sensors is determined more by form factor issues than by cost. Since sensors on plastic are desirable for various consumer packaging applications including product content monitoring, there is generally interest in development of sensors that are printed directly on plastic. In particular, there is interest in gas / vapor / fluid sensor for monitoring of product quality within a package. In turn, this defines materials / device needs including the need for printed sensing elements as well as printed circuitry for processing the signals produced by the sensing elements.

RFID Tags

The size of RFID tags operating at 13.56MHz is largely determined by the antenna size. As a result, printed electronics may fit well in this space from an economic perspective. To realize printed RFID tags, materials / device requirements include printed transistors and potentially also printed diodes, printed capacitors, and printed high-Q inductors. This in turn defines material needs including printed low-resistance conductors, dielectrics with low-dispersion at 13.56MHz, and high-mobility printed semiconductors.

TECHNOLOGY STATUS

Based on the applications above, we have been pursuing a program aimed at developing materials, devices, and circuits targeted at the above needs. Here, we review our general progress towards development of fully-printed low-cost electronic systems.

Printed conductors and passive components

As discussed above, there is a need for printed conductors for interconnection in all the aforementioned applications, as well as for formation of passive components for printed RFID applications. To achieve low-resistance, we have exploited the substantial melting point reduction that is exhibited in metallic nanoparticles. Organic-encapsulated metallic nanoparticles are synthesized and solubilized to form an ink. This is printed and sintered at plastic-compatible temperatures. Through appropriate design of the nanoparticle, this temperature is sufficient to fuse the nanoparticles to form high-conductivity metals. Using this process, we have fabricated both interconnects and passive components, as shown below.

1-4244-0181-X/06/$20.00 ©2006 IEEE

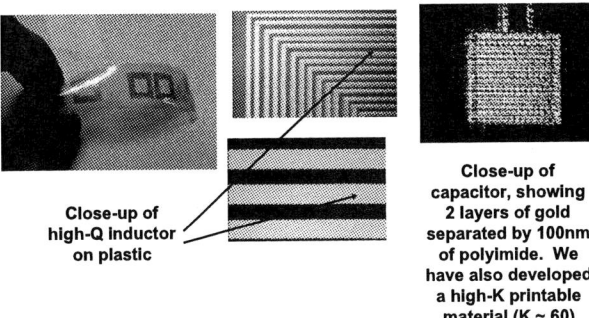

Close-up of high-Q inductor on plastic

Close-up of capacitor, showing 2 layers of gold separated by 100nm of polyimide. We have also developed a high-K printable material (K ~ 60)

FIGURE 1: PRINTED PASSIVE COMPONENTS FORMED USING NANOPARTICLE INKS

Printed transistors

Printed transistors are fabricated using the metal nanoparticles above in conjunction with various printed semiconductors and dielectrics. A conventional bottom-gated transistor structure is typically use. All layers are printed.

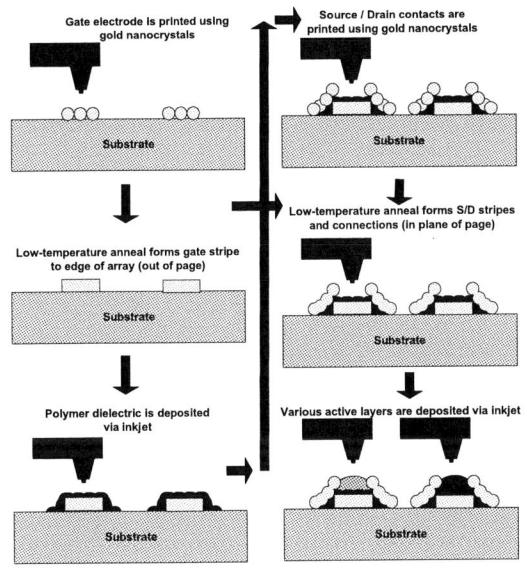

FIGURE 2: PROCESS FLOW USED TO FABRICATE PRINTED TRANSISTORS

Various types of printable semiconductors are used, including various printable organic semiconductors and nanoparticle-based inorganic semiconductors. A key feature of our process is that all printed layers are converted during the process into insoluble form. This allows multiple layers to be printed over each other without causing significant solvent-interaction problems. Thus, the nanoparticles are converted into insoluble metal, polymer dielectrics are crosslinked to make them insoluble, and organic and inorganic semiconductors are thermally converted into insoluble forms.

Among organic semiconductors, we have focused on various oligothiophene and pentacene derivatives. Mobilities as high as $0.2 cm^2/V$-s have been achieved. Operating voltages as low as 10V have been achieved by printing thin gate dielectrics (dielectrics as thin as 20nm have been achieved with high yield and no pinholes).

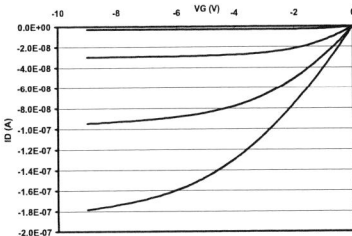

FIGURE 3: ELECTRICAL CHARACTERISTICS OF PRINTED TRANSISTORS FORMED USING A SOLUBLE PENTACENE PRECURSOR

Devices with inorganic semiconductor channels have also been realized using ZnO nanoparticles. Mobility as high as $0.2 cm^2/V$-s have been achieved. ZnO has the additional advantage of being transparent, making it highly attractive for use in displays.

Gas sensors

By exploiting the chemical sensitivity of organic materials, we have also realized gas sensors. Arrays of individual sensing transistors are deployed on the same substrate. This is conveniently achieved using inkjet printing. By functionalizing individual chemicals to be sensitive to different materials and deploying them into an array, highly specific sensors may be realized. The response of a sample array sensor to different analytes is shown below.

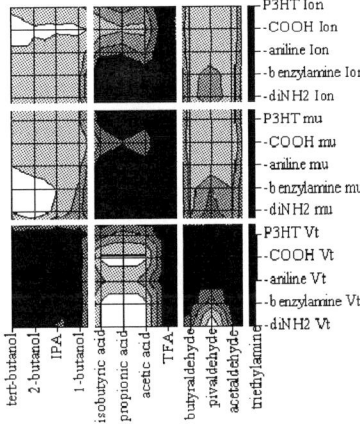

FIGURE 4: ARRAY RESPONSE OF VARIOUS PRINTED SENSORS (HORIZONTAL AXIS) TO VARIOUS VAPORS (VERTICAL AXIS)

CONCLUSIONS

In this paper, we have discussed our development of various printed electronics technologies targeted at meeting the needs of various optimal applications for printed electronics. While several concerns remain, particularly related to stability and reliability, the potential for printed electronics is strong, and progress will continue, driven by development of new materials and processes designed to exploit the opportunities that exist at the intersection of economics and engineering through the benefits of printing.

2006 International Symposium on VLSI Technology, Systems, and Applications

Future Prospects of Flexible, Large-Area Sensors and Actuators with Organic Transistor ICs

Takao Someya[1*], Takayasu Sakurai[2], and Tsuyoshi Sekitani[1]

[1] Quantum-Phase Electronics Center, School of Engineering, the University of Tokyo, Tokyo, Japan
[2] Center for Collaborative Research, the University of Tokyo, Tokyo, Japan
* Email: someya@ap.t.u-tokyo.ac.jp, Phone: +81(Japan)-3-5841-6820, Fax: +81(Japan)-3-5841-6828

ABSTRACT

We report recent progress and future prospects of organic field-effect transistors and their applications to flexible, large-area sensors and actuators. In particular, we describe stretchable electronic artificial skins (e-skins) for robots, pocket image scanners suitable for mobile applications, and Braille sheet display with plastic actuator arrays. We also present recent progress of reliability and stability issues; encapsulation techniques to suppress chemical degradation and annealing techniques to reduce the effect of DC bias stress..

INTRODUCTION

Over the past several years, organic thin film transistors (TFTs) and their integrated circuits have attracted considerable attention (1–4) since organic TFTs possess attributes that complement silicon-based LSI devices, which are high-performance, but expensive. Organic TFTs can be manufactured on plastic films at ambient temperatures; therefore, they are mechanically flexible and potentially inexpensive to manufacture. Recent studies organic TFTs are based on two major applications. The first application is flexible displays, such as paper-like displays or e-paper, in which electronic inks or other media are driven by organic TFT active matrices. The other is radio frequency identification (RFID) tags. The printable features of organic transistors should facilitate the implementation of RFIDs on packages.

As the third application, we have proposed and demonstrated flexible, large-area sensors and actuators in which organic TFT active matrices are used for data readout from area-type sensors or to drive large-area actuators. In this paper, we report the recent progress, issues, and future prospects of organic TFT-based flexible, large-area sensors and actuators.

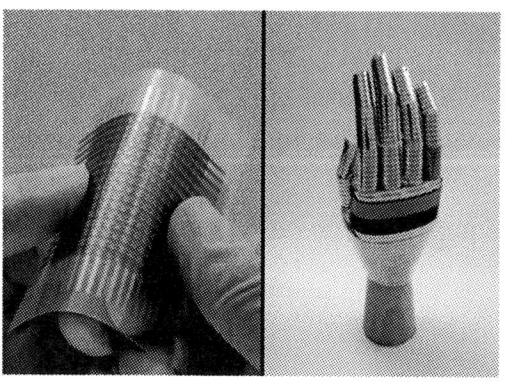

Fig. 1: A picture of an electronic artificial skin (e-skin), a large-area, flexible pressure sensor.

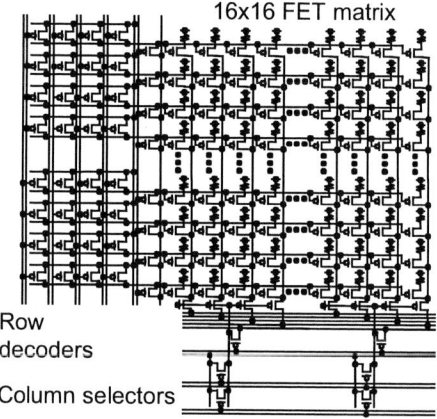

Fig. 2: (a) A circuit diagram of an artificial skin system. A 16×16 organic transistor active matrix, a row decoder and a column selector are manufactured on plastic films separately and them assembled to make integrated circuits for data readout.

ELECTRONIC ARTIFICIAL SKINS

The first application of large-area sensors is an electronic artificial skin (e-skin) (5). An e-skin is a flexible pressure sensor (Fig. 1), which will be used in next-generation robots. Although the mobility of organic semiconductors is approximately two or three orders of magnitude less than that of poly- and single-crystalline silicon, the slower speed is tolerable for most applications of large-area sensors. In particular, for the fabrication of E-skins, the integration of pressure sensors and organic peripheral electronics avoids the drawbacks of organic transistors, while taking advantage of their mechanical flexibility, large area, low cost, and relative ease of fabrication.

Figure 2 shows a circuit diagram of an artificial skin system. An organic TFT-based 16×16 active matrix, a row decoder, and a column selector are assembled by a physical cut-and-paste procedure to manufacture integrated circuits for data readout. Three functional films — an interconnection layer, a pressure-sensitive rubber sheet, and a top electrode for power supply— are then laminated together with the organic ICs. Pressure images were obtained by a flexible active matrix of organic transistors whose mobility is as high as 1.4 cm^2/Vs. Organic TFTs can be bent to a radius of 0.5 mm, which is sufficiently small for the fabrication of human-sized robot fingers.

However, human skin is more complex than transistor-based imitations demonstrated thus far. It performs certain functions including thermal sensing. Furthermore, without conformability, the application of e-skins to three-dimensional surfaces is impossible.

1-4244-0181-X/06/$20.00 ©2006 IEEE   70

Fig. 3: An image of a pocket *Braille sheet display*. It was manufactured on a plastic film by integrating the active matrix of organic TFTs with a plastic sheet actuator array based on a perfluorinated polymer electrolyte membrane. The device is mechanically flexible, very thin, and lightweight. One character is displayed by a 3 × 2 array of rectangular actuators (4 mm in length and 1 mm in width). A semisphere is attached to each actuator, which bends and lifts the semisphere.

Based on an organic semiconductor, we have developed conformable, flexible, wide-area networks of thermal and pressure sensors. A plastic film with organic transistor-based electronic circuits was processed to form a net-shaped structure that allows the e-skin films to be stretched by 25%. The net-shaped pressure sensor matrix was attached to the surface of an egg and pressure images were successfully obtained in this configuration.

Moreover, a similar network of thermal sensors was developed using organic semiconductors. A possible implementation of both pressure and thermal sensors on various surfaces is presented. By using laminated sensor networks, the distributions of pressure and temperature are simultaneously obtained.

SHEET-TYPE BRAILLE DISPLAYS

Organic transistors are also suitable for applications to large-area plastic actuators. We have fabricated a novel, flexible, lightweight *Braille sheet display* that is fabricated on a plastic film by integrating high-quality organic TFTs with plastic actuators (7). A small hemisphere that projects upwards from the rubber-like surface of the display is attached to the tip of each rectangular actuator (Fig. 3).

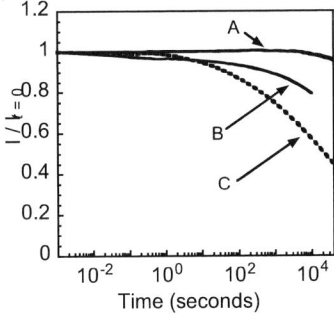

Fig 4: Normalized source-drain current (I_{DS}) under continuous DC voltage biases ($V_{GS} = V_{DS} = -40$ V) as function of time t. $I_{t=0}$ represents I_{DS} at the start of the measurement ($t=0$). Measurements are performed on (A) the annealed and (B) nonannealed TFTs on polyimide base films (solid line), and on (C) nonannealed TFTs on Si/SiO_2 gate dielectric layers (dashed line), for comparison.

DC BIAS STRESS

We report organic TFTs that exhibit a very small degradation in performance under a continuous DC bias stress (10). When the pentacene TFTs are annealed at 140 °C for a duration of 12 h in a nitrogen environment, the change in I_{DS} is 3 ± 1% even after the application of continuous DC voltage biases of $V_{DS} = V_{GS} = -40$ V for 11 h (Fig. 4).

ISSUES AND PROSPECTS

We describe issues of organic TFTs from the viewpoint of large-area electronics. First, stability and reliability are the main concerns for the organic transistors. Degradation may be induced by oxygen and/or moisture like electroluminescent (EL) devices; therefore, it can be suppressed drastically by appropriate encapsulation. Although plastic films usually exhibit high gas penetration, applications that require mechanical flexibility, such as electronic skins and sheet image scanners, also require flexible substrates with low gas permeability. Thus, it is very important to develop flexible base films with low gas permeability. The second issue is the reduction in operation voltage. This can be achieved by reducing the device dimensions. Reduction in the thickness of the gate dielectric layers and/or of those with a higher gate dielectric constant is very effective in decreasing the operation voltage.

It is expected that organic TFT-based ICs would play an important role in large-area electronics wherein the manufacturing cost per area must be very low. Undoubtedly, one of the most important directions for future electronics is ambient intelligence or wireless sensor networks. In order to realize such networks, one of the key technologies is a sensor to detect physical or chemical information distributed over a large area. We believe that the large-area features of the organic TFTs would be instrumental in realizing such large-area sensors. A new class of applications that require large-area detections has gained importance in future electronics. It should be tough for organic TFTs to compete with silicon in terms of the cost per function. However, they are very suitable for applications that require low-cost features in a large area. Therefore, organic TFTs are appropriate for large-area electronics.

ACKNOWLEDGEMENTS

This study is partially supported by IT program, MEXT and CREST, JST.

REFERENCES

[1] A.Tsumura, et al., APL 49, 1210 (1986).
[2] C. D. Dimitrakopoulos, et al., Adv. Mater. 14, 99 (2002).
[3] A. Dodabalapur, et al., Science 268, 270 (1995).
[4] C. J. Drury, et al., APL 73, 108 (1998).
[5] T. Someya, et. al., IEDM, 203 (2003); PNAS 101, 9966 (2004); PNAS 102 (35) (2005); ISSCC, 288 (2004); H. Kawaguchi, et. al., IEEE JSSC 40, 177 (2005); Y. Kato, et. al., APL 84, 3789 (2004).
[6] T. Someya, et. al., IEDM #15.1, 580 (2004); H. Kawaguchi, et. al., ISSCC, 365 (2005).
[7] Y. Kato et. al., #5.1, IEDM (2005).
[8] T. Sekitani, et. al., APL 86, 073511 (2005); T. Sekitani, et al., APL 87, 073505 (2005).
[9] S. Iba, et. al., APL 87, 023509 (2005).
[10] T. Sekitani, et. al., APL 87, 073505 (2005).

2006 International Symposium on VLSI Technology, Systems, and Applications

TUBES, RIBBONS AND WIRES FOR FLEXIBLE ELECTRONICS

Yugang Sun and John A. Rogers

Department of Materials Science and Engineering, Seitz Materials Research Laboratory, Beckman Institute for Advanced Science and
Technology, University of Illinois at Urbana-Champaign, Urbana, Illinois 61801, USA

ABSTRACT

Solution processable conductors, dielectrics and semiconductors represent enabling materials for electronic circuits that can be fabricated on plastic sheets by continuous, high speed printing techniques. It is generally believed that these types of systems, which can cover large areas, will be important for new applications in consumer electronics. This talk describes the operational aspects of flexible transistors and circuits that use printable semiconductors based on carbon nanotubes, and on ribbons and wires of single crystal silicon, gallium arsenide (GaAs), indium phosphide (InP) and gallium nitride (GaN). High mobilities, optical transparency, GHz switching speeds and mechanical bendability and even stretchability represent a few of the unusual characteristics that can be achieved in these systems. These and other aspects, including the soft lithographic printing techniques used to form the devices and some costing studies for the materials and patterning methods will be discussed.

INTRODUCTION

Flexible electronics on unconventional low-cost substrates (e.g., plastics, rubbers, papers, etc.), often referred to as macroelectronics when they cover large areas, represents a rapidly expanding area of modern electronics. These kinds of systems might find applications in a wide range of areas including wall-size displays as well as RF (radio frequency) surveillance systems for military facilities.[1] Various materials, such as small organic molecules, conjugated polymers, amorphous (a-) and polycrystalline silicon with mobilities from one thousandth to tens of cm^2/V·s, have been explored to serve as channel materials for thin film transistors (TFTs) on flexible substrates that are suitable as backplanes for driving large-area displays. On the other hand, high-end applications which require high operating speeds can be achieved by using semiconductor materials with high mobilities, such as single crystalline silicon, GaAs, InP, GaN and single-walled carbon nanotubes (SWNTs). These building blocks exhibit mobilites in the range of hundreds to 10,000 cm^2/V·s. Herein, we report some of our recent work on flexible as well as stretchable devices (e.g., TFTs) that are fabricated with the aforementioned high-performance materials using printable elements with lateral dimensions on the scales of tens of nanometers to several micrometers.

RESULTS AND DISCUSSIONS

Semiconductor building blocks with high mobilities have been prepared through various approaches.[2-10] Figure 1 shows typical images. SWNTs were synthesized through the well-developed chemical vapor deposition (CVD) process catalyzed with nanoparticles. Networks of SWNTs exhibit different densities and electrical transport properties when ferritin and a mixture of ion acetate, cobalt acetate and molybdenum acetate are used as catalysts for growing samples shown in A (low density and semiconducting behavior) and B (high density and metallic behavior), respectively. Because nanotubes are often formed on silicon substrates covered with SiO$_2$ layers, networks of them can be released from the growth substrates by dissolving the underneath SiO$_2$ in HF solution.[2] This

process is general to fabricate free-standing building blocks from wafers with multiple layers, such as silicon-on-insulator (SOI) wafers and GaN layers on Si wafers buffered with AlN layers. Frames C and D show the as-fabricated ribbons of Si and GaN.[7,10] Anisotropic etching provides the means to generate appropriate profiles to directly undercut the uniform bulk wafers into wires with small lateral dimensions. Frames E and F give the scanning electron microscopic (SEM) images of GaAs wires and InP wires with triangular cross sections.[9] The curved profiles of some of these wires/ribbons provide evidence of their mechanical flexibility. The characteristic makes them well suited for fabricating flexible electronics formed by transfer printing these elements onto plastic or rubber substrates using techniques developed in our group.[7,8,11,12]

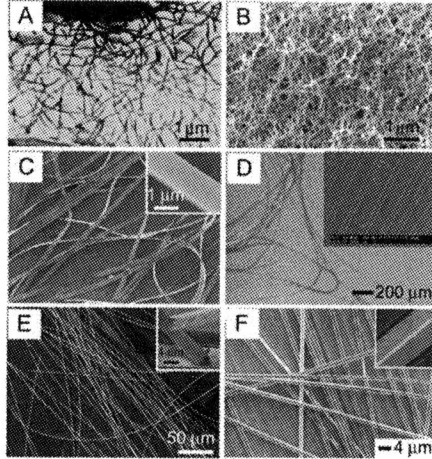

FIGURE 1. Images of various structures used for the fabrication of flexible electronics: (A, B) SWNTs with low and high densities; (C, D) ribbons of Si and GaN with rectangular cross sections; (E, F) wires of GaAs and InP with triangular cross sections.

These printed wires, ribbons and tubes can be used to build electronic devices (e.g., TFTs discussed in this paper) with high performance. For example, arrays of GaAs wires with integrated ohmic contacts maintain the order defined by the lithography process used to create them after they were transfer printed onto a poly(ethylene terephthalate) (PET) substrate. Depositing source/drain electrodes (overlay of the ohmic contacts) and Schottky contacts as gate electrodes forms thin-film metal-semiconductor field-effect-transistors (MESFETs), as shown in the inset of Figure 2B. Figures 2A and 2B present the DC characterization results of a MESFET with channel length of 50 μm, channel width of 150 μm and gate length of 2 μm. The results indicate that the current flows from source to drain are well modulated by the voltages applied to the gate electrode; the ON/OFF current ratio is around 10^6 in the saturation region. The transistor also exhibited promising response to RF signals. The unity current gain frequency was determined 1.55 GHz, which is much higher than flexible TFTs formed in other ways. The flexibility of the devices was evaluated by measuring the variation of device performance during bending the substrates with the bending stage shown as the inset of Figure 2D. The data show that the

1-4244-0181-X/06/$20.00 ©2006 IEEE

transistors still work well even with surface strains as high as 0.8% and the change of saturation current is less than 20%.[13,14] Silicon ribbon devices also exhibit similarly good electrical and mechanical performance.[7,15]

SWNTs exhibit low absorption coefficient in the visible region and they can serve as ideal material for fabricating transparent electronics with integrating transparent substrates (e.g., PET sheet) and dielectrics (Al_2O_3, epoxy and PDMS).[4] In this case, the low density tube networks serve as channels, while the high density tubes serve as electrodes due to their metallic behavior (see, inset of Figure 3A). Figure 3A shows the *I-V* curves of an all-tube transistor with 750 µm channel width and 225 µm channel length, indicating the good modulation over current. Figure 3B presents transmission spectra in the visible range for the PET substrate and the source/drain region of a completed TFT. The data indicate that the transmittance of the source/drain region of the completed TFT is ~80% or larger in the visible region, which is only slightly less than the 85% transmittance of the bare PET substrate. The all-tube devices exhibit extreme bendability due to the ultra-small diameter of SWNTs and their excellent mechanical properties.

In addition to plastic sheets, rubber substrates (e.g., PDMS) can also provide platforms to support the building blocks as shown in Figure 1 to prepare electronics with stretchability and compressibility.[11] In the fabrication process, silicon ribbons were first transferred to the surface of a pre-stretched PDMS stamp, and relaxing the stamp forced the ribbons to form buckles (i.e., periodical wave) along their longitudinal directions (Figure 4A), i.e., the direction of forces applied to stretch PDMS. The folded wavy geometry of ribbons on PDMS surfaces provides stretchability and compressibility. Completed ribbon devices with integrated electrodes and dielectrics can also form the similar wavy profiles as shown in Figure 4B. The wavy devices could be stretched and compressed with external applied strains up to 10% and the mechanical operations do not significantly degrade the device performance (Figure 4C).

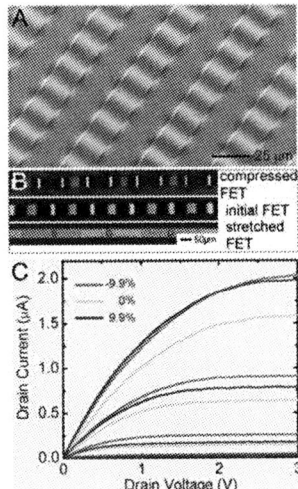

FIGURE 4. (A) SEM image of periodically wavy Si ribbons on a PDMS substrate. (B) Optical image and (C) electrical characterization of a wavy Si-ribbons transistor on a PDMS stamp before and after it had been applied with external strains.

In summary, various semiconductor building blocks with small lateral dimensions have been used to fabricate high performance electronics with flexibility and stretchability/compressibility. Further integration of these TFTs, diodes and passive elements can yield circuits with complex functionalities.

We thank the financial supports from DARPA, DOE and NSF and contributions from all members of Rogers' group at UIUC.

REFERENCES

[1]   R. H. Reuss, et al. *Proc IEEE*, vol. 93, pp. 1239-1256, 2005.
[2]   S.-H. Hur, O. O. Park, J. A. Rogers, *Appl. Phys. Lett.*, vol. 86, no. 243502, 2005.
[3]   S.-H. Hur, C. Kocabas, A. Gaur, O. O. Park, M. Shim, J. A. Rogers, *J. Appl. Phys.*, vol. 98, no. 114302, 2005.
[4]   Q. Cao, S.-H. Hur, Z.-T. Zhu, Y. Sun, C. Wang, M. A. Meitl, M. Shim, J. A. Rogers, *Adv. Mater.*, vol. 18, pp. 304-309, 2006.
[5]   C. Kocabas, S.-H. Hur, A. Gaur, M. A. Meitl, M. Shim, J. A. Rogers, *Small*, vol. 1, pp. 1110-1116, 2005.
[6]   C. Kocabas, M. A. Meitl, A. Gaur, M. Shim, J. A. Rogers, *Nano Lett.*, vol. 4, pp. 2421-2426, 2004.
[7]   E. Menard, K. J. Lee, D.-Y. Khang, R. G. Nuzzo, J. A. Rogers, *Appl. Phys. Lett.*, vol. 84, pp. 5398-5400, 2004.
[8]   Y. Sun, J. A. Rogers, *Nano Lett.*, vol. 4, pp. 1953-1959, 2004.
[9]   Y. Sun, D.-Y. Khang, F. Hua, K. Hurley, R. G. Nuzzo, J. A. Rogers, *Adv. Funct. Mater.*, vol. 15, pp. 30-40, 2005.
[10] K. J. Lee, J. Lee, H. Hwang, Z. J. Reitmeier, R. F. Davis, J. A. Rogers, R. G. Nuzzo, *Small*, vol. 1, pp. 1164-1168, 2005.
[11] D.-Y. Khang, H. Jiang, Y. Huang, J. A. Rogers, *Science*, vol. 311, pp. 208-212, 2006.
[12] M. A. Meitl, Z.-T. Zhu, V. Kumar, K. J. Lee, X. Feng, Y. Y. Huang, I. Adesida, R. G. Nuzzo, J. A. Rogers, *Nat. Mater.*, vol. 5, pp. 33-38, 2006.
[13] Y. Sun, S. Kim, I. Adesida, J. A. Rogers, *Appl. Phys. Lett.*, vol. 87, no. 083501, 2005.
[14] Y. Sun, H.-S. Kim, E. Menard, S. Kim, G. Chen, I. Adesida, R. Dettmer, R. Cortez, A. Tewksbury, J. A. Rogers, *Appl. Phys. Lett.*, submitted.
[15] Z.-T. Zhu, E. Menard, K. Hurley, R. G. Nuzzo, J. A. Rogers, *Appl. Phys. Lett.*, vol. 86, no. 133507, 2005.

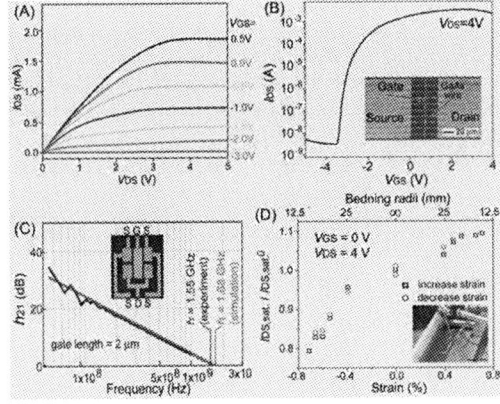

FIGURE 2. Characterization of GaAs-wire MESFETs on PET substrates: (A,B) ac characteristics; (C) microwave measurement; (D) bending test.

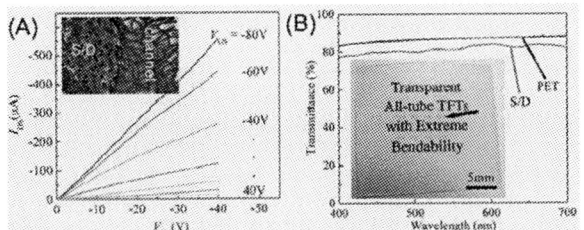

FIGURE 3. Electrical (A) and optical (B) characterization of all-tube TFTs fabricated on a PET substrate.

2006 International Symposium on VLSI Technology, Systems, and Applications

Printed Electronics for System Application

Z.Pei, C.P. Kung, P.Y.Lo, J.J.Chang, C.A.Chung. Stanley H. Huamg and Y.J.Chan

ERSO/Industrial Technology Research Institute
195-4, Sec. 4, Chung Hsing Road
Chutung, Hsinchu
Taiwan, R.O.C.

ABSTRACT

Printed electronics are of highly interested in recently years for its flexible and low cost. Extensively results on the transistor, light-emitting diode, memory and diode were reports. However, rarely works on the integration of those components into an electronic system and the architecture for the system are discussed. In this article, we discussed the design and performance of transistor, memory and diode based on the integrated process consideration and the system architecture for a printed RFID.

INTRODUCTION

Two major applications account for printed electronics in the future 20 years [1]. One is the display and lighting, another is the memory and logic circuit. For display and lighting, the major points are brightness, uniformity of process and lifetime of the devices. The system and circuit are less complexity. For logic and memory, major product is a radio-frequency identification (RFID) tags. A RFID tags should integrated several types of devices (Antenna, capacitor, resistor, transistor, diode and memory), received RF signal from reader, extract DC power, demodulator the RF signal, data processing the order from the reader and emit a RF signal to give a response to the reader. The total circuit would counts for several thousand of transistor for EPC or ISO standard. Therefore, RFID has much system issue than display and other applications. The general requirement is the uniform and high speed transistors. However, for printed electronics, the material is not stable as Si technology and the resolution for printing technology could not easily less than 1 um as in the Si technology. Printed RFID tag may not fulfill the full function of EPC and ISO standard in the recent year. Besides item level tracking by using EPC and ISO standard, there are some other applications could be applied, such as access control, ticketing, visitor control and home management or personal filing. These applications talk to reader one by one and need very less memory volume that could largely simplify the system. In addition, these applications need very low cost and capability of custom designed, is a niche area for printed RFID. In article, we propose a protocol for 13.56MHz RFID that can remove the speed and stability limitations of printed electronics alleviating the performance requirement on transistors. Moreover, a printed process for all device was also printed that could have low production cost for not only RFID tag but other logic circuits.

PROTOCOL AND ARCHITECTURE

For these simple applications, proposed protocol and architecture put much effort of RFID operation on the reader part which do not manufacture by printed process. For the signal, the Manchester codes were use to define "1" or "0" in a write once read many architecture, as in the Fig. 1(a). To read a data in the RFID tag, the data in the tag will send to reader once they were wake by the reader. The replied signal would be a series of Manchestor code with a preamble signal

added in front of the data and a code "0" followed to denote the end of the data as depicted in the Fig.1 (b).

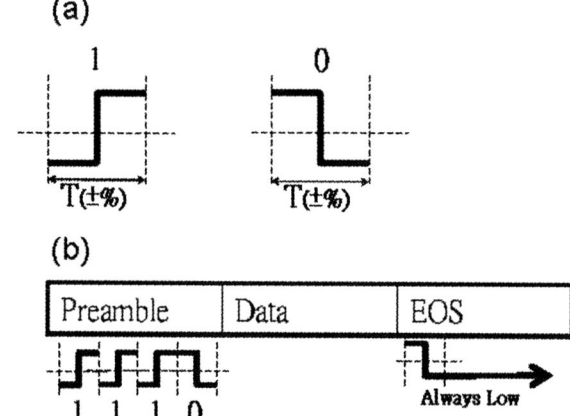

FIGURE 1. The protocol of proposed RFID tag, (a) Monchester code, (b) preamble read-out sequence.

For this protocol, the architecture for operation was depicted in the figure 2. An antenna was used to couple the signal from reader into the tag; a rectifier was used convert AC signal to DC power output. Once the power generated, a ring oscillator will be turned on generating a reference clock for control circuit and Manchester encoder. The data or the code pattern was then read and sent to the reader by the load modulator and the antenna.

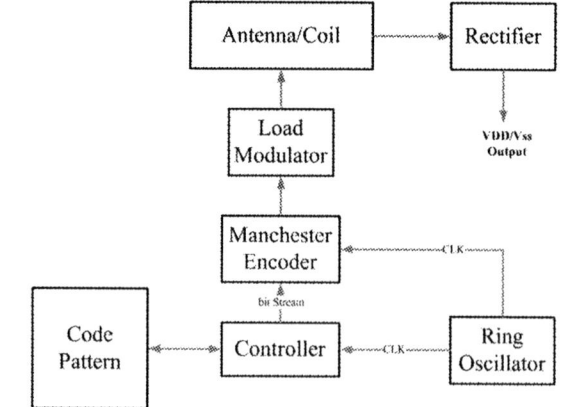

FIGURE 2. The operation architecture for proposed protocol

In this architecture, the code pattern read-out circuit is the most "complicated" part. The data was read one-by-one by using the shift register. The tags do not response for the demodulation the order from the reader. Therefore, this architecture has two advantages. First, there is no demodulation circuit for 13.56 MHz, the data rate

1-4244-0181-X/06/$20.00 ©2006 IEEE

could be as low as possible to match the general properties of printed devices. Second, the generated clock is for internal used only for each tag, different tag could have different clock rate. The decoding of each tag could rely on the phase-locked loop process of the reader. These two advantages just match the properties of printed electronic devices: slow and not so uniform as Si-based devices. Therefore, the most important devices are the diode and memories.

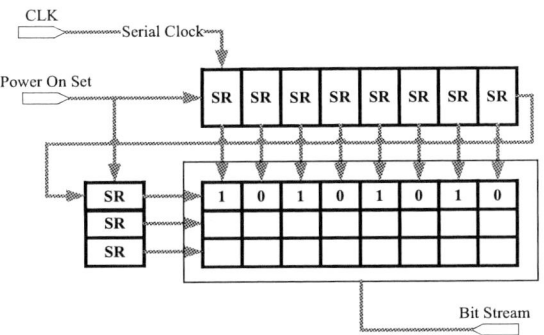

FIGURE 3. The read architecture of the code pattern (memory).

PLATFORM AND DEVICES
Platform

Although the architecture is simple enough, it still rely on the production process to realize a RFID tag in a very low cost As mentioned above, devices including transistors, diode, antenna, capacitor, and memory should be integrated as a whole system. In general, these devices are produced by different process to have highest performance. For example, Antenna coil are printed by Ag ink to have low cost and low curing temperature. Transistors generally using top gate structure by using Au or PEDOT: PSS as the Source/Drain material to match the requirement of low contact resistance. Different materials and printing method would cause a high cost process. An integrated printing process was proposed that printing all devices with same method and same material to reach the low cost requirement. This integrated structure was demonstrated in the Fig. 4.

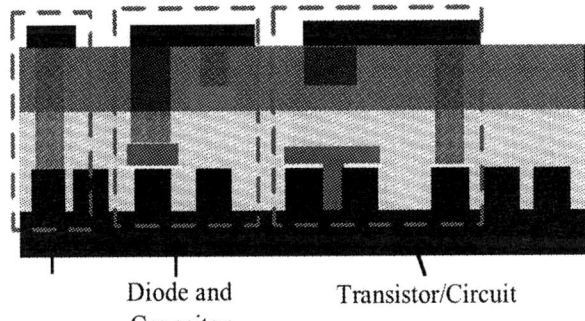

Diode and   Transistor/Circuit

FIGURE 4. The schematic cross-section view for proposed integrating platform of process.

This integrated platform has a common substrate. The substrate could be plastic or paper depends on the requirement of application. On top of the substrate is the first conductor material. Ag paste could be printed by ink-jet printer, screen printer or pad printer. This conductor layer was used as the coil, bottom electrode of the diode and capacitor and the source/drain of the transistors. After Ag paste,

conducting polymer such as PEDOT: PSS selectively printing on the diode's bottom electrode and source/drain area of transistor used as the surface and the work function modifier for the following semiconductor layer. A layer of P3HT was then printed on the diode and transistor as the semiconductor layer. As the gate dielectric layer, polymer insulator was coated entirely act as the electrical isolator between devices and the dielectric for capacitor as well. The second electrode, second insulator, via hole and third conductor layer were processed sequentially to complete the circuit. For the memory, the material is unique and for memory use only, do not plot in the schematic diagram. However, it still can be integrated into the platform and will be discussed in the later section.

Diode

The diode that could integrated the capacitor and resistor to form a rectifier is the most important device in a simplified RFID tag; it is the only active device should be operated as fast as the carrier signal to convert the AC signal to DC power. The requirements for the diode are fast, low forward turn on voltage, low reverse leakage current and high rectifying ratio. Although either diode connected or Schottky diode rectifier has been reported operating at 13.56MHz [2,3] , the Schottky type rectifier has advantages of high speed and larger forward current due to essential properties of majority carrier transport.

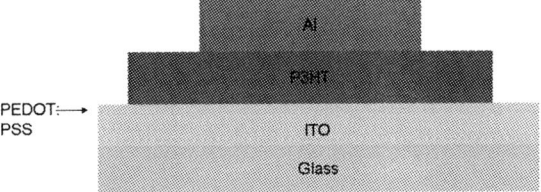

FIGURE 5. The structure of a organic diode

To explore the performance of the diode, an ITO coated glass was used as the substrate, on top of ITO, a layer of PEDOT: PSS was spun on as the interface smoother and the work function modifier to form the ohmic contact. A P3HT layer was then also spun on. A thermal evaporated Al was used as the Schottky contact to the P3HT. The Al patterning was accomplished by a shadow mask. Although this diode was fabricated on the glass, its process flow is compatible to the proposed platform. The ITO could be replace by Ag ink because its act just as a conductor. The top Al can also replace by Ag ink for they has similar work function to Al. The cross section view of the diode was shown in figure 5 schematically.

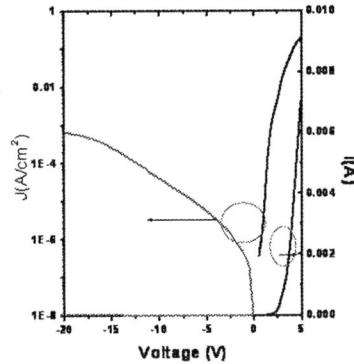

FIGURE 6. The electrical behavior of a organic diode

The diode exhibit well diode behavior as expect. The electrical property was shown in the Figure 6. The forward turn on voltage is around 2.5 V and has soft breakdown behavior at reverse bias. The reverse breakdown voltage is larger than 30V. The forward current density is around 0.2 A/cm^2 at 5 V, and the reverse leakage is around 2×10^{-4} A/cm^2 at -15V. The rectifier ratio is therefore about 1,000. The forward current could be further increased by modify the process in PEDOT: PSS layer.

Memory

Two type of organic memory are generally reported. One is the three terminal ferroelectric memories (transistor type) another is the two terminal bistable (diode type) memories [4, 5]. The process flow of ferroelectric memory is similar to organic transistor except the gate dielectric is a ferroelectric material. The operating voltage is rather high and speed is slow that is limited by the thickness of ferroelectric layer and the length of the source/drain. In contrast, two terminal bistable devices have very simple architecture. The operating voltage is relative low due to its thin memory material.

Bistable Memory

The structure of diode type bistable memory is shown in the Figure 7 (a). The memory material is simply sandwiched by two electrodes. The memory material consist three elements; electron acceptor, electron donor and host material. The memory material is behavior like an insulator film at initial. The applied voltage will then force the electrons transport to the acceptor from the donor. Once the acceptors are filled, a transport channel is formed, the current is increased abruptly. This is the write process. The channel won't disappear unless the electrons in the acceptor back to the donor. Therefore, for read, a small voltage will have large current. For erase process, a reverse voltage was applied; force the electrons back to the donor. The memory device was then back to its initial states. Au nanoparticles were used as the acceptor and the donor is 8HQ (8-hydroxyquinoline) with polystyrene as the host the in this experiment. The electrical properties of a memory device was shown in the Fig. 7(b). The write voltage could be as low as 1.5V. The on/off ratio could as high as 10^7.

(a)

(b)

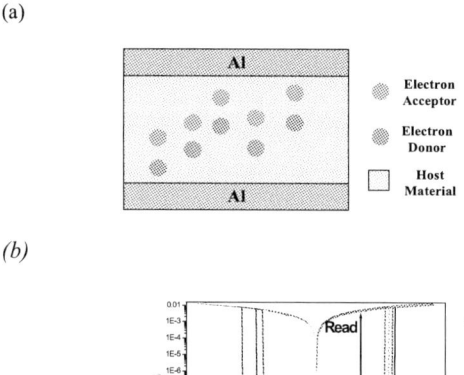

FIGURE 7. The schematic structure of bistable memory (a) and the electrical behavior of the memory. (b)

Flexibility

To further explore the flexibility of the memory, it was fabricated on an Al foil. The rest process is the same as the device fabricated on glass. The memory devices are almost invisible as shown in the Figure 8(a). The electrical property before bending shows almost nine order of magnitude of On/Off memory window. After bending, the behavior shows slightly increase on the write voltage. This experiment indicates the memory has good flexibility.

(a)

(b)

(c)

FIGURE 8. The picture of memory device fabricated on Al foil (a), electrical behavior before bending (b) and after bending. (c)

Au Nanoparticle Memory

From the operation mechanism of the bistable memory, filling Au nanoparticles with electrons is a main step to set up the conducting channel. This might not necessary accomplished by electron donors which surround the acceptor. The electrons can be supplied by the electrode as well. Figure 9 (a) shows the electrical properties of a bistable memory in which its memory material consist only Au nanoparticles and the host material. This device shows memory behavior as the device with electron donor. The on/off ratio of this device is around 10^4 which is lower than memory with acceptor-donor pair. The retention time of the device was tested for "1" state at 2V. The memory keeps in the "1" state last more than 5 hours in the ambient environment.

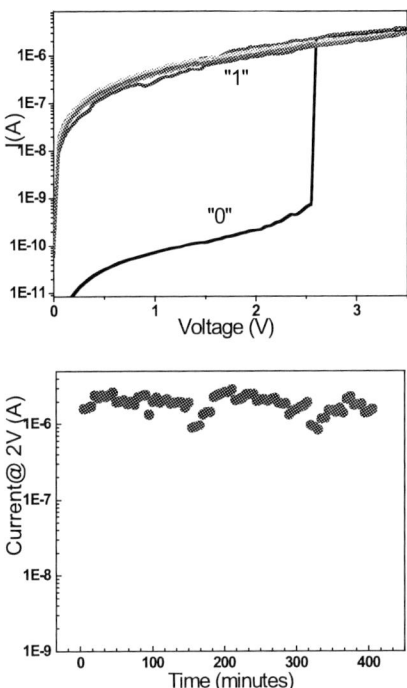

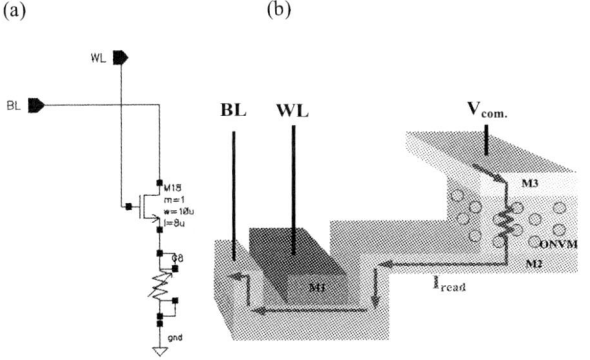

FIGURE 9. The electrical properties of Au nanoparticle memory (a) and the retention measured at 2V(b) .

Integrated Memory

In the process platform, the memory device was not sketched. The reason is the material is not common to other devices such as diode and transistor. But this not indicates the memory can not be integrated into this platform.

(a)   (b)

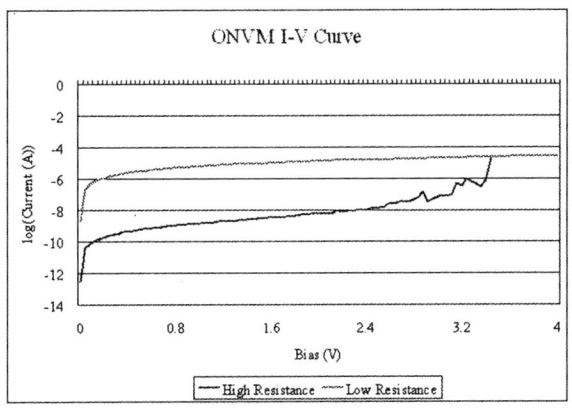

FIGURE 10. The circuit of transistor derived memory(a) and the schematic cross section structure

By adding a printed memory material layer after gate dielectric formation onto pre-defined memory area, the memory could be formed with only one extra printing process. Before this process was integrated, the transistor-drive memory was demonstrated first. The circuit and the device structure are sketched in the Fig. 10 (a) and (b), respectively. Once the word line (WL) was selected, the bit line (BL) was used to control the memory state. Figure 10 (c) show the electrical behavior of "0" state and "1" state. The transistor is a low temperature polysilicon transistor in this experiment.

FIGURE 11. The electrical properties of a transistor drive bistable memory

SUMMARY

In summary, we proposed a protocol and architecture for printed RFID tag. This protocol can be applied to many simple and unique applications such as ticketing, access control and personal items control. The architecture takes the inherent drawbacks of unstable and low mobility of printed electronics into consideration. Based on this architecture, an integrated process platform was also proposed. This platform could have lowest process and investment cost. Besides RFID tag, other product such as display backplane and some logic circuit for simple application can be designed based on this platform as well. For device part, diode based on organic material show good behavior, memory device has very large memory windows.

REFERENCES

[1]. Peter Harrop and Raghu Das, Organic Electronics Forecasts, Players, Opportunities 2005-2025, IDTechEx Press, 2005.

[2]. Soren Steudel, *et.al.*, *50MHz rectifier based on an organic diode*, Nature Materials, 2005, vol.4, 597-600

[3]. Robert Rotzoll et.al., *13.56MHz Organic Transistor Based Rectifier Circuit for RFID tags*, MRS Proc. Vol.871E, I11.6.1-I11.6.6, 2005

[4]. Jianyoung Ouyang et.al., *Proggramable Polymer Thin Film and Non-volatile Memory Device*, Nature Materials, Vol.3, 918-922,2004

[5]. Ronald C.G. Naber et.al., *High-performance solution-processed Polymer Ferroelectric Field-effect Transistors*, Nature Materials, Vol.4, 243-248, 2005

Advanced Substrate Engineering for the Nanotechnology Era

Carlos Mazure
SOITEC
Parc technologique des Fontaines, Bernin,
38926 Crolles Cedex
France

ABSTRACT

Engineered substrates is one of the most important innovations of the nanotechnology era driven by the vanishing boundary between substrate design and device architecture. SOI substrates, the first engineered substrates of its kind, have made possible an efficient optimization of MOSFET current drive while minimizing the leakage and reducing parasitic elements, thus enhancing the overall IC performance. Mobility enhancing substrates have added new handles to traditional scaling further improving device and IC performance. An overview of the advances in Smart Cut engineered substrates and the impact on device performance will be given.

INTRODUCTION

IC density increase through device geometry shrinking no longer results in an IC performance increase if the scaling is not coupled to the introduction of new materials. The addition of new materials has always been present in the CMOS world, but it has become the dominant "scaling" approach beyond the 90nm technology node with substrate engineering playing a central role in it.

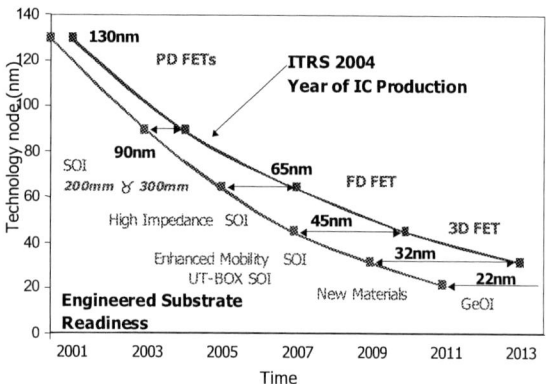

FIGURE 1. Substrate readiness roadmap compared to the ITRS 2004 roadmap.

The adoption of SOI technology [1] illustrates this very well. SOI substrates constituted a paradigm shift in the 90's that made possible to increase current drive while reducing leakage, to improve IC performance while keeping a relatively low power consumption level. Since competitive solutions have been developed to address smart power for automotive, very low power IC for consumer applications and high performance logic with microprocessors as the main technology driver.

While the 90nm node is characterized by the adoption of SOI for high performance applications [2], the future nodes are driving numerous substrate solutions.

The ITRS roadmap [3] predicts a 3 year cycle for the node generations but the industry has accelerated and the period is down to 2 years for the most aggressive IC makers. The substrate industry is thus forced to anticipate faster with an increasing number of options, the final choices of the IC industry. The challenge for the substrate industry is illustrated in figure 1. The advanced substrate readiness roadmap follows a 2 year cycle with an increasing gap to the ITRS production forecast, which highlights the early implication of the substrate design in the device architecture development.

SUBSTRATE ENGINEERING

Epitaxy and single crystal layer transfer are the two most critical processes of substrate engineering that allow the tailoring of the composite substrate to the targeted application. For thin layer transfer, Smart Cut^{TM} technology [4] is today the industrial standard, a powerful tool that applies to many materials, making it possible to create a wide range of composite substrates and its tailoring to the requirements of the application by properly choosing the active layer, the buried dielectric and the base substrate.

High Impedance SOI

High impedance SOI is an evolutionary SOI approach where the handle or base wafer is a high resistivity (HR) substrate (>1 kOhm-cm) [5-7]. SOI technology provides complete oxide isolation cutting off direct paths of substrate injection noise and a high resistivity substrate reduces the capacitive coupling, thus further reducing the substrate related RF loss. Because of the SOI nature of the high impedance substrate, latchup is of no concern. CMOS SOI constitutes a cost effective alternative to GaAs and BiCMOS technologies [7]. Moreover, layer transfer offers the unique possibility to engineer SOI HR, high impedance substrates, to optimize RF gain, while reducing noise with no significant change in the fabrication process.

Embedded Memory

Capacitor-less one transistor DRAM cells are a new development which takes advantage of the floating body effect in SOI MOSFETs [8-11].The generation of excess negative or positive charge in the body can be used to store data states. In an n-channel device an excess of positive charges leads to an increase of the current drive, defining state "1". The removal of positive charges from the body decreases the channel current, defining state "0". The strong industrial potential of the floating body cell is due to the fact that very dense embedded memory blocks are realized with current SOI processes with footprints as small as $4F^2$, where F is the minimal feature size [11].

Mobility Enhancing Substrates: sSOI

The most important innovation at the transistor level is the introduction of mobility engineering through Si substrate straining techniques. Process-induced stress has been known for many years.

For example, silicidation, STI isolation, and cap layers all induce a certain level of device geometry dependent stress that can be beneficial or detrimental to transistor performance. The challenge is to optimize all CMOS process modules, i.e. their uniaxial stressor components, in order to maximize beneficial effects while minimizing the negative contributions. The IC industry has intensively developed the stressors techniques making it possible for uniaxial strained silicon to make its way into 90nm IC manufacturing.

At the substrate level a similar development has occurred. Strained silicon on insulator (sSOI) has been developed as the solution that offers higher carrier mobility, combining the advantages of SOI with those of strained silicon [12]. The fundamental difference between the uniaxial and the wafer level (biaxial) strained Si is that the former is introduced during CMOS processing while the latter is built into the substrate. As a consequence, sSOI wafers offer tensile strain that is device-layout and gate-pitch independent, giving IC design a high degree of freedom. The sSOI evaluation by IC manufacturers highlights its potential as a low power, high performance solution. On-going work with IC makers and research institutes shows that the combination of uniaxial stressors and sSOI amplifies the mobility enhancement of both n- and p-channel devices. The scalability of sSOI has been thoroughly investigated for ultrathin body sSOI with the fabrication of short channel devices [13], as well as the impact of 40 to 50nm thick sSOI, super critical strained Si, on PD MOSFETs [14], showing a 30% reduction in gate oxide leakage and a 60% improvement of the SRAM write margins.

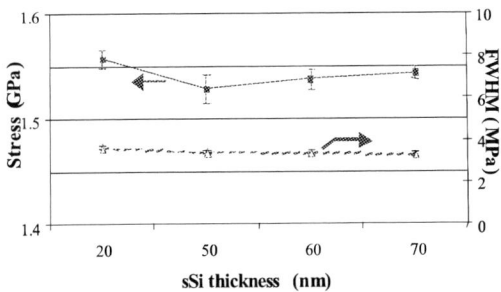

FIGURE 2. UV Raman stress measurement of sSOI as a function of sSi thickness. Laser: 363.8nm.

Typical sSi film thicknesses range from 10-20nm for fully depleted (FD) designs and up to 70nm for full compatibility with existing partially depleted (PD) architectures. Fig. 2 shows UV Raman stress measurements of sSOI for different sSi film thicknesses. No significant effect of the sSi thickness is observed on neither the strain Si peak, i.e. the stress level, nor a broadening of the full width half maximum (FWHM) Raman signal.

Ultra thin buried oxides

A factor of 3 improvement in thermal conductance can be achieved by reducing the BOX thickness from 150nm to 20nm [15]. It is an effective manner to reduce local self-heating for devices that are used in the on-state most of the time or for circuits with a high duty cycle. The tradeoff is, of course, increasing the parasitic capacitances and reducing the overall IC performance. Another approach is to introduce a high thermal conductivity material as the buried dielectric. There are several options [15] but silicon nitride appears to be the most attractive one. It is an industrially mature material, it exhibits an order of magnitude higher thermal conductivity than SiO_2 and it is a well characterized insulator. It has

been shown that a composite nitride/oxide buried dielectric is a viable approach for an improved thermal conductivity substrate [16].

In contrast, if the focus is low power consumption, ultra-thin BOX is very advantageous. It offers the possibility to easily form buried n and p regions in the handle substrate as a back gate for low voltage operation. By applying a back bias, the off current can be reduced while a forward bias lowers the device V_T resulting in a current drive increase [17]. It is a particularly interesting concept for SOC integration.

Multi gate FETs

FinFETs are a major innovation in the device architecture that will help nanoscaling. A key element in the manufacturability of these new devices is the SOI substrate. The Si layer thickness defines in part the transistor width, and assures its reproducibility. The buried oxide is used as an etch stop. If a composite nitride/oxide dielectric is used it will improve the etch stop efficiency and will avoid dielectric undercut during H_2 smoothing of the fins lateral roughness.

CONCLUSION

Engineered substrates have become a critical element of transistor scaling for nano technologies. Demands of future circuits, regardless of specific application, will continue to drive the development of specialized engineered substrates. Mobility enhancing substrates will constitute one of the key innovations of the 65 and 45 technology nodes. SOI will continue to gain interest for RF applications, in particular due to high resistivity SOI substrates, and for SOC applications. The ultra thin buried oxides are a feature that allow the integration of a back-gate for dynamic threshold voltage control. The integration of embedded one transistor floating body memories on SOI in the future will result in ICs that will not only have speed and power advantages, but should also be significantly lower cost. Moreover, SOI appears also as a key element in the development of 3D multi-gate FETs and its manufacturability.

REFERENCES

[1] G.K. Celler and S. Cristoloveanu, J. Appl. Phys., vol. **93**, p.4955 (2003).
[2] D. Greenlaw et al.; IEDM Tech. Dig., pp. 277-280 (2003).
[3] http://public.itrs.net/
[4] C. Maleville & C. Mazure, Solid State Elect., vol. **48-6**, p. 855 (2004).
[5] C. Raynaud et al.; *Silicon-On-Insulator Technology and Devices XII*, ECS Proc., vol **2005-03**, pp. 331-344 (2005).
[6] A. Viviani, et al.; IEDM Tech. Digest, pp. 713-716 (1995).
[7] T. Matsumoto et al.; IEDM Tech. Digest, pp.219-222 (2001).
[8] P. Fazan, IEEE Int. SOI Conf., p. 10 (2002).
[9] T. Oshawa et al., Proc. ISSCC, p. 152 (2002).
[10] T. Shino et al., Proc. Symp. VLSI Circuits, (2004).
[11] S. Okhonin et al.; IEEE EDL, **23**, p. 85 (2002).
[12] C. Mazure & I. Cayrefourcq; IEEE Int. SOI Conf., p.1 (2005).
[13] K. Rim et al.; IEDM Tech. Dig., p. 49 (2003).
[14] A. Thean et al.; Proc. Symp. VLSI Tech, p. 134 (2005).
[15] N. Bresson et al.; IEEE Int. SOI Conf., p. 62 (2004).
[16] O. Rayssac et al.; *Silicon-On-Insulator Technology and Devices X*, ECS Proc., vol **2001-03**, p. 39 (2001).
[17] R. Tsuchiya et al.; IEDM Tech Dig., pp. 631-634 (2004).

2006 International Symposium on VLSI Technology, Systems, and Applications

A CMOS Bulk-Micromachined Thermal Imager

D.-H. Liu[2], L.-S. Zheng[2], C.-Y. Hsu[3], D.-J. Yao[3], and M. S.-C. Lu[1,2]
[1]Department of Electrical Engineering, [2]Institute of Electronics Engineering, and
[3]Institute of Microelectromechanical System
National Tsing Hua University
Hsinchu 30013, Taiwan, R.O.C.

INTRODUCTION

Most uncooled thermal imager arrays are integrated with CMOS detection circuits to facilitate the routing, multiplexing, and processing of the sensed signals. Thermal imaging by using MOS transistors [1, 2] provides a low-cost solution for the application, for example, Reay [1] placed a diode-connected sense transistor inside the suspended n-well and measured the voltage across it with respect to a temperature change. Liu [2] also reported an 8×8 infrared imager fabricated by using a PMOS sensing transistor operated in the saturation region. In this paper, a MOS transistor operated in the sub-threshold region is used for the sensing purpose. The experimental result shows that it has a high temperature coefficient of resistance (TCR) at -2.7%/°C, showing a good potential in making sensitive thermal imagers.

DEVICE FABRICATION

The TSMC 0.35-μm two-polysilicon four-metal CMOS process is adopted for device fabrication. The cross-sectional view in Fig. 1 illustrates how a sensing transistor within a suspended membrane can be preserved after CMOS micromachining steps. An anisotropic deep silicon reactive etch is first performed from the backside to thin down the substrate to about 20 μm. Next a front-side reactive ion etch is performed to remove the dielectric layers with the top metal as the etch-resistant mask, followed by a front-side anisotropic silicon etch to release the microstructure. An isotropic silicon etch is performed afterward to remove the silicon underneath the support beams in order to enhance thermal isolation from the sensing plate to the supports. The stacked metals and vias as shown in the suspended plate are connected to the substrate surrounding the sensing transistor to enhance thermal conduction of the absorbed heat from the top. Fig. 2 is the micrograph of the thermal imager array after post micromachining.

SENSOR DESIGN

The output resistance of a PMOS transistor operated in the sub-threshold region is used for thermal sensing. Its resistance is expressed by,

$$R_{SD} = \left(\frac{\partial I_{SD}}{\partial V_{SD}} \right)^{-1} = \left[\frac{I_{s0}}{V_T} exp\left(-\frac{V_{SD}}{V_T} \right) \cdot exp\left(\frac{V_{SG} - |V_{th}|}{nV_T} \right) \right]^{-1} \quad (1)$$

where V_{th} is the threshold voltage, V_T is the thermal voltage, and,

$I_{s0} = \mu_0 \frac{W}{L} V_T^2 \sqrt{\frac{q\varepsilon_{Si}N_{ch}}{4|\phi_F|}}$. The parameter μ_0 is the carrier mobility,

W and L are the transistor width and length, ε_{Si} is the permittivity of silicon, ϕ_f is one half of the built-in junction potential, and N_{ch} is the channel doping concentration. Simulation of a PMOS transistor with $L = 4$ μm, $W = 1$ μm, and a source-to-gate voltage of 0.53 V shows an average TCR value of -3.3%. The reason is attributed to the dominant effect of the reduction in the threshold voltage over the

other temperature-dependent factors, such as the carrier mobility. The sensing pixel size is 70 μm by 70 μm. The sensing circuit in Fig. 3 comprises of a PMOS source follower and the sense transistor M1 operated in the sub-threshold region. The capacitor and the M1 transistor form a high-pass filter with $f_{-3dB} = \frac{1}{2\pi R_{SD} C}$. For detection, an a.c. voltage with a frequency f_m close to f_{-3dB} is applied, followed by demodulation after the pre-amp.

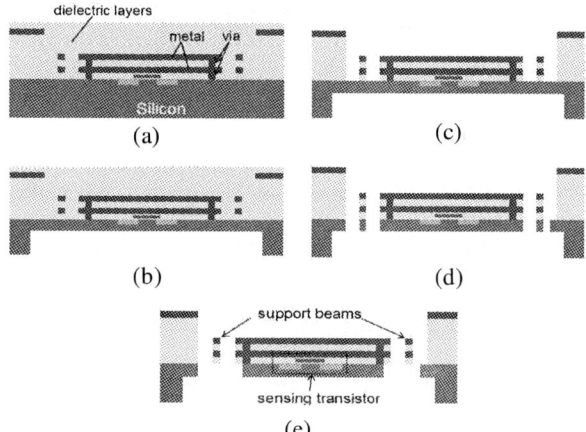

Figure 1. The post CMOS micromachining process.

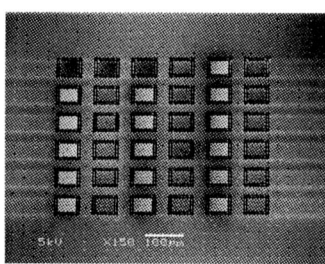

Figure 2. SEM of the 6×6 pixel array.

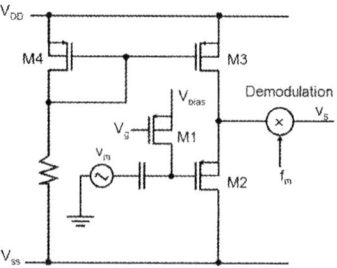

Figure 3. Schematic of the sensing circuit.

1-4244-0181-X/06/$20.00 ©2006 IEEE

EXPERIMENT

For device characterization, an on-chip polysilicon heater was used to provide the temperature change. The high-pass characteristic of the sensing amplifier was measured at different heating powers as shown in Fig. 4. The pole frequency moves from 294 to 333 Hz with an increasing power to 7.3 mW. This is equivalent to a change of MOS resistance from 108 MΩ to 90 MΩ. Fig. 5 shows the measured and the simulated pole frequencies with respect to the transistor bias V_{SG}, illustrating an offset in V_{SG} about 0.05 V. Temperature distribution of the sensing membrane was measured by an infrared scope (QFI, Inc.), which measured a heating efficiency of 0.84 °C/mW with the polysilicon heater. Thus a 7.3 mW power gives a total 6.1°C increase. We thus obtain the TCR of the PMOS resistance at -2.7%/°C, which is close to the -3.3%/°C from simulation. In another experiment the circuit input was provided by an a.c. voltage at 325 Hz, and the output was measured by a spectrum analyzer. The amplitude is 0.814 V with no applied power as shown in Fig. 6. The noise spectral density of the pre-amp is close to 7 μV/$\sqrt{Hz}$. With an increasingly applied power, the corresponding pre-amp output was measured. The sensor signal is the difference between the measured and the initial signals. The signal and the signal-to-noise ratio with respect to the temperature change are plotted in Fig. 7, in which a 73°C rise gives a signal of 34.2 mV and a SNR of 4900. Thus the calculated equivalent noise temperature is 14.9×10^{-3} K/$\sqrt{Hz}$.

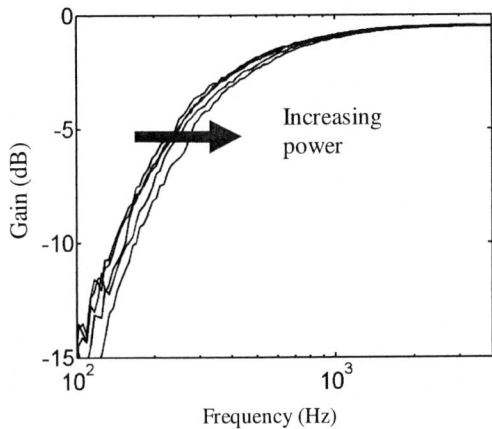

Figure 4. Measured high-pass curves at different powers.

CONCLUSION

The thermal imager has been fabricated by successive dry etching steps after completion of CMOS. We are the first to successfully demonstrate thermal sensing by transistors operated in the sub-threshold region. With no need for deposition of sensing films as in conventional micro bolometers, this CMOS-based approach provides a low-cost sensing solution for fabrication of thermal imaging arrays.

ACKNOWLEDGMENT

The authors would like to thank the National Science Council, Taiwan R.O.C., for support of this research effort (NSC94-2215-E-007-020), the National Chip Implementation Center for support of chip fabrication, and the National Center for High-Performance Computing for support of simulation tools.

REFERENCES

[1]   R. J. Reay et al., "Thermally and electrically isolated single silicon structures in CMOS technology," *IEEE Elec. Dev. Lett.*, vol. 15, no. 10, pp. 399-401, Oct. 1994.

[2]   C.-C. Liu and C. H. Mastrangelo, "A CMOS uncooled heat-balancing infrared imager," *IEEE J. Solid-State Circuits*, vol. 35, no. 4, pp. 527-535, Apr. 2000.

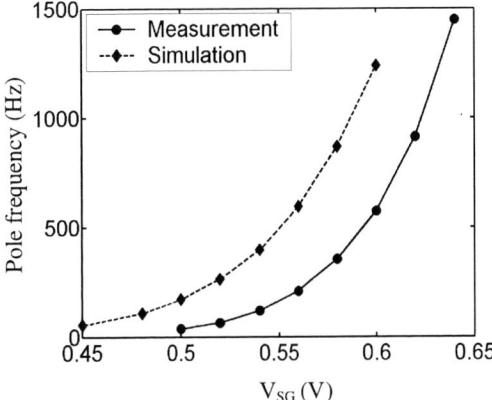

Figure 5. The pole frequency versus the transistor bias.

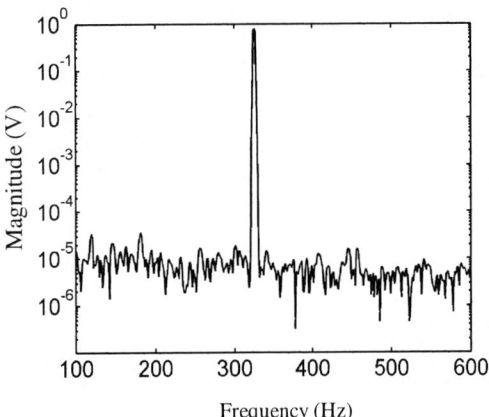

Figure 6. Measured signal spectrum from the pre-amp output at no heating power.

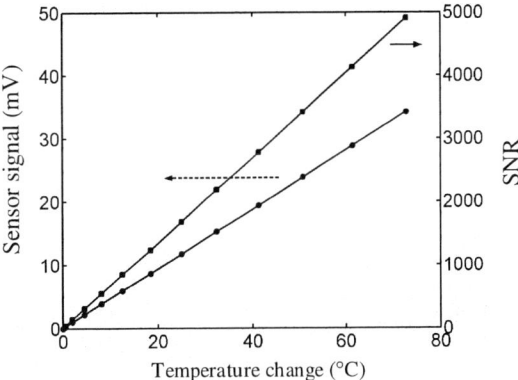

Figure 7. The sensor signal and the SNR with respect to the temperature change.

2006 International Symposium on VLSI Technology, Systems, and Applications

Impact of Back Gate Bias on Hot-Carrier Effects of n-channel Tri-Gate FETs (TGFETs)

Chia-Pin Lin (*Student Member, IEEE*) and Bing-Yue Tsui (*Senior Member, IEEE*)
Department of Electronics Engineering & Institute of Electronics, National Chiao Tung University
Room ED630, No.1001, Ta-Hsueh Road, Hsinchu, Taiwan, R.O.C.
Tel: 886-3-5712121-54112   Fax: 886-3-5724361   e-mail: cplin.ee90g@nctu.edu.tw

ABSTRACT

The hot-carrier effects of non-planar tri-gate SOI FET (TGFET) with back-gate bias were investigated. Negative back gate bias could raise the influence of buried oxide defects and then degrade the device quickly. For TGFETs with ultra-narrow fin width and side gate extension, the smaller buried oxide interface area and more obvious screening effect terminate the field lines to obviate the back gate bias efficiently. The extrapolated hot-carrier lifetime encourages the TGFETs as promising sub-10nm devices.

INTRODUCTION

The TGFET is attractive for sub-10 nm gate lengths CMOS technologies because of the superior scalability and conventional CMOS compatible process [1, 2]. The requirement of ultra-thin fin width (W_f) in double gated FinFET was efficiently relaxed to approximately the electrical gate length and kept the superior fully depleted (FD) behaviors at the same time in TGFET. As the device dimension scales down, the hot-carrier effects must be evaluated carefully. Recently, the hot-carrier effects of TGFETs were reported that the devices with narrower W_f could balance the gate electric field effectively and then have better hot-carrier immunity (HCI) [3]. Moreover, the effects of buried oxide quality on HCI of FD SOI devices were also reminded, especially for SIMOX wafer [4, 5]. In this work, the relationship between the HCI and the device dimensions and bias conditions, including back-bias effect, of TGFETs are investigated firstly.

EXPERIMENTS

The detailed fabrication process of the TGFETs used in this work has already been reported previously [6]. They were fabricated using SIMOX SOI wafers with 150 nm thick buried oxide and patterned by e-beam lithography. The gate length (L_g) of device is fixed at 130 nm, the fin thickness (T_{Si}) is fixed at 40 nm, and the fin width (W_f) is down to 40 nm. The gate dielectric in all devices is 4 nm thick thermally grown SiO_2. This work primarily focused on the HCI of n-channel TGFETs with different W_f stressed at different front-gate (V_f) and back-gate (V_b) bias conditions. The cross sectional TEM images of the Si fin with $W_f = T_{Si} = 40$ nm is shown in Fig.1(a). The schematic cross-sectional structure is shown in Fig.1 (b). The side gate extension (D_{ext}) is about 8 nm from the bottom of Si fin.

The worst bias condition for hot-carrier stress at $V_b = 0$ V was determined at first. The maximum transconductance (G_m) degradation and V_t shift were found to be the largest when the front gate voltage equals to the drain voltage. Therefore, devices with different W_f were stressed at $V_g = V_d = 3.2V$ and biased at various V_b. 3D numerical simulator was also used to understand the electric field distribution [7].

RESULTS AND DISCUSSIONS

Fig.2 shows the G_m degradation and the threshold voltage shift (V_{th}) of the TGFETs induced by hot-carrier stress at $V_b = 0V$ and $V_g =$ $V_d = 3.2V$. It is known that the energetic electrons trapped at the gate oxide/Si interface or in the gate oxide generated by impact ionization near drain is the major reason of positive V_{th} shift and G_m degradation. The DC hot-carrier lifetime is projected in Fig.3. The failure criterion used here is 10% change of G_m. The TGFETs with narrower W_f could balance the electric field in Si fin induced by the front gate and result in better HCI. On the other hand, for FD SIMOX SOI devices, the back interface degradation can also influence the front channel operation owing to the poor buried oxide quality and interface-coupling effect. The degradation of G_m and V_t shift of the TGFETs induced by hot-carrier stress at various V_b from 5 to -30 V is shown in Fig 4 (a) and (b), respectively. For devices stressed at $V_b = 0V$, G_m degrades and V_{th} shifts to positive direction, which indicates that hot electrons are trapped in the front gate oxide and interface. Nevertheless, for devices stressed at negative V_b, the hot holes are accelerated toward the buried oxide; then, the G_m and V_t change toward the reverse direction relative to those of devices stressed at $V_b = 0$ V.

The effects of back gate bias and W_f on HCI are investigated in this work. Fig.5(a) shows the variations of G_m of TGFETs stressed at $V_b = -20V$. For the devices with $W_f = 40$nm and stressed at $V_b = -20V$, the maximum variations of positive G_m shift are smaller and turn around faster than those of devices with $W_f = 200$nm and stressed at the same conditions. This observation indicates that the effect of V_b for narrower devices is slighter than that for wider ones. Fig.6 shows the simulated electric field of devices with $W_f = 40$ nm and 200nm. For devices with 40 nm W_f, the influence of electrical field induced by back gate is smaller than that of 200 nm devices. In the narrower devices, not only the reduction of the ratio between the buried oxide interface to front-gate oxide interface but also the strong shielding effect attributed to the D_{ext} obviously terminate the penetration of field lines from the back gate under the Si fins. Therefore, the maximum positive G_m variations for the devices with different W_f as shown in Fig.5(b) also demonstrates that the narrower devices have greater ability than the wider ones to restrict the back bias effect during hot-carrier stress.

CONCLUSIONS

The hot-carrier effects of n-TGFETs with different W_f, V_b, and D_{ext} were investigated in this work at first time. The TGFETs with narrower fins have better HCI and could more efficiently restrict the influence of buried oxide quality induced by the V_b. It is not only attributed to the decreasing of the ratio of buried oxide interface to front gate oxide interface but also attributed to the screening effect by the extension of side gate.

ACKNOWLEDGMENT

This work is supported by the NSC of the Republic of China (No. NSC-93-2215-E-009-004).

REFERENCES

[1] R. Chau, et. al., *Int. Conf. Solid State Dev. & Mat.*, p.68, 2002.

[2] B. S. Doyle, et. al., *IEEE Elec. Dev. Lett.*, Vol. 24, p.263, 2003.
[3] C. P. Lin, et. al., *IEEE Elec. Dev. Lett.*, Vol. 26, p.394, 2005.
[4] M. Bruel, et. al., *Proc. IEEE Conf. SOI*, p.178, 1995.
[5] S. H. Renn, et. al., *Proc. IEEE Conf. SOI*, p.81, 1998.

[6] C. P. Lin, et. al., *Proc. VLSI-TSA-Tech*, p.118, 2005.
[7] "ISE TCAD Rel. 10.0 Manual," DESSIS, 2004.

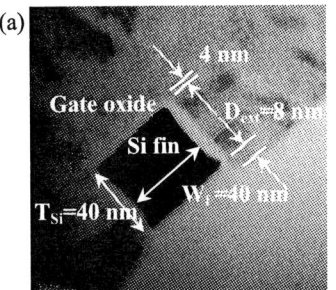

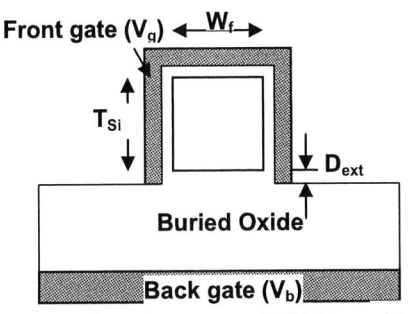

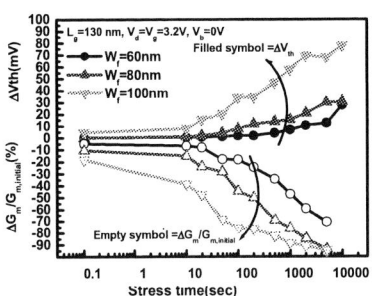

FIGURE 1. The cross-sectional (a) TEM picture and (b) schematic structure of TGFET.

FIGURE 2. The G_m degradation and V_{th} shift of TGFETs stressed at $V_g=V_d=3.2V$ and $V_b=0V$.

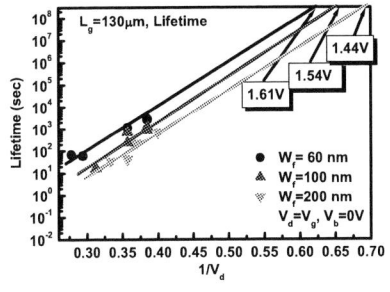

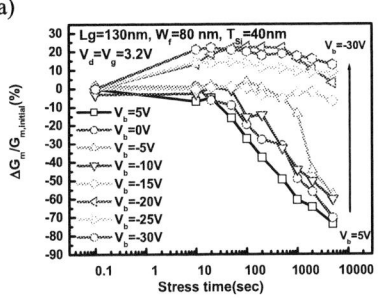

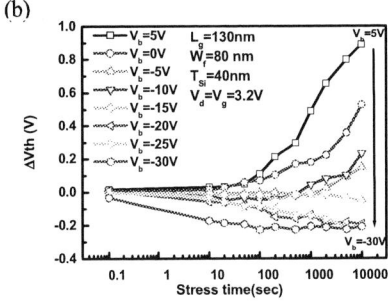

FIGURE 3. The DC hot-carrier lifetime of TGFETs with different W_f stressed at $V_g=V_d=3.2V$ and $V_b=0V$.

FIGURE 4. (a) The G_m degradation and (b) V_{th} shift of TGFETs stressed at $V_g=V_d=3.2V$ and different V_b.

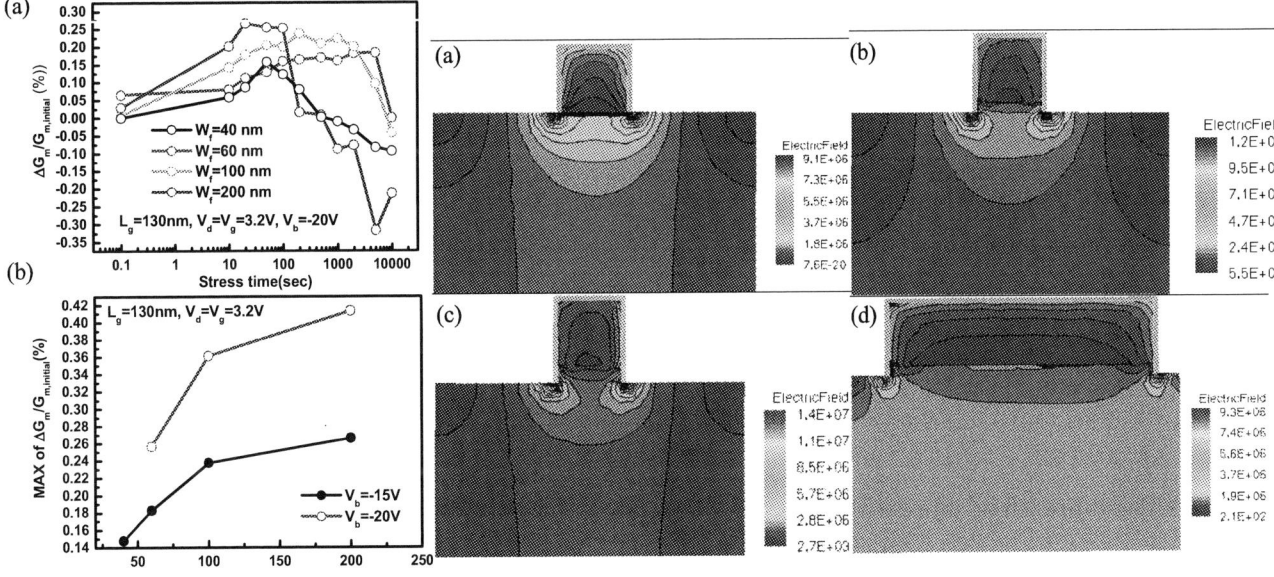

FIGURE 5. (a) The G_m degradation and (b) the maximum positive variations of G_m of TGFETs with different W_f stressed at $V_g=V_d$ and $V_b=-15$ and -20 V.

FIGURE 6. The simulated electric field of TGFETs with (a)$W_f=40nm$, $D_{ext}=0$, (b) $W_f=40nm$, $D_{ext}=8nm$, (c)$W_f=40nm$, $D_{ext}=16nm$, and (d) $W_f=200nm$ and $D_{ext}=8$ nm biased at $V_g=V_d=3.2V$ and $V_b=-20V$.

2006 International Symposium on VLSI Technology, Systems, and Applications

Ultra-Thin SOI CMOS Using Laser Spike Anneal

Zhibin Ren, J. Sleight, *J. M. Hergenrother, *D. V. Singh, O. Gluschenkov, O. Dokumaci,
§L. Black, §J. Pan, *K.-L. Lee, *J. Ott, P. Ronsheim, J. Lee, *W. Haensch, M. Ieong and C.Y. Sung
IBM Semiconductor Research and Development Center (SRDC), Hopewell Junction, NY 12533
*IBM T J Watson Research Center, Yorktown Heights NY 10598
§AMD Inc., 2070 Rte. 52, Hopewell Junction, NY 12533
e-mail: zhibinr@us.ibm.com, phone: 845-892-1735, fax: 845-892-3039

Abstract

We have investigated the impact of Laser Spike Anneal (LSA) on the performance of Ultra-thin SOI MOSFETs. LSA was found to significantly reduce the parasitic external resistance in UTSOI devices. Reduced external resistance in conjunction with improved gate activation resulted in a substantial improvement in nFET performance. A conventional spike RTA followed by LSA at 1300C enhances nFET drive current, I_{on}, by 20% and the effective drive current, I_{eff}, by 30 % (at $I_{off} = 1\mu A/\mu m$). The RTA+LSA approach was found to have a smaller impact on pFET performance. This is attributed to boron loss due to segregation into the buried oxide (BOX) during the RTA. The RTA+LSA process also resulted in improved AC performance (~ 10% improvement in ring oscillator stage delay at fixed leakage current) compared to an RTA-only process. We have found that an LSA-only process significantly suppresses boron segregation and increases dopant activation, resulting in a 50% reduction in the p-type sheet resistance when compared to a conventional high temperature RTA-only process. The introduction of LSA provides a path for high performance UTSOI CMOS.

Introduction

The potential advantages of fully depleted UTSOI CMOS architectures have been extensively studied through both simulation and experiment [1,2]. However the successful integration of high performance UTSOI devices is very challenging. One of the key issues is the parasitic external resistance. While a raised source/drain architecture alleviates contact resistance issues, the thin extension region represents a resistance bottleneck. The high resistance in the extension region can arise due to (1) the ultra-shallow conducting layer (2) dopant loss to the BOX and (3) limited dopant solid solubility. Unless the parasitic external resistance issues in UTSOI can be effectively controlled, it will not be possible to leverage the performance advantages offered by UTSOI. In this work we demonstrate for the first time that improved dopant activation achieved using the RTA+LSA approach, significantly improves both the AC and DC performance of UTSOI CMOS. Additionally, the LSA-only approach is shown to be extremely promising for UTSOI PFET devices.

Fabrication

The process starts with Si thinning, shallow trench isolation and gate oxide growth. Si thickness under the gate varies from 13nm to 23nm. After the poly-silicon gate etch and disposable spacer formation, the raised source and drain (RSD) is grown epitaxially. A buffer implant is done to improve the electrical connection between the extension and source/drain regions. This is followed by disposable spacer removal, halo implants, offset spacer formation and extension implant. A wide final spacer is then formed followed by deep source/drain implant and activation by a 1085C spike anneal (RTA) or RTA+1300C LSA. A conventional Ni salicide process is then implemented, followed by a low-temperature back-end of line process. A TEM cross-section of a fabricated device is shown in Fig. 1.

Results

Figures 2, 3 compare the DC IV characteristics of UTSOI devices with i) a 1085C spike RTA and ii) a 1085C spike RTA followed by a 1300C LSA. Figure 2 shows the I_{on}-I_{off} and I_{eff}-I_{off} curves for these devices. Significant I_{on} enhancement (~20%) is observed for the nFETs, whereas improvement for pFETs is relatively small (~5%). I_{eff} which is more relevant to AC performance [3] than I_{on} exhibits a larger enhancement of ~30% for the nFETs and 8% for the pFETs. Parameters describing the short channel behavior of these devices are shown in Fig. 3. Negligible change is seen in the DIBL and subthreshold slope for devices with and without LSA indicating minimal dopant redistribution due to the LSA. Excellent control of SCEs is demonstrated down to an L_{eff} of ~20nm. High frequency split CV measurements on long channel nFETs show that the threshold voltage of forming the inversion layer decreases by ~70mV for devices with LSA (Fig. 4a). The threshold voltage shift is likely due to the t_{inv} reduction

associated with higher dopant activation in the poly-Si gate. The enhanced dopant activation reduces the t_{inv} from 1.95nm to 1.75nm (~10%). Fig 4(b) shows the V_{tsat} roll-off curves for devices with and without LSA. The V_{tsat} reduction of ~ 70mV is consistent with the reduction of threshold voltage defined by the CV curves. Figure 5(a) plots the R_{on} vs. L_{eff} characteristics for nFETs. R_{on} is the device resistance in the linear regime (V_{ds}=50mV, V_{gs}-V_t=0.7V). The slope of R_{on} vs. L_{eff} is related to t_{inv}, while the y-intercept represents the external parasitic resistance. The change in slope is not clearly observed due to the presence of non-linear halo effects, however a reduction in nFET parasitic resistance of ~30ohm-um is evident. The decrease in nFET linear resistance can be due to both improved sheet resistance in the extensions regions and lower contact resistance. The resistance reduction impacts drive current (and gm_{sat}) in saturation region more strongly, as observed in the superior I_{on}-I_{off} slope for nFET devices with LSA (Fig. 2a). The nFET performance improvement results in a decrease in the ring oscillator stage delay of ~10% as shown in Fig. 5b. The ring oscillator RC delay reduction is primarily caused by resistance reduction, an improvement in t_{inv} would not directly contribute to an RC delay improvement.

Discussion

In UTSOI devices, the SOI layer thickness (T_{soi}) strongly impacts the extension sheet resistance. Reducing T_{soi} decreases the dopant sheet concentration, increasing the sheet resistance. We have found that RTA induced dopant loss is significant in UTSOI structures, particularly for boron doped samples. SIMS analysis indicates that the boron loss can be as high as ~70% in a 13nm SOI layer after a 1085C spike RTA (Figs. 6a, 6b). While As segregation effects are much less severe than in B, the As sheet resistance in an ultra-thin layer is limited by activation (Fig. 6b). LSA is effective in addressing the resistance issues. For As doped samples, we found a high temperature LSA (>1300C) enhances dopant activation, reducing Rs by >30% (Fig. 7a). We also found that LSA followed by a conventional RTA or a LSA-only approach provides similar Rs reduction for As doped samples. These results indicate that Rs is limited by As solubility. To further improve Rs, a different species or higher temperature is needed. For boron doped samples, Rs is strongly degraded by RTA related dopant loss. Although a conventional RTA followed by LSA enhances dopant activation, it does not prevent B loss through segregation. An LSA-only approach minimizes B loss, and simultaneously enhances dopant solubility, significantly reducing the p-type sheet resistance (50% of Rs of the RTA processed sample, Fig. 7b). The sheet resistance results are consistent with our observations of MOSFET performance: an RTA followed LSA is very effective in enhancing nFET drive current, but boron loss makes it less effective in improving pFETs. These results clearly indicate that a high-temperature LSA-only process is a key enabling element to a high-performance UTSOI CMOS technology.

V. Conclusion

We found that an RTA+LSA approach significantly enhances dopant activation resulting in improved t_{inv} and lower parasitic external resistance in UTSOI MOSFETs. Using an RTA+1300C LSA process we have demonstrated ~20% I_{on} improvement for UTSOI nFETs, and 10% ring oscillator delay reduction. The improvement in pFET performance using the RTA+LSA approach is less than observed for the nFETs due to significant RTA-driven boron segregation to the BOX. We demonstrated that an LSA-only process eliminates the RTA-driven boron loss to the buried oxide decreasing the sheet resistance by 50% compared to an RTA-only process and is a very promising approach for fabrication high performance UTSOI pFETs.

Acknowledgements

We wish to thank G. Shahidi and T.C. Chen for supporting this work.

References:
[1] H.-S. P Wong, *et al.*, *IEDM*, p.407, 1997.
[2] D. V. Singh *et al.*, IEEE Intl. SOI conference, p.178, 2005.
[3] M. Na *et al.*, *IEDM*, p.121, 2002.

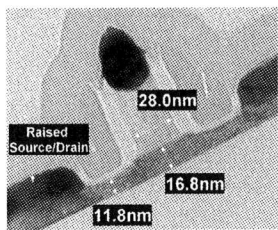

Fig. 1 A TEM cross-section of a UTSOI MOSFET. A raised source-drain structure is used to reduce resistance.

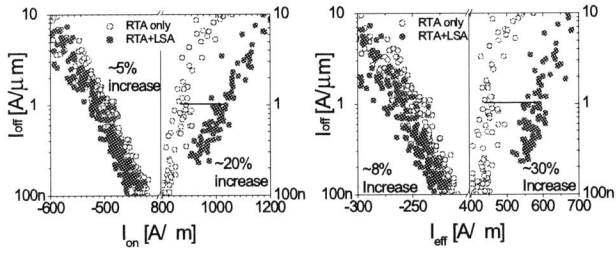

a)   b)

Fig. 2 a) I_{on} vs. I_{off}. At a power supply supply of 1.0V, an LSA (1300C) following the regular RTA improves nFET I_{on} by ~20% and improves pFET Ion by ~5%. b) I_{eff} vs I_{off}. More improvement is seen for I_{eff}, which is a derived DC parameter that is much more closely correlated to AC performance than I_{on} [3]. A power supply of 1.0V is used.

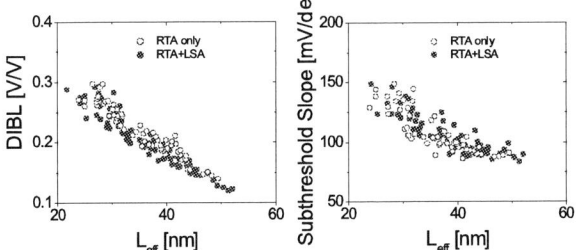

Fig. 3 The LSA process following a regular RTA has negligible impact on DIBL (left) and subthreshold slope (right).

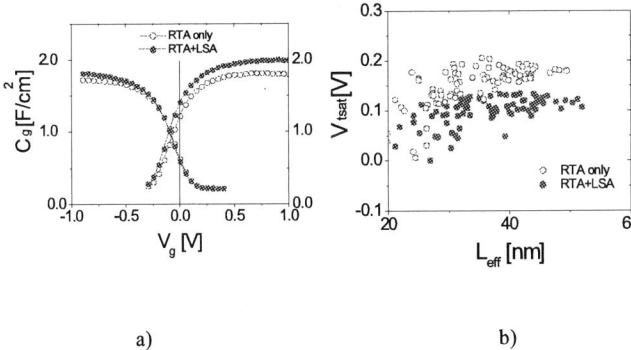

a)   b)

Fig. 4 a) High frequency split CV. The LSA process reduces poly-silicon gate depletion and it also causes a small Vt shift of for nFET. The impact on pFET CV is smaller. b) nFET V_{tsat} vs L_{eff}. The LSA following a regular RTA reduces nFET Vt by ~70mV. Negiligible Vt shift is seen for pFET.

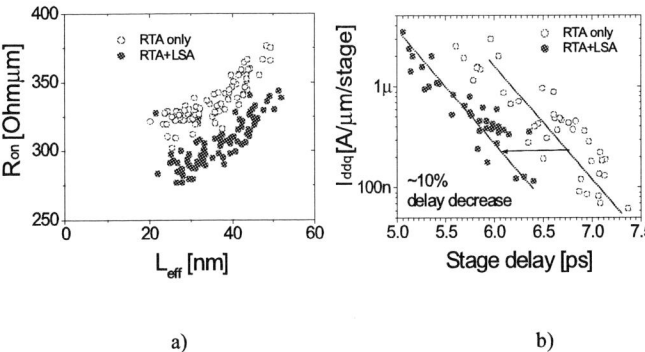

a)   b)

Fig. 5 a) nFET linear region (V_d=0.05V, V_{gs}-V_t=0.7V) resistance vs. L_{eff}. The LSA process reduces the resistance by ~30 ohm-um. The resistance reduction shows weak dependence on channel length, indicating it is related to parasitic resistance. b) Ring oscillator performance comparison. The LSA following a regular RTA (1085C spike) improves the oscillaor speed by ~10% at a V_{dd} of 1.0V.

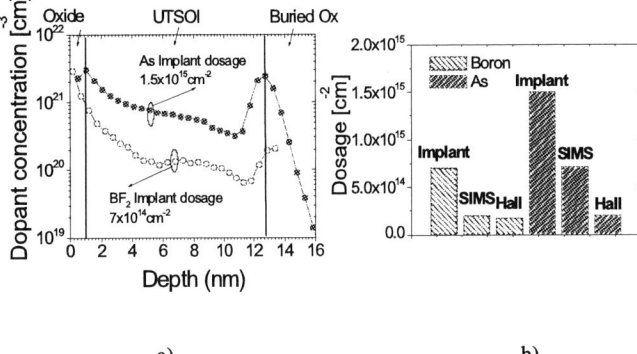

a)   b)

Fig. 6 a) SIMS dopant profiles of UTSOI samples after a 1085C, spike RTA. b) Comparison of dosage: i) Implant dosage, ii) SIMS dosage after a RTA and iii) Activated dosage obtained from Hall effect analysis. SOI thickness is ~13nm. The RTA induced boron loss is significant in UTSOI samples.

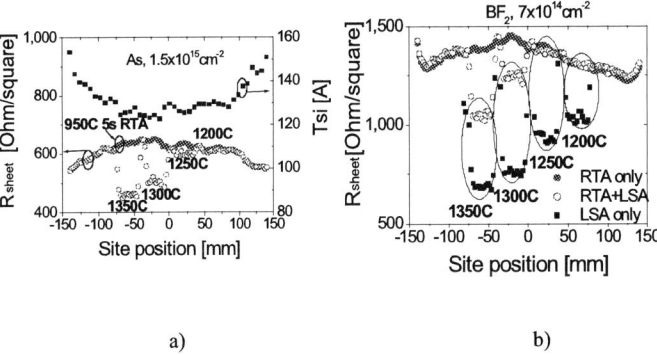

a)   b)

Fig. 7 a) The LSA impact on sheet resistance of an As doped UTSOI sample. The same wafer was processed by a 950C, 5s RTA (red), then by LSA with stripes at 4 different temperatures (blue). SOI thickness is also plotted (along the measurement line across the wafer). A strong correlation between sheet resistance and SOI thickness is seen. b) The LSA impact on sheet resistance of a Boron doped UTSOI sample. The same wafer was processed by a 950C, 5s RTA (red), then by LSA striping at 4 different temperatures (blue). The results for LSA-only case (black) are taken on a different wafer with a similar SOI thickness of ~13nm.

2006 International Symposium on VLSI Technology, Systems, and Applications

Effect of Oxygen Absorption on Contact Resistance between Metal and Carbon Nano Tubes (CNTs)

Bing-Yue Tsui, Chien-Li Weng, Chih-Lien Chang, Jeng-Hua Wei*, and Ming-Jinn Tsai **

Department of Electronics Engineering & Institute of Electronics, National Chiao Tung University
1001, Ta-Hsueh Road, Hsinchu, Taiwan, R.O.C.
Tel: 886-3-5712121-54170, Fax : 886-3-5724361, e-mail : bytsui@mail.nctu.edu.tw.
* Department of Electrical Engineering, Ching Yun University, Jung-Li, Taiwan, R.O.C.
**ERSO/ITRI, Hsinchu, Taiwan, R.O.C.

Abstract

Contact resistance (R_c) between metal and carbon nanotubes (CNTs) is studied extensively. Metal oxide formation at interface due to oxygen absorption plays very important role. Chemically inert metals such as Pt and Au results in the lowest R_c. Adding Ta into Pt can solve the adhesion issue while keeps low R_c. Those metals can break metal oxide and form metal carbide are also preferred.

Introduction

Since carbon nanotube was observed in 1991 [1], it attracts much attention due to its small geometry, high current carrying capability, high thermal conductivity, and high mechanical strength. Recently, high performance CNT field effect transistors (CNTFETs) and logic units have been demonstrated [2-5]. On the other hand, the effective resistivity of metallic CNT would be lower than Au line as the diameter becomes smaller than 2nm [6]. Therefore, metallic CNT is proposed to be nano-interconnection. No matter what kind of applications, CNT must communicate with real world through metals. Therefore, contact resistance (R_c) between metal and CNT becomes very important.

In this work, we developed an aligned process to form multi-contact test structure. Resistance of metal/CNT contact with different metals is extracted. Effect of post metal deposition thermal treatment and binary alloy are also studied.

Test Structure Fabrication

Semiconductor process compatible techniques were used to fabricate the aligned metal-on-CNT test structure. Alignment mark was patterned at first. Single-wall CNTs (SWNT) were then spread on oxidized silicon surface. SEM inspection was employed to find isolated CNTs and determine their coordinates relative to the nearest alignment marks. According to the coordinates, metal electrodes were formed by e-beam lithography and metal lift-off technique. Since the CNTs have been pre-selected, multiple contacts structure can be fabricated with almost 100% yield. Nine kinds of metal including Al, Ti, Cr, Cu, Ni, Ta, W, Au, and Pt were studied. Fig.1(a) and (b) show the AFM image of CNTs and the SEM micrograph of the multi-contact test structure, respectively. The diameter of CNT in Fig.1(a) is 1.8 nm and is believed to be SWNT. The end-resistance (R_e) but not contact resistance was extracted by 3-terminal or 4-terminal method as indicated in Fig.2. Although R_e is always lower than R_c, the trend of R_e can reflect the trend of R_c [7].

Results and Discussion

Fig.3 shows the typical 2-terminal I-V characteristics of Pt/CNT structure at room temperature. The linear characteristic indicates that the metallic type CNT and ohmic contact were obtained. is metallic and the bundle is dominated by metallic SWNT. Fig.4 shows the mean value of the extracted R_e with different metals. Pt and Au result in the lowest R_e among the 9 metals. Al and Cr result in the highest R_e.

Contact resistance is determined by the work function difference (ϕ_{mc}) as well as the interfacial properties mainly. Fig.5 compares the work function of metals with the extracted R_e. The work function of metallic SWNT is about 4.3eV. Pt and Au have the highest ϕ_{mc} but result in the lowest R_e, while Al has the lowest ϕ_{mc} but results in the highest R_e. It is clear that the R_e is inconsistent with the ϕ_{mc}. On the other hand, Fig.6 shows that the extracted R_e is consistent with the heat-of-formation (ΔH_f) of metal oxide. Except Cu, the higher the ΔH_f, the higher the R_e. It is thus postulated that metal may react with oxygen absorbed by CNTs to form a thin metal oxide layer at interface. This oxide layer not only changes the effective ϕ_{mc} but also serves as a tunneling barrier [8].

Although Pt results in the lowest R_e, the adhesion of Pt to SiO_2 and Si_3N_4 is poor. We tried to add a small amount of Ta into Pt. Stud pull test shows that the adhesion of Pt:Ta to SiO_2 is excellent. Fig.7 shows that the 2-terminal total resistance of Pt:Ta/CNT and Pt/CNT test structures are in the same range.

It is known that Ti can react with C to form TiC and the contact resistance can be reduced [9]. Fig.8 shows the R_e values of Ti/CNT structure after rapid thermal annealing in N_2 ambient for 30 sec. The R_e increases to almost open circuit after 400°C annealing while it decreases dramatically after 800°C annealing. It is postulated that TiO_x insulator forms at 400°C while Ti breaks TiO_x layer to form TiC at 800°C. To verify this postulation, Ti was deposited on gathered CNTs followed by an 800°C annealing. The Ti layer was then sputtered out by Ar ions in a XPS system for 200 sec, 220 sec, and 240 sec. Fig.9 shows that the closer to the Ti/CNT interface is, the stronger the Ti-C/Ti-O intensity ratio is. This result supports the postulation that Ti breaks TiO_x to form TiC at 800°C.

Conclusions

Contact resistance of metal/CNT contact is studied extensively. Metal oxide formation at interface plays very important role. Chemically inert metals such as Pt and Au results in the lowest R_c. Adding Ta into Pt can solve the adhesion issue while keeps low R_c. Those metals can break metal oxide and form metal carbide are also preferred. Oxygen absorption must be controlled to obtain reliable and low resistance contact to CNT.

Acknowledgment

This work was supported by ERSO/ITRI, UST CNST, and NSC (No. NSC 94-2120-M-009 -018), Taiwan, R.O.C.

References

[1] S. Iijima, *Nature*, vol.354, p.56, 1991.
[2] S. J. Tans, et al., *Nature*, vol.393, p.49, 1998.
[3] V. Derycke et al., *Nano Letters*, vol.1, p.453, 2001.
[4] T. Durkop, et al., *Nano Lett.*, vol.4, p.35, 2004.
[5] J. Chen, et al., *IEEE Conf. on Nanotech.*, p.576, 2005
[6] F. Kreupl et al., *Microelec. Eng.*, vol.64, p.399, 2002.
[7] J. Chern and W. G. Oldham, *IEEE Electron Device Lett.*, vol. 5, p.178, 1984.
[8] N. Park and S. Hong, Phy. Rev. B, vol.72, p.045408, 2005.
[9] Y. Zhang et al, *Science*, vol.285, p.1719, 1999.

1-4244-0181-X/06/$20.00 ©2006 IEEE

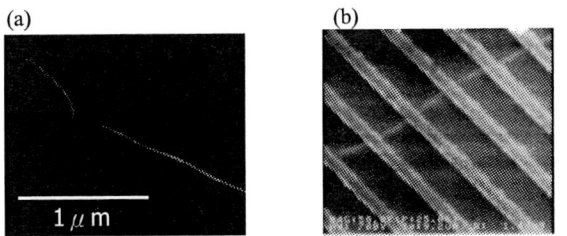

Fig.1 (a) AFM image of CNTs and (b) SEM micrographs of the multi-contact test structure. The diameter of CNT is about 1.8nm.

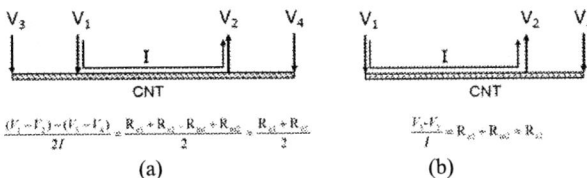

(a)   (b)

Fig.2 Test structure for contact resistance extraction. (a) 4-terminal structure and (b) 3-terminal structure.

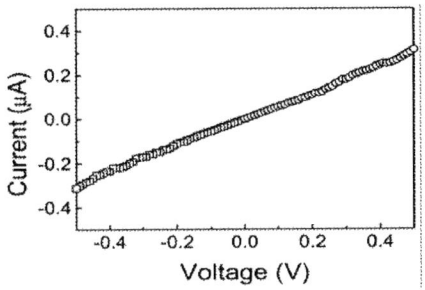

Fig.3 Typical I-V characteristic of 2-terminal Pt/CNT structure.

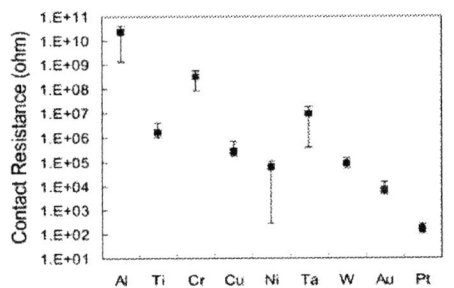

Fig.4 Mean value and range of the extracted metal/CNT contact resistance.

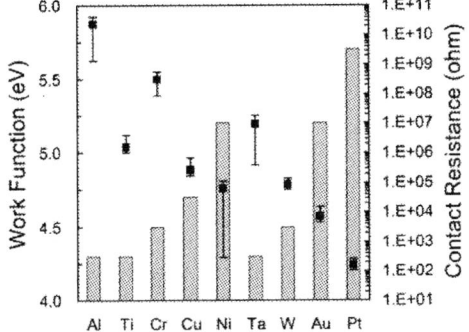

Fig.5 Correlation between metal work function and metal/CNT contact end-resistance.

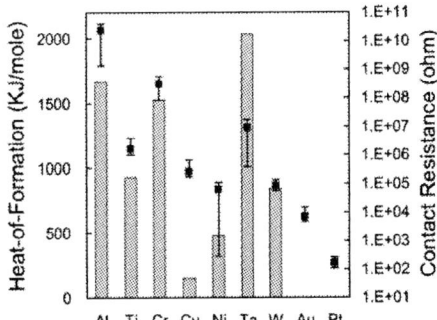

Fig.6 Correlation between Heat-of-formation of metal oxide and metal/CNT contact end-resistance.

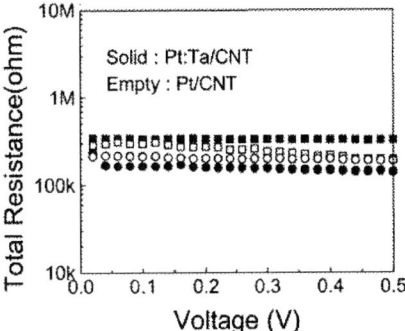

Fig.7 Two-terminal total resistance of Pt:Ta/CNT and Pt/CNT structures.

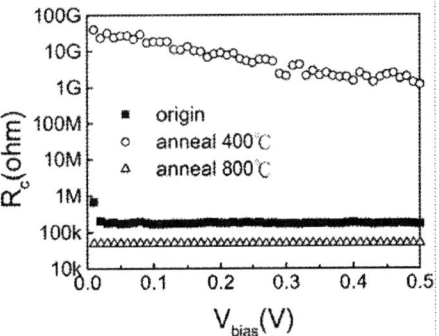

Fig.8 Contact end-resistance of Ti/CNT after annealing at 400℃ and 800℃.

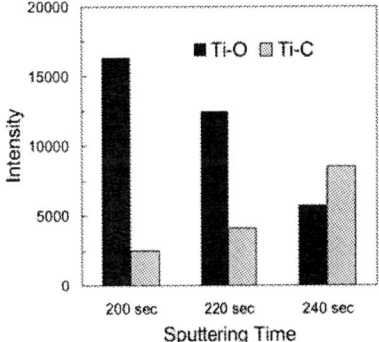

Fig.9 XPS signal intensity of Ti-O and T-C bonds after surface sputtering.

2006 International Symposium on VLSI Technology, Systems, and Applications

A Novel Deep Trench Isolation Featuring Airgaps for a High-Speed 0.13μm SiGe:C BiCMOS Technology

L.J. Choi , E. Kunnen , S. Van Huylenbroeck , A. Piontek , A. Sibaja-Hernandez [a], F. Vleugels, T. Dupont,
P. Leray, K. Devriendt, X.P. Shi, R. Loo, S. Vanhaelemeersch and S. Decoutere
IMEC Interuniversity Microelectronics Center, Kapeldreef 75, 3001 Leuven, Belgium
(a) K.U.Leuven, ESAT, Kasteelpark Arenberg 10, 3001 Leuven, Belgium
Choi.Lijen@imec.be, Tel. Phone: +32 16 281139, Tel. Fax: +32 16 281844

ABSTRACT

A novel scheme for deep trench isolation is presented, which uses an airgap as insulator. When incorporated in our 0.13μm SiGe:C BiCMOS technology, the peripheral substrate parasitics decrease with an order of magnitude to a record value of 0.02fF/μm, which significantly improves the device RF performance.

INTRODUCTION

Over the past few years, SiGe:C HBTs have made a tremendous improvement in RF performance. Reduction of device parasitics is key for obtaining even higher speed. Smart device architectures have been conceived to reduce device parasitics, including raising the extrinsic base regions [1], cutting away unnecessary active device regions [2] or using double poly architectures with selectively grown base epitaxial layers [3]. Ever since deep trench isolation (DTI) was invented, a strong reduction of the collector-substrate capacitance (C_{cs}) was obtained through a reduction of the collector area and the perimeter specific collector-substrate capacitance density ($C_{cs,p}$). The latter is usually obtained by partially filling the deep trench with oxide, which has a low κ-value, and the remainder with polysilicon, which acts as a stress-relief buffer. In this work however, we use an airgap to isolate devices. The integration scheme is fully compatible with standard STI processing and does not require novel materials or complex processing steps. Integrating the airgap DTI module in a high-speed SiGe:C BiCMOS technology, we will show the isolation properties and the impact on device performance.

DEEP TRENCH ISOLATION PROCESSING

The fabrication of the deep trench isolation module is illustrated in Figure 1. On top of the STI oxide-nitride stack, a thick oxide layer is deposited, which is used as a hardmask to etch the deep trench (a). The trench is 1μm wide and 6μm deep. The trench is filled with oxide and polysilicon. Then the wafer is planarized using CMP, stopping on the oxide layer. After creation of a recess in the polysilicon plug (b), D-shape oxide spacers are created inside the deep trench. Subsequently, the underlying polysilicon is completely removed by dry etch using an isotropic SF_6 chemistry (c). The purpose of the spacers is to narrow down the trench opening, which allows an easy trench closing by depositing oxide. Then the wafer is polished again, stopping on the nitride layer (d). Figure 2 shows a cross-section SEM picture after this second planarization step. After this step, the original state of the wafer is restored and the shallow trench isolation (STI) module can be processed. One of the key features of this DTI module is that the STI module can be processed without any modification of the processing steps. Figure 3 shows a TEM picture of the complete HBT at the end of processing.

ISOLATION PROPERTIES

The collector-substrate isolation is illustrated in Figure 4, where the leakage current is plotted versus bias. The leakage level is low and comparable with the case where no deep trenches are present. This means that the presence of the deep trench did not introduce additional leakage. The collector-collector isolation is depicted in Figure 5. In this kind of structures where two n-type regions in a p-type substrate are separated by a deep trench, a parasitic conduction channel around the deep trench is present [4]. Due to the absence of a polysilicon plug in the deep trench, there is no gate that can turn on this channel. With a collector-substrate and collector-collector breakdown voltage exceeding 100V, the airgap DTI is superior to classical DTI schemes using an oxide/polysilicon filling and is also suitable for high voltage applications. The collector-substrate capacitance associated with the airgap isolation is strongly reduced due to a reduction in $C_{cs,p}$. Figure 6 compares $C_{cs,p}$ for different isolation schemes. At reverse bias, a value of 0.02fF/μm is obtained for the airgap DTI, which is an order of magnitude lower than for classical DTI schemes.

DEVICE RESULTS

To demonstrate the impact of the airgap DTI on device performance, the module is incorporated in our high-speed 0.13μm SiGe:C BiCMOS technology. The architecture for this technology is based on a low-complexity quasi-self-aligned integration scheme, which was recently scaled down vertically and laterally towards high-speed operation [5]. A typical Gummel plot is shown in Figure 7 and shows only a slight impact of the airgap isolation on DC device characteristics. Despite the lateral confinement of the heat-flux by the airgap, the devices are thermally very stable. The thermal resistance R_{TH} for a 0.13μmx2μm device is about 12.6kK/W, which is only slightly higher than the value for classical DTI schemes (Figure 8). Figure 9 shows the relative increase of C_{cs} with decreasing device length in comparison with the base-emitter and base-collector capacitances C_{be} and C_{bc}. It is clear that using the airgap DTI allows controlling the substrate parasitics as device dimensions are scaled down. Figure 10 shows the cut-off frequency f_T and the maximum oscillation frequency f_{max} versus collector current. Peak f_T/f_{max} values exceeding 200GHz are obtained. An excellent figure-of-merit to study the impact of the airgap DTI is the recently introduced available bandwidth f_A [6], which gives a more realistic idea of the device RF performance in real circuit blocks like for instance emitter-coupled pairs. In comparison with classical DTI schemes, the airgap isolation improves the peak f_A value with 17% (Figure 10). At low power, the device RF performance is heavily impacted by parasitic capacitances. The airgap isolation boosts up the low-power f_A value with 75%.

CONCLUSIONS

We have demonstrated the integration of a novel airgap DTI module into a BiCMOS process, which is fully compatible with standard STI processing. Without affecting the thermal stability of the devices, the airgap strongly reduces the substrate parasitics, allowing significant improvement in device RF performance.

ACKNOWLEDGEMENTS

The authors would like to thank the IMEC pilot line, the Flemish IWT and Philips Research Leuven for supporting the BiCMOS programs.

REFERENCES

[1] Khater et al., IEDM Tech. Dig., 2004, pp. 247-250
[2] Heinemann et al., IEDM Tech. Dig., 2004, pp. 251-254
[3] Böck et al., IEDM Tech. Dig., 2004, pp. 255-258
[4] Parthasarathy et al., IEDM Tech. Dig., 2002, pp. 459-462
[5] Van Huylenbroeck et al., Proc. BCTM, 2004, pp.229-232
[6] Hurkx et al., IEEE Trans. on Electron Devices, 2004, vol. 51, pp. 2121-2128

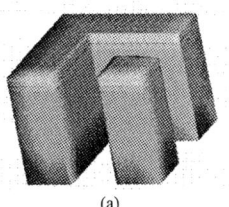  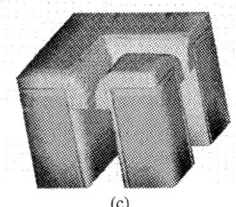 

(a)　　　　　　　　　(b)　　　　　　　　　(c)　　　　　　　　　(d)

Fig.1: Overview of the airgap processing: (a) after trench etch and liner oxidation, (b) after trench fill and recess formation in the polysilicon plug, (c) after spacer formation and trench clearing and (d) after trench closing and planarization.

Fig.2: Tilted cross-section SEM picture at the end of the airgap DTI module.

Fig.3: TEM picture at the end of processing, the airgap deep trench isolates the HBT.

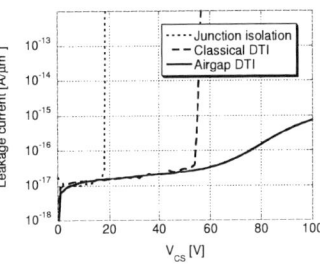

Fig.4: Collector-substrate leakage for different isolation schemes.

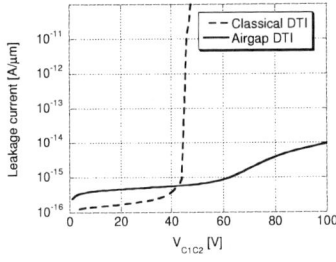

Fig.5: Collector-collector leakage for different types of DTI. Measured between two collectors separated by a deep trench.

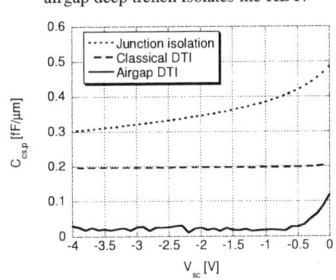

Fig.6: Perimeter specific collector-substrate capacitance density $C_{cs,p}$ for different isolation schemes.

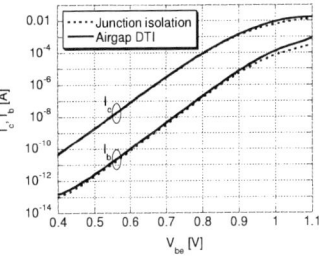

Fig.7: Typical gummel plot for a device with $A_E=0.13\mu m \times 2\mu m$. Measured at $V_{cb}=0V$.

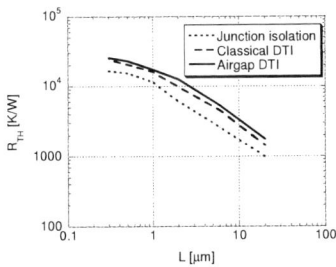

Fig.8: Thermal resistance for devices with a minimum emitter width of $0.13\mu m$ and various lengths.

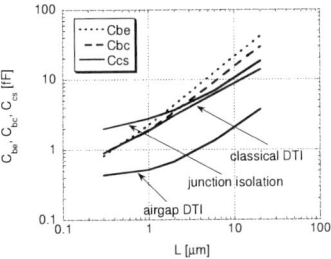

Fig.9: C_{be} and C_{bc} at zero bias and C_{cs} at 1V reverse bias for devices with a minimum emitter width of $0.13\mu m$ and various lengths.

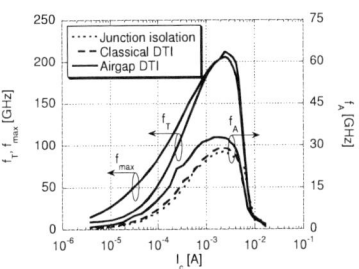

Fig.10: F_T, f_{max} and f_A vs. I_c for a device with $A_E=0.13\mu m \times 2\mu m$. Measured at $V_{cb}=0.5V$. F_A is the available bandwidth for a DC gain of 10.

2006 International Symposium on VLSI Technology, Systems, and Applications

Beyond Scaling – Realizing Value Through the Integration of Memory and Autonomic Chip Features

Subramanian S. Iyer

IBM Corporation, Systems & Technology Group,

2070 Route 52, Hopewell Junction, NY 12533. USA.

email: ssiyer@us.ibm.com

ABSTRACT

As we scale to 45nm and beyond, the expectations of increased functional value per unit cost may only be met with a more holistic technology integration based on systems needs. We consider two approaches: Leveraging embedded dense memory. We examine the scalability and performance issues of embedded memory and the autonomic chip – where a judicious use of circuit innovation, material and physical phenomena are employed to yield self diagnosing and self healing and perhaps even self reconfiguring chips extending both the lifetime and scope of the chip

INTRODUCTION

The technological challenges of scaling into the 45nm node are well known. These include the lithographic challenges being addressed by hyper NA immersion lithography, device performance challenges being met through the use of advanced strain engineering, and not the least the challenges of interconnect performance which dominate at the chip level especially for complex System-on-Chip (SOC) applications. We address two issues that promise to deliver increased value at both the chip and system level. One has to do with integrating dense dynamic memory on chip, and the other has to do with building autonomic features where self diagnosis and self repair are possible at various times during the chips lifetime.

Embedded Memory

From a functional perspective, the trend to integrate increasingly large amounts of memory on chip (embedded memory or eMemory) continues unabated. eMemory dominates the area of mid- and high-end processors chips as well as most Application Specific Integrated Circuits (ASICs) and in many cases, memory can occupy up to 70% of the die area. In the high end applications, the key attributes of on-chip memory are performance, area, power, soft error rate (SER) and cost. eMemory is also an important attribute in consumer ICs especially for portable applications. Here power (especially standby power), area, cost and SER are important considerations.

Static RAMs (SRAMs) have been the workhorse of embedded memory applications but are becoming increasingly difficult to scale. As SRAMs scale, the close proximity of N- and P-channel devices makes isolation more complex. Additionally, the need for performance (as measured by read current), the inherent threshold mismatches (dominated by statistical dopant fluctuation) cause read instabilities that manifest themselves as increasing number fails as the operating voltage is decreased (see reference 1). Standby power in SRAMs is also not scaling because of the leakier nature of the devices used. Finally, SER has reached very high values as both capacitance and stored charge at the internal SRAM nodes has decreased as a result of scaling. Finally, for large memory sizes, time of Flight (TOF) is a significant delay component that detracts from performance.

DRAMs address several of the issues above. IBM has pioneered the use of deep trench embedded DRAM for logic applications for over four logic generations. The trench based eDRAM is extremely logic friendly and guarantees logic equivalency since the DRAM specific processes are established well before the logic devices are fabricated. There is no impact on the middle of the line and backend processing ensuring timing and library equivalency. DRAM cells are typically about 5 - 8 times smaller than corresponding SRAM cells. At the usable memory macro level this advantage is eroded by various circuit overheads but still is about 3.5 to 4 times as dense as SRAMs with the advantage increasing with memory size. From a power perspective DRAMs containing only one low leakage device typically have about 6 to 8 times lower standby power per Mb compared to 6 transistor SRAMs. Finally, the larger capacitance of the DRAM cell makes it immune to SER events and SER fail rates are typically 5000 times better compared to SRAMs.

Scaling of eDRAM in logic technology has two goals: to reduce size consistent with the scaling factor and trench based eDRAM has met this approximately 50% scaling per generation rather well; the other is to improve performance – in this it differs from the more conventional commodity DRAM. Performance improvement comes from three factors: logic performance improvements; cell performance improvements and finally DRAM architecture improvements. As a result of these three factors, it is expected that eDRAMs in the 45 nm generation will have latencies of the order of 1-2ns and random cycle times of about 2-4 ns. This coupled with their smaller size and consequent lower TOF delays will make

1-4244-0181-X/06/$20.00 ©2006 IEEE

eDRAMs an excellent replacement for the large on chip level 2/level 3 caches on processors. This has special significance in the context of multi-core processor chips, where processor speed has saturated and performance is expected to come from the multi-core architecture and an optimized memory subsystem with large caches that are electrically close and dedicated to the processors.

Finally, we must address the issue of eDRAM cost and complexity adders. The trench based eDRAM adds 3-4 extra mask levels to a typical logic flow. Two to three of these masks are used to optimize the standard I/O device for a low leakage transfer device and are block level masks. The other is a critical deep trench level. The overall complexity adder in 90nm is about 15% and becomes a smaller fraction as the device menu increases and backend interconnect complexity increases. A simple calculation shows that if a chip allocates about 25% of the die area to SRAM and this SRAM can be replaced functionally by eDRAM, the resultant die will in fact be more economical. Clearly, the more complex technologies in 45 nm and certainly Silicon-on-insulator (SOI) technology make this tradeoff more favorable for eDRAM. There will obviously gains in power and SER resistance. From an area scaling perspective, replacing standard SRAM with eDRAM is almost equivalent to scaling the memory an additional two generations! Finally, from a performance perspective, improvements in eDRAM cell and circuit architecture as well as judicious optimization of the memory sub-system to leverage the larger eDRAM cache will allow for a system level performance improvement of over one generation.

The reader is encouraged to consult reference (2) which explores these ideas in greater detail.

Autonomic Chips

Another important development is the building of autonomic capabilities into the chip. Most advanced SOCs employ Built-in Self Test (BIST). BIST engines are dedicated engines of increasing versatility that can apply a wide range of test vectors to functionality test a wide variety of IP blocks including memory, logic and analog functions. In the case of memory, BIST can be combined with Redundancy Allocation Logic (RAL) to fix the defects using redundant elements. Advances in the development of electrical fuses (eFUSE) that can be programmed on chip allow us to combine BIST, RAL and eFUSE to form a Built-in Self Repair (BISR) system. Additional sophistication is possible through the use of Fuse string compression, hierarchical repair where repair solutions are augmented at wafer level, module level, and even multiple instances of field repair.

Realization of this BISR requires the use of a reliable fuse methodology. We have employed electromigration a hither-to-fore reliability problem constructively to build a highly reliable fuse. Unlike rupture mode fuses, electromigration does not create debris and thus fuse healing does not occur (reference 3). This system has been employed not just for repair but even more extensively for chip reconfiguration, radio frequency (rf) tuning, and die by die yield optimization but more innovatively to store on-chip the chip Bill of Materials, the power frequency characteristics, chip temperature – frequency characteristics etc. These can be effectively used at the system level to manage performance and power.

Summary

While technology scaling is now more challenging, there are some relatively easy ways to extract more value for the technology. We have described the use of embedded DRAMs and eFUSEs to leverage additional functional value beyond scaling alone.

REFERENCES

1. C. Wann et al, "SRAM Cell design for stability methodology" proc. VLSI-TSA (2005) p 21
2. S.S. Iyer et al " eDRAM the technology platform for the Blugene /L chip", IBM Jour, Res. Dev. 49 (2/3) p333 (2005)
3. C. Kothandaraman et al " Electrically programmable fuse (eFUSE) using electromigration in silicides" IEEE Electron Device Letters EDL (23) p 523 (2002)

2006 International Symposium on VLSI Technology, Systems, and Applications

Trench DRAM Technologies for the 50nm Node and Beyond

W.Mueller, G.Aichmayr, W.Bergner, M.Goldbach, T.Hecht, S.Kudelka, F.Lau*, J.Nuetzel, A.Orth,
T.Schloesser, A.Scholz, A.Sieck, A.Spitzer*, M.Strasser*, P.-F.Wand, S.Wege, R.Weis

Infineon Technologies, D-01099 Dresden, Germany
*Infineon Technologies, D-81609 Munich

ABSTRACT

This paper reviews the DRAM technology challenges for overcoming the 50nm barrier. First the product requirements and barriers to shrink the DRAM cell beyond 50nm will be addressed. Then the technology solutions for DRAM cell capacitor, cell transistor, and support transistors will be presented. Key enablers are high aspect ratio cell capacitor structures, new capacitor materials, 3-dimensional cell transistor schemes and high performance LSTP support device technologies.

INTRODUCTION

The DRAM product roadmap down to the year 2010 is shown in Table 1.

TABLE 1. DRAM PRODUCT ROADMAP

	DDR3	DDR4	DDR5
Year	2006	2008	2010
Data rate (Mbps)	1066+	3200+	6400+
V_{dd} **(V)**	1.5	(1.35)	(1.2)
V_{blh} **(V)**	1.2	1.1	1.0
t_{WR} **(ns)**	12	10	8

Higher data rates, and reduced voltages will drive the performance of the support transistors. Shorter array timings will drive the performance of the array transistor as well as the reduction of the parasitic RC for the array path. As a consequence of the reduced array voltage V_{blh}, the cell capacitance C_s has to stay constant at about 30fF in order to ensure a constant cell signal [1].

TRENCH CELL

Cell Capacitor

To support a constant cell capacitance of 30fF the aspect ratio (AR) of the trench cup capacitor has to increase as shown in Fig. 1, and at the same time high k dielectric materials with metal electrodes have to be implemented in order to reduce the equivalent oxide thickness CET of the storage capacitor.

Beyond 70nm high k dielectric materials such as AlO (70nm and 60nm) [2] and HfSiON (50nm and 40nm) [3] have been developed. In order to reduce capacitance loss due to the gate depletion effect, metal electrodes (TiN) will be utilized as MIS or MIM capacitor structures. Atomic layer deposition (ALD) processes have become mainstream for capacitor dielectric and metal electrode formation due to their excellent step coverage.

Trench capacitor aspect ratios around 90 have been demonstrated for a 70nm technology (Fig. 2). The achievable aspect ratio is not

limited by the Si trench etch capability, but by the trench etch hard-mask capability.

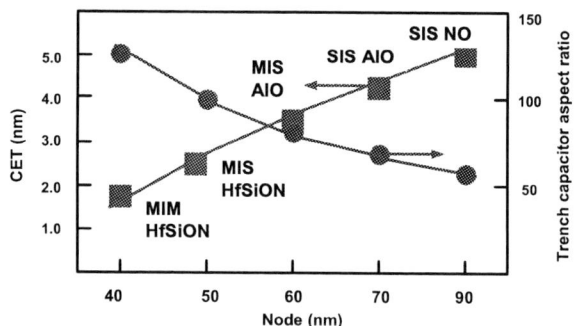

FIGURE 1. CET AND CAPACITOR ASPECT RATIO ROADMAP

For a Si-trench aspect ratio of 100 a SiO hard-mask thickness of about 2um and a corresponding AR=25 are required, reaching the capability limit for SiO etch. Thus new hard-mask technologies become an key enabler [4].

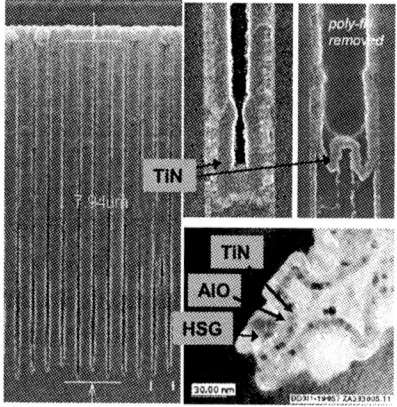

FIGURE 2. 60NM TRENCH CAPACITOR STRUCTURE WITH AR=90 (LEFT), HSG/MIS ALO CAPACITOR (LOWER RIGHT) AND TiN METAL CONNECTION FROM THE CAPACITOR TO THE POLY SI STRAP CONTACT (UPPER RIGHT).

By implementing for the 60nm node a metal (TiN) connection from the capacitor to the cell transistor (Fig. 2) the parasitic resistance is reduced by 2x: Thus the write back time is reduced by 1.5nsec compared to the prior generation.

Cell Transistor

Scaling the cell transistor below 90-70nm the doping level required to meet the off current criterion of $I_{off} < 1fA$ gets so high, that the electric field at the node junction may exceed 0.5 MV/cm, thus initiating trap assisted tunneling leakage [5]. Vertical and U-shape

1-4244-0181-X/06/$20.00 ©2006 IEEE

structures have been developed in order to ensure sufficient device length at the given cell footprint thus keeping the electric field below the critical level [6, 7]. However, besides the I_{off} current requirement also an I_{on} >20-30μA has to be achieved for fast array timings. Different structures with side-wall gates have been developed in order to realize a large effective width, such as the Extended U-shape Device (EUD) [1], FinFET cell transistor [1, 8], vertical double transistor [9], and saddle transistor [10].

For the trench DRAM the planar cell transistor can be scaled down to the 70nm node [11]. For the 60nm and 50nm node EUD or FinFET are candidates for the cell transistor structure. Compared to the U-shape device [6], the EUD device improves the sub-threshold slope (Fig. 3), the I_{on}, and eliminates the sensitivity of V_t on the bottom U-shape curvature.

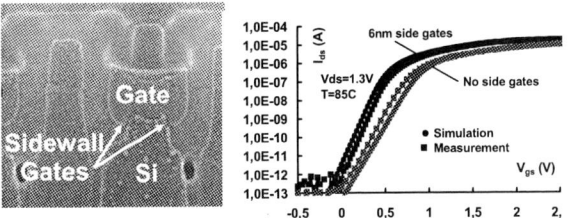

FIGURE 3. EUD CROSS SECTION IN WIDTH DIRECTION AND IDS-VGS CHARACTERISTICS FOR EUD (BLACK) AND U- DEVICE (BLUE).

The integration of a local damascene FinFET cell transistor into a 90nm 512M product [1] is shown in Fig. 4.

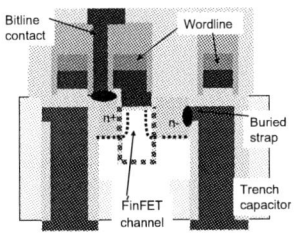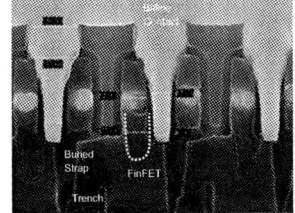

FIGURE 4. FINFET INTEGRATION SCHEME (LEFT), AND VERTICAL SEM VIEW FOR THE FABRICATED TRENCH FINFET CELL.

Measured I_{ds}-V_{gs} characteristics of the 90nm FinFET cell devices (Fig. 5, left side), demonstrate a steep sub-threshold slope (77mV/dec), high Ion (40uA), small DIBL and negligible body effect. Results of 3d TCAD simulations for a 40nm trench FinFET structure calibrated to the experimental data are given in the right side of Fig. 4.

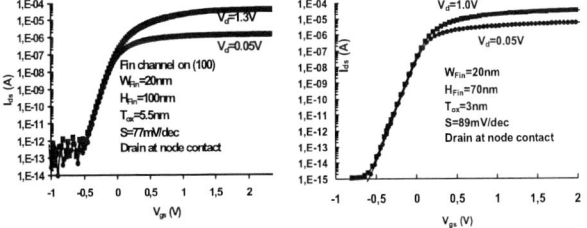

FIGURE 5. MEASURED 90NM FINFET I_{DS}-V_{GS} CHARACTERISTICS (LEFT), AND SIMULTED 40NM FINFET I_{DS}-V_{GS} CHARACTERISTICS (RIGHT).

For 40nm an I_{on} <1fA, Ioff =40μA, a sub-threshold slope of 89mV/dec, and small DIBL is predicted, showing the excellent scalability of the FinFET cell transistor.

SUPPORT CIRCUIT TRANSISTOR

To support high data-rates up to 6400Mbps in DDR5 technologies high performance LSTP DWF support devices [11] are needed. The internal voltage has to be reduced to accompany a low-power design. For SiON gate dielectric the EOT cannot be reduced below 1.5nm without compromising the gate leakage current. High k materials are in development to support the further EOT scaling, these materials will not be ready for production prior to the 40-30nm node however. The thermal budget has to be reduced and new technologies such as junction co-implants and ultra–short annealing have to be implemented in order to realize steep junctions and better gate doping activation respectively. Salicidation of source and drain will enable a low parasitic transistor resistance and a low contact resistance.

SUMMARY

The technical challenges for scaling the trench DRAM cell to 40nm have been reviewed. Concepts to overcome the technical hurdles have been developed and evaluated on larger ground rules. Projections to the target feature sizes have been done via simulation. The main challenges for scaling the capacitor to 40nm will be the implementation of the high aspect ratio capacitor structures required, together with new capacitor materials. For the cell transistor EUD and FinFET are candidates for the 60nm and 50nm node, with the scaling potential of the FinFET reaching well beyond 40nm.

REFERENCES

[1] W.Mueller et al., "Challenges for the DRAM cell scaling to 40nm", *IEDM 2005*.

[2] M.Gutsche et al., "Capacitance enhancement techniques for sub-100nm trench DRAM", *IEDM 2001*.

[3] J.Heitmann et al., "HfAlO and HfSiO based dielectrics for future DRAM applications", *209th ECS Meeting, Denver, May 2006*.

[4] S.Wege et al., "High aspect ratio deep trench Si etching for technologies below 70nm", *SEMI-ECS ISTC*, March 2005.

[5] A.Weber et al., "Data retention analysis on individual cells of 256Mb DRAM in 110nm technology", *ESSDERC 2005*.

[6] H.S.Kim et al., "An outstanding and highly manufacturable 80nm DRAM technology", *IEDM 2003*.

[7] R.Weis et al., "A highly cost efficient 8F² DRAM cell with a double gate vertical transistor for 100nm and beyond", *IEDM 2001*.

[8] R.Kamatsumata et al., "Fin-array FET on bulk silicon for sub-100nm trench capacitor DRAM", *Symp.VLSI Technology 2003*.

[9] T.Schloesser et al., "Highly scalable sub-50nm vertical double gate trench DRAM cell", *IEDM 2004*.

[10] K.H.Park et al., „Highly scalable Saddle MOSFET for high density and high performance DRAM", *IEEE Trans. Electron Device Letters, Vol. 26, No. 9, September 2005, p.690-692*.

[11] J.Amon et al., "A highly manufacturable deep trench based DRAM cell layout with planar device in a 70nm technology", *IEDM 2004*.

2006 International Symposium on VLSI Technology, Systems, and Applications

Overview and Future Challenge of Floating Body Cell (FBC) Technology for Embedded Applications

Akihiro Nitayama, Takashi Ohsawa, and Takeshi Hamamoto

SoC Research and Development Center, Toshiba Corp.

8, Shinsugita-Cho, Isogo-Ku,Yokohama 235-8522, Japan

Phone: +81-45-770-3207, Fax: +81-45-770-3510, E-mail: akihiro.nitayama@toshiba.co.jp

ABSTRACT

A one-transistor memory cell on silicon-on-insulator, called Floating Body Cell (FBC), has been developed for high density embedded DRAM applications. The functionality of a 128Mb FBC DRAM using fully compatible 90nm CMOS technology has been successfully demonstrated. The memory cell design, such as fully-depleted (FD) operation with substrate-bias, and the process integration, such as well and Cu wiring, are reviewed. The scalability and future challenge of FBC technology are discussed as well.

INTRODUCTION

Some kinds of the capacitor-less DRAM cell have been proposed and developed. FBC (Floating Body Cell) is a promising candidate from the viewpoint of its simplest structure and its scalability [1]-[5]. For the purpose of the verification of the extendibility of FBC for a embedded macros, a 128Mb SOI DRAM with FBC has been designed and fabricated. The key concerns of FBC for high density embedded macros is how to guarantee the memory cell characteristics such as the signal margin and the retention characteristics with keeping the process compatibility with CMOS devices. From such a viewpoint, two new technologies, the optimized well structure and Cu wiring, have been introduced. As a result, 36bit-fail chip at 32 Mb area has been obtained.

FBC STRUCTURE FOR 128Mb SOI DRAM

Figure 1 summarizes the fabricated FBC structure. The technologies, which have been already developed, such as substrate plate cell, FD operation and salicide for array device are adapted. Newly implemented technologies are (1) the well design optimized for both the array device and the peripheral circuit, (2) Cu wiring for Bit Line (BL) and Source Line (SL).

Applying the substrate plate bias voltage is very important for increasing cell signal and realizing good retention time. But, the back surface leakage of the support device occurs if the same bias is applied to the peripheral circuit. In order to eliminate this leakage, twin tub with deep n-well has been formed beneath the peripheral circuit. Figure 2 schematically shows the well structure and bias conditions, in which the suitable back biases for NFET and PFET of the peripheral circuit are applicable. Figure 3 shows the operation principle of FD FBC [4]. Data "1" is written by creating holes by impact ionization and pushes up body potential to high level. The created holes are accumulated to the back surface by negative plate bias. On the other hand, data "0" is written by extracting holes from the body to BL and pulls down the body potential to low level.

The array unit has 512 local BLs and 512 local WLs. Each S/A is shared among eight units in the hierarchical BL architecture [6]. In this array architecture, high parasitic resistances of BL and SL reduce signal. The low sheet resistance of Cu wiring can eliminate this signal reduction. Cu wiring also realizes the full compatibility with CMOS BEOL process.

DESIGN AND FABRICATION

The memory cell layout and the design parameters are shown in Fig.4 and Table1, respectively. The unit cell area is $6.2F^2$ ($0.17\mu m^2$). To reduce the parasitic resistance, Co salicide is formed on source, drain and gate. M1 (Cu wiring) is used both for SL and the pad for BL contact. M2 (Cu wiring) is used for BL. Fabrication process is based on the standard 90nm CMOS process (CMOS4). The additional process is only formation of deep n-well. Figure 5 shows the cross-sectional picture of FBC in the fabricated 128Mb SOI DRAM.

EXPERIMENTAL RESULTS

Before starting the fabrication of 128Mb SOI DRAM, the impact of the plate bias on signal and the retention characteristics were clarified by previous ADM. Figure 6 and Figure 7 show threshold voltage and signal (V_T and ΔV_T) as a function of the plate bias. The threshold voltage of "0" cell significantly increases by applying the plate bias, indicating it is operated in FD mode [4]. In our estimation, the threshold voltage difference of 0.4V is sufficient for obtaining fully functional die with the redundancy system of 128Mb SOI DRAM.

Figure 8 shows the retention distribution of partially depleted (PD) FBC (Tsi=150nm) with parameter of channel dose. In the case of PD FBC, signal and GIDL leakage dominate the retention characteristics. Increase of dose enlarges the retention time due to increase of signal. Too much dose, however, degrades the retention time due to GIDL current increase by the strong electric field. As a result, the relationship between the signal and the retention time has bell shape, shown in Fig. 9. The best peak time is 160ms. In the case of FD FBC (Tsi=55nm), also shown in Fig.9, deeper plate bias boosts up the retention time longer than peak of PD FBC. Important to note is that the electric filed is larger than that of PD FBC of the same LDD dose (Fig.10). The deeper plate bias eliminates the back surface leakage and thin silicon thickness is contributing to the reduction of leakage source. Figure 11 shows the actual retention distribution of FD FBC. The median retention time of 250ms, which is 50% longer than PD FBC, has been achieved at the plate bias of –1V.

Impact of the wiring resistance on signal and write speed has been estimated by device simulation. If the same BEOL structure as previous work is used for the 128Mb SOI DRAM, the sheet resistance of BL (W) and SL (CoSi2) of the worst bit are estimated to 4500Ω and 830Ω, respectively. In the case of Cu wiring, those resistances are 800Ω and 13Ω. Significant improvement of signal and write speed are summarized in Fig.12. IR drop of BL at "0" write makes the severest effect on both signal and write speed. Cu wiring, therefore, is indispensable for realizing full functionality of high density FBC DRAM.

A chip microphotograph and a fail bit map are shown in Fig.13. 36bit-fail chip at 32 Mb area has been obtained.

FUTURE CHALLENGE

We've simulated the device operation by using device structure with 45nm CMOS (Leff=50nm) shown in Fig.14. It is verified that the signal margin is still kept 0.4V, and it leads to cell size of 0.031um2, which is less than half of that of embedded DRAM. If shrinking the thickness of SOI and BOX properly, FBC is scalable even down to 32nm ground rule. Table 2 shows the comparison among embedded memories. FBC has the potential that 1cell/bit operation will replace the e-DRAM or even commodity DRAM for high density use, and 2cell/bit operation will replace e-SRAM for high speed applications.

CONCLUSIONS

128Mb SOI DRAM with FBC has been successfully fabricated. FBC is a promising structure for embedded memory for future SoC.

ACKNOWLEDGEMENTS

The authors would like to thank Dr. T.Furuyama and Dr. S.Fujii for their fruitful discussions and supports throughout this work.

REFERENCES

[1] T. Ohsawa, et al., ISSCC Tech. Dig., pp152, 2002
[2] K. Inoh, et al., Symp. on VLSI Technology, pp.63, 2003.
[3] T. Shino, et al., Symp. on VLSI Technology, pp.132, 2004.
[4] T. Shino, et al., IEDM Tech. Dig., pp.281, 2004.
[5] Y. Minami, et al., IEDM Tech. Dig., pp.317, 2005.
[6] T. Ohsawa, et al., ISSCC Tech. Dig., pp.458, 2005.

1-4244-0181-X/06/$20.00 ©2006 IEEE

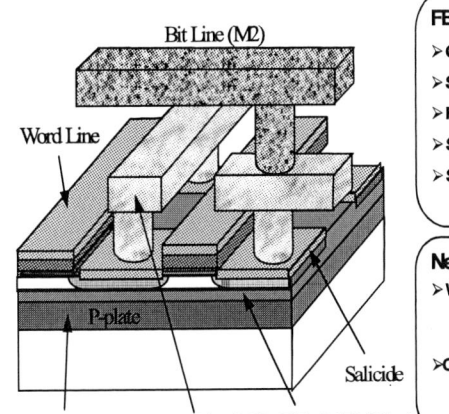

Bit Line (M2)

Word Line

P-plate

Substrate Plate   Source Line (M1)   Thin BOX (25nm)

Salicide

Fig.1 Schematic view of FBC cell structure.

FBC Technologies already implemented
➢ Cell Concept (ISSCC2002)
➢ Substrate Plate Cell (VL2004)
➢ FD Operation (IEDM2004)
➢ Salicide for Array Device (IEDM2004)
➢ S/A designed with countermeasure
   for Charge pumping (ISSCC2005)

New technologies for 128Mb SOI DRAM
➢ Well Design Optimized for Array & Support Device
   ✓ Improve ΔV_T & Retention Time
   ✓ Eliminate Back Surface Leakage at Support Device
➢Cu Wiring is used for SL and BL
   ✓ Reduce Parasitic Resistance
   ✓ Fully Compatible w/ LOGIC Device

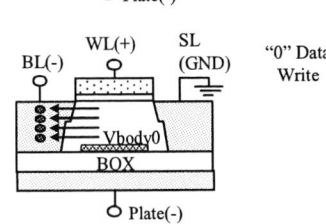

"1"Data Write

Hole Storage Region ("Volume" for PD, "Plane" for FD)

"0" Data Write

Fig.3 Operation of FD FBC.

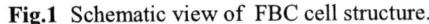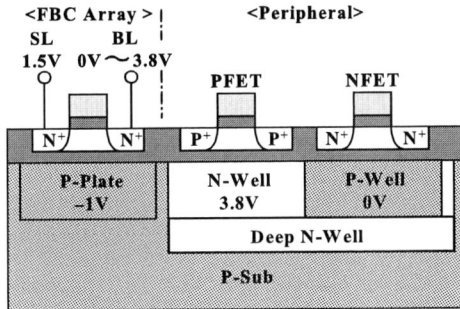

Fig.2 Well structure and typical bias condition of 128Mb SOI DRAM.

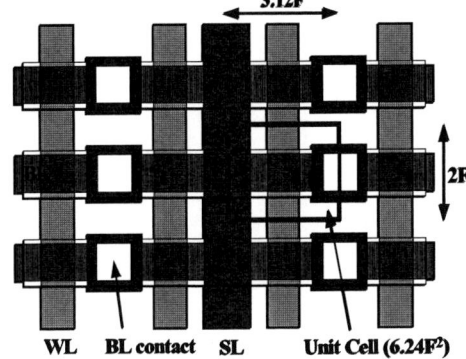

3.12F

2F

WL   BL contact   SL   Unit Cell (6.24F²)

Fig.4 Layout of 128Mb FBC array.

	WL Pitch (μm)	0.515
Cell Array	BL Pitch (μm)	0.33
	Cell Size (μm²)	0.17
	Cell Size (F²)	6.24
	AA Width (μm)	0.18
	Lpoly (μm)	0.145
	Tox (nm)	6
Process	Platform	90nm(CMOS4)
	Aditional process	(Only) Deep Nwell
	Silicide	CoSi₂
	Wiring	Cu(6layers)

Table 1 Design parameters of FBC.

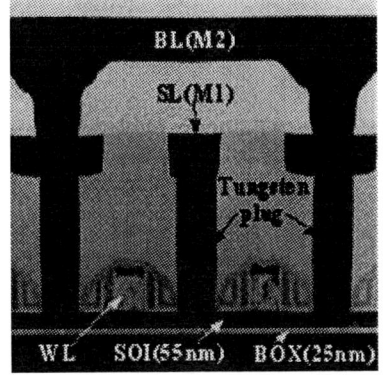

BL(M2)

SL(M1)

Tungsten plug

WL   SOI(55nm)   BOX(25nm)

Fig.5 TEM cross–section along Bit-line of fabricated 128Mb FBC.

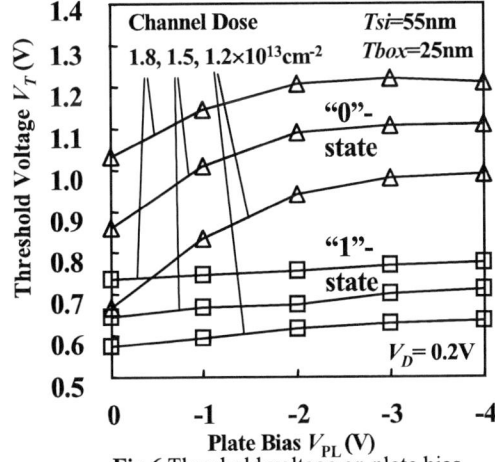

Fig.6 Threshold voltage on plate bias.

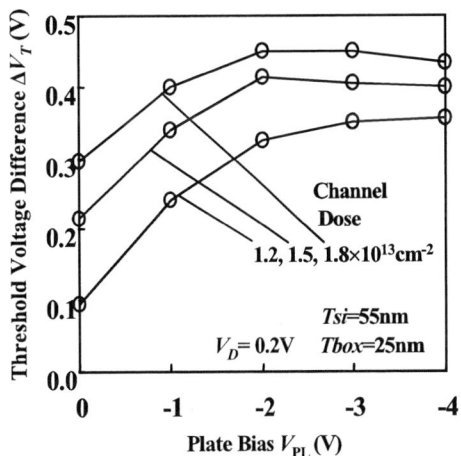

Fig.7. Difference of threshold voltage on plate bias.

95

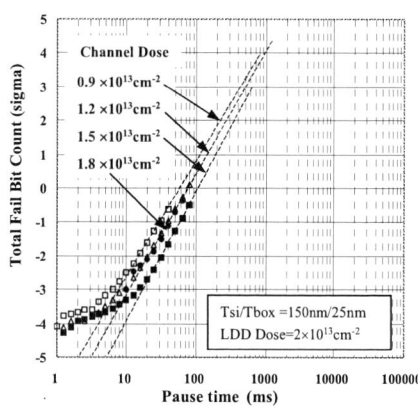

Fig.8 Retention characteristics of PD FBC.

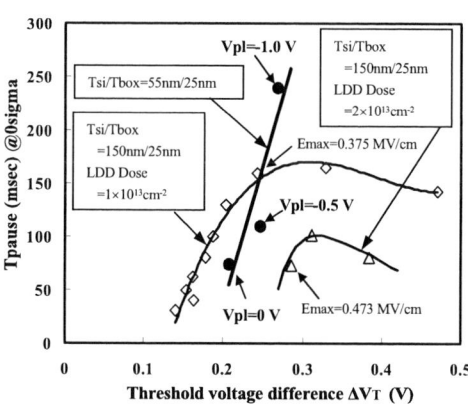

Fig.9 Relationship between ΔV_T and Tpause

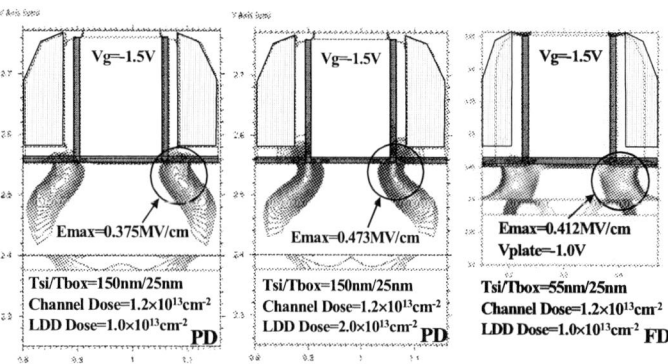

Fig.10 Electric potential at "0" hold.

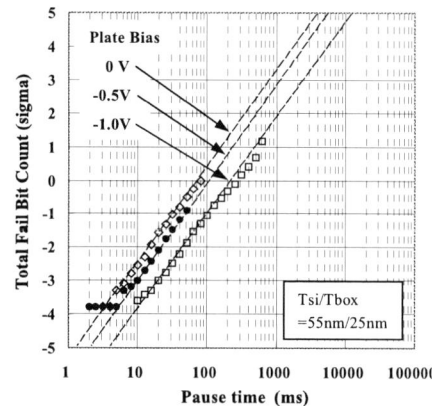

Fig.11 Retention characteristics of FD FBC.

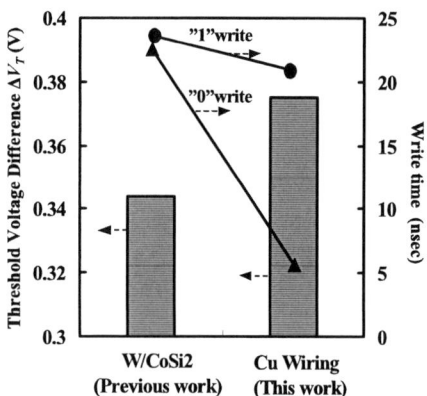

Fig.12. Estimated effect of Cu wiring.

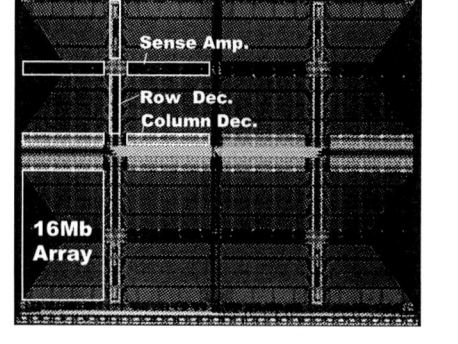

Fail Bit Count =36

Fig.13 Chip photograph of fabricated 128Mb SOI DRAM and Fail Bit Map of 32Mb area.

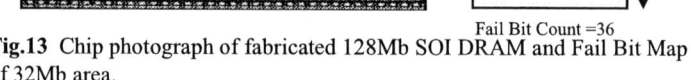

Fig.14 Device structure of 45nm CMOS.

	DRAM	eDRAM	FBC (1Cell/Bit)	FBC (2Cell/Bit)	eSRAM
Cell Size	6 - 8	12 - 20	6	12	80
Process	CMOS + Capacitor	CMOS + Capacitor	CMOS		CMOS
tRC (ns)	20	5	20	3	1
Median Retention (s)@85C	>10	>1	>0.1		-
Scalability Limits	Storage Capacitance	Storage Capacitance	Small Hole Density		Signal Stability
Issue to shrink to <45nm	Capacitor Transistor	Capacitor Transistor	Transistor		Transistor
Advantage	High Density	High Performance	Low Cost		High Performance
Basis of memory	Charge Storage in Capacitor	Charge Storage in Capacitor	Charge Storage in Floating Body		F-F

Table 2 Comparison among embedded memories.

96

2006 International Symposium on VLSI Technology, Systems, and Applications

Stack DRAM Technologies for the Future

Donggun Park, Wonshick Lee, and Byung-il Ryu
Semiconductor R&D Center, Samsung Electronics, Co.
San #24 Nongseo-dong, Giheung, Yongin, Gyeonggi-Do, KOREA 449-711

Abstract

Newly developed stack DRAM technologies such as RCAT, FinFET, MIM capacitor, improved support transistors, and litho-friendly patterning are introduced. Using these technologies, we expect that the scaling of DRAM can be continuously achieved until 2010.

Introduction

To meet the requirements such as cost, density, and performance for DRAM, aggressive scaling trend of ~30% per year has been achieved. However, since the technology has entered into sub 100nm scale, several barriers have appeared. In this paper, we will review the barriers and introduce our newly developed technologies to overcome the barriers for the future DRAMs by 2010. Among them are RCAT (Recessed Cell Array Transistor), Body-tied Fin CAT, MIM& MIS capacitor, and high performance peripheral transistors. Consideration for the scaling trend will also be discussed.

Technology Trend

Since the design rule of DRAM was reduced below 100nm in 2003, the scaling has been continued to show the bit growth of ~50% every year, while the DRAM price also dropped 30~40% per year as shown in Fig. 1. In addition, the DRAM characteristics such as data retention time and data rate also improved dramatically with several innovative technologies for the data storage capacitor and the cell array transistor as shown in Fig. 2 [1]. There are several key issues to scaling the stack DRAM as illustrated in Fig. 3. Three dimensional cell capacitor for 30 fF per cell, self-align contact to minimize the cell size, extremely low leakage cell transistor, sophisticated cell layout minimizing parasitic capacitance and resistance of bit-line and word-line, and high performance support transistors for high speed operation are required.

Cell Array Transistor Technology

Recently, using 3 dimensional cell array transistor is well accepted in order to meet the data retention requirement entering into the sub-100 nm technology node. The shorten gate length scaling the cell array size caused severe loss of the stored charge through transistor leakage (dynamic data retention) and junction leakage due to the high electric field using heavy channel dose to prevent the short channel effect of cell array transistor. The first attempt on this approach is RCAT, which digs the channel substrate below junction depth to make the gate (channel) length long. As a more improved shape, SRCAT is expected to be used for further shrink and reduced threshold voltage[2]. Then, finally, FinFET cell array transistor will be used for its excellent short channel effect and large current drivability. Especially, using body-tied FinFET, the floating body concern of SOI FinFET can be overcome. This cell array transistor trend is shown in Fig. 4 [3].

Support Transistor Technology

W/Poly-gate is replacing the WSi/Poly gate to meet the requirement of low word line resistance. Using this W/Poly-gate, poly-si depletion effect is reduced resulting in the inversion capacitance increase with smaller equivalent oxide thickness (EOT) as shown in Fig. 5. Also, dual poly-Si process is adapted even to the commodity DRAMs to provide the low threshold voltage p-ch MOSFET for low Vdd operation of DDR2 and/or DDR3 [4].

Cell Capacitor Technology

Cell capacitor is the device which requires the most innovative technology. Since the cylindrical structure was adapted, the most effort to have enough capacitance in smaller area has been focused on the high-k dielectric materials and metal electrodes. This approach seems to be of great success. In order to increase the height of the cylinder, MESH-Cap is also introduced as shown in Fig. 6 [5]. Fig.7 shows the cell capacitor technology trend of new dielectric and electrode materials.

Lithography and Interconnection Technology

In addition to ArF, OAI(Off Axis Illumination), PSM(Phase Change Mask), and Model based OPC(Optical Proximity Correction), Litho-Friendly Layout and Design for Manufacturing is needed for the sophisticated patterns of the sub-50 nm feature as shown in Fig. 8. To overcome the small contact issues, bar type contact is introduced to avoid the transistor current penalty as shown in Fig. 9. Storage node may need metal plug to reduce the contact resistance as shown in Fig. 10 [6].

Integration Technology

Cell size is shrinking not only by reducing the design rule but also by reducing the cell scheme from $8F^2$ to $6F^2$ [7]. $6F^2$ cell array is illustrated in Fig. 11.

1-4244-0181-X/06/$20.00 ©2006 IEEE

$4F^2$ cell will be the ultimate goal of DRAM cell as NAND Flash memory. Fig.12 shows the full integration scheme for sub-50 nm DRAM.

Conclusions

DRAM scaling below 50nm technology node has been reviewed. Several innovative technologies such as FinFET, MIM capacitor with MESH structure, metal/poly stacked gate, metal plug, cell array schemes of $6F^2$ and $4F^2$, and dual-poly process for low Vdd operation are expected to be adapted to meet the scaling below 50nm. Table1 summarizes the technology roadmap of future DRAM until 2012.

References

[1] K. Kim, IEDM p.333, 2005
[2] J.Y. Kim, et al., VLSI, p.34, 2005
[3] C. Lee, et al., IEDM, p.61, 2004
[4] N. Son, et al., Trans. on ED, Vol.51, p 1644, 2004
[5] D.H. Kim, et al., IEDM, p69, 2004
[6] C. Cho, et al., VLSI, p32, 2004
[7] C. Cho, et al., VLSI, p36, 2005

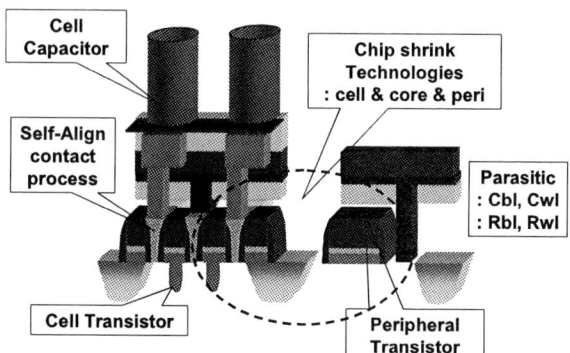

Fig. 3 Key technologies for DRAM scaling

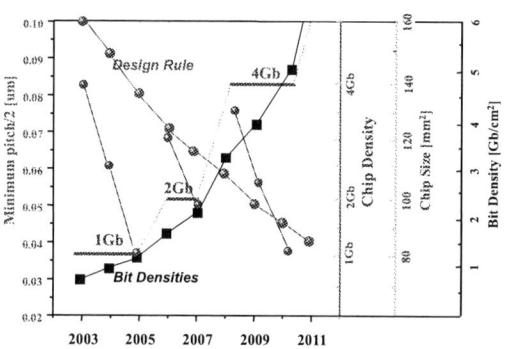

Fig. 1 DRAM Technology Trend in ITRS Roadmap

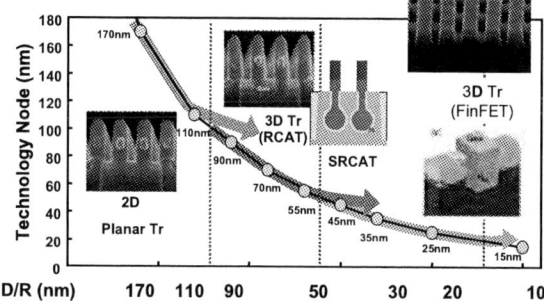

Fig. 4 3-dimensional cell array transistors, RCAT and body-tied FinFET, are required for the scaling in order to avoid short channel effect and achieve current drivability.

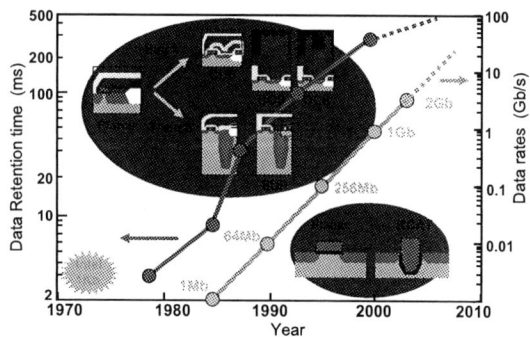

Fig. 2 Several innovative technologies are required for the continuous scaling of DRAM.

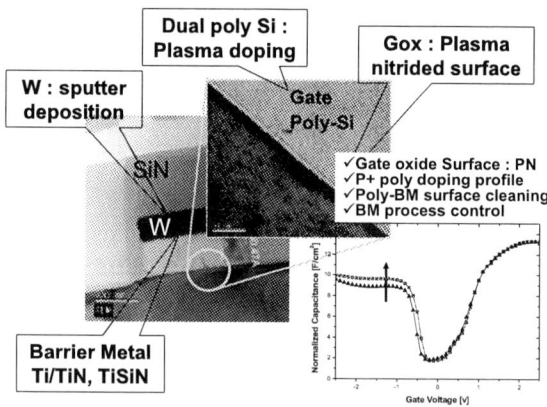

Fig. 5 Dual-Poly Gate and W/Poly-Si gate stack technologies are needed for the low voltage operation of DDR2 and DDR3 operation and high density of 2G bits.

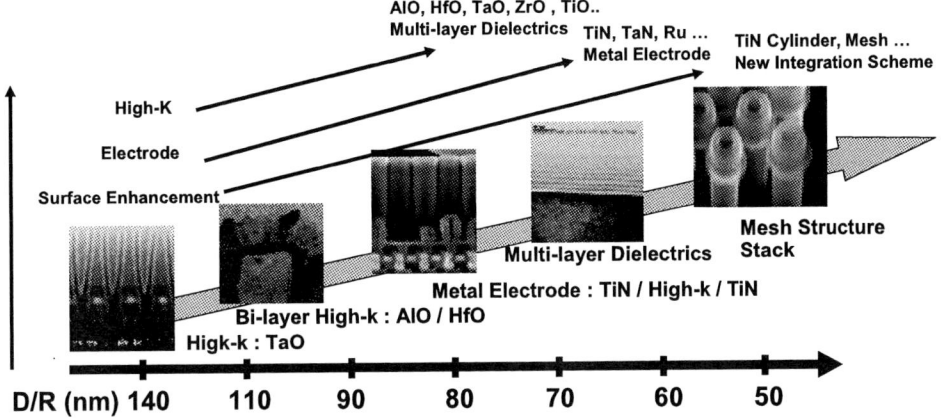

Fig. 6 In addition to new high-k dielectric materials, metal electrode and new structural innovation are needed for the scaling of stack-cell capacitor.

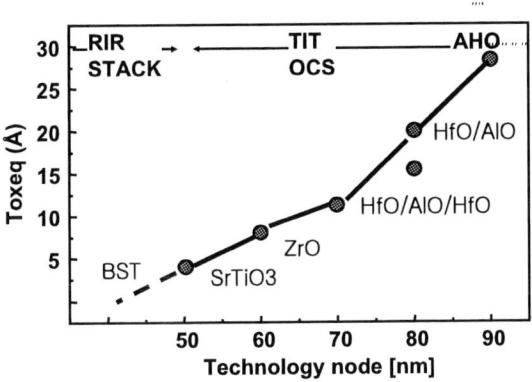

Fig. 7 High-k dielectric materials such as ZrO, SrTiO3, BST may be needed with proper electrode metal electrodes.

	70nm	sub-50nm
	ArF	ArF, immersion EUV
	OAI PSM OPC LFL	OAI PSM OPC LFL/DFM Self-align

Fig. 8 For sub-50 nm DRAM fabrication, Litho-friendly layout , Design For Maunfacturing, and self-align techniques should be used with ArF immersion lithography.

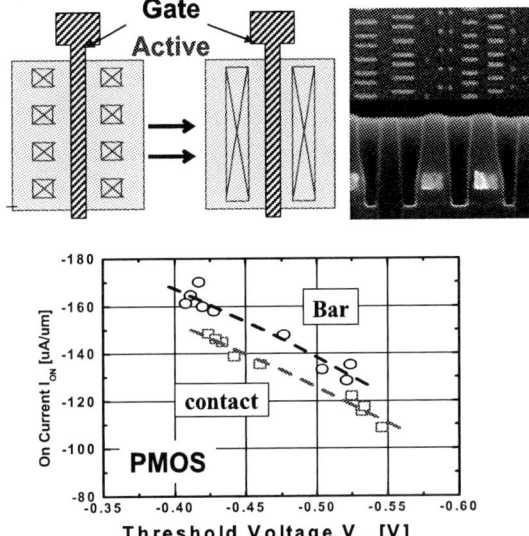

Fig. 9 Improving contact resistance with bar type contact, transistor current has been increased more than 7% in P-ch MOSFET.

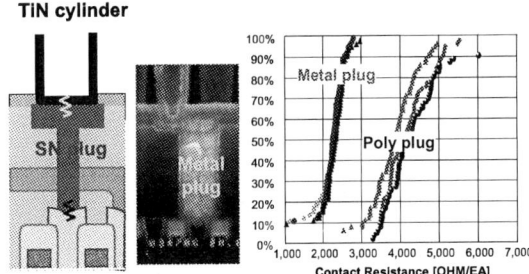

Fig. 10 To provide small enough contact resistance for the charge storage to the capacitor, metal plug is needed.

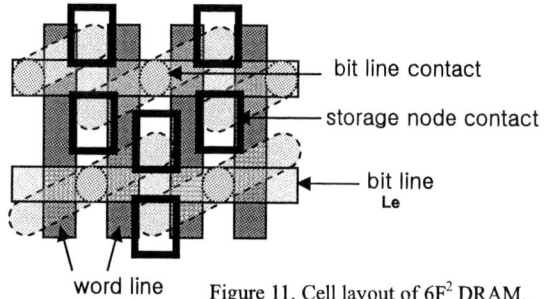

bit line contact

storage node contact

bit line
Le

word line

Figure 11. Cell layout of 6F^2 DRAM.

Table 1. Technology Roadmap of future DRAM

Year	04	06	08	10	12
Node (nm)	80	65	55	48	40

Data rate	800	1066	3200		6400
Vdd (V)	1.8	1.5	1.3		1.2
Density	1Gb	2,4Gb	4,8Gb		16Gb
Gate	WSix		W + Poly-Si		metal
Metal IMD	Al			Cu, Low-k	
Cell Tr.	RCAT		FinFET		
Capacitor	MIS	MIM (HfO, ZrO)		RIR (BST)	

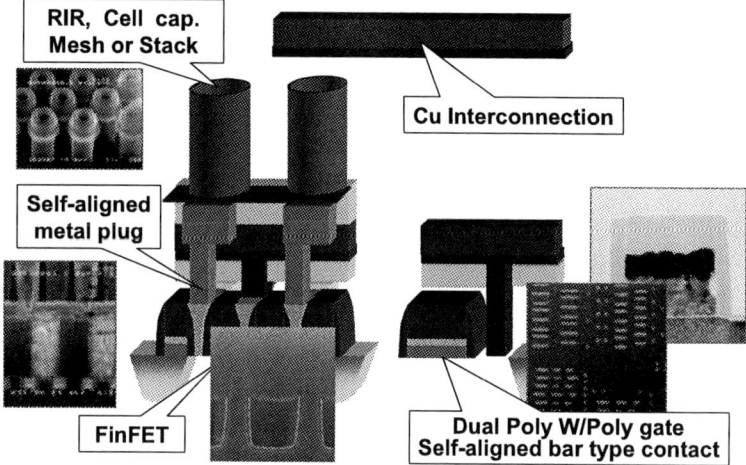

RIR, Cell cap. Mesh or Stack

Cu Interconnection

Self-aligned metal plug

FinFET

Dual Poly W/Poly gate
Self-aligned bar type contact

Fig.12   Schematic structure for sub-50nm Stack DRAM

2006 International Symposium on VLSI Technology, Systems, and Applications

A NOVEL DRAM CELL DESIGN AND PROCESS FOR 70NM GENERATION

C.H. Chung, T. Chien, J.S. Hsiao, C.H. Chu, W.S. Kuo, CC Cheng, F. Li, S. Nieh, S. Wu, B. Wang, C. Wang, T. Hu, G. Hsiao, M. Che, R.Y. Hon, H.M. Chen, G. Chou, G. Chang, L. Chou, H.C. Shu, K.Y. Huang, V. Tsai

Core Memory Tech. Development Div. ProMOS Technologies Inc.
No19, Li Hsin Road, HsinChu Science Park
HsinChu
Taiwan, R.O.C.

ABSTRACT

A new DRAM cell design for 70nm generation is demonstrated. We develop a novel cell design which is different from that previously published for trench DRAM technology and can be shrunk into 70nm generation. Some innovative processes are also introduced to successfully demonstrated the cell function.

INTRODUCTION

Big challenge we are facing for trench DRAM technology to scale down to below 100nm generation had been discussed widely [1]. The top one issue would be how to keep the cell capacitance to meet retention requirement. In terms of trench technology the CKB (Checker Board) DT pattern design is for the purpose to enlarge the DT dry etch process window for deeper trench depth and without neighboring DT-DT's short. CKB DT cell pattern also provides the possible space for following WB (Wet Bottle) process to enhance the surface area of the bottom part of the trench. HSG and high k dielectric material (Al_2O_3) are also implemented for 70nm process to enhance more cell capacitance by over 30% [3].

A new cell design is needed to couple with CKB DT due to the overall cell is different from previous generation and had better to keep $8F^2$ [3]. We develop a brand new memory cell and have successfully demonstrated a fully function product. The new cell design has opposite BS (Buried Strap) direction for each adjacent row of DT. Some novel processes like small STI, STI HDP voids free process and gate oxide spacer are also discussed in this paper. The new cell architecture and process can extend the trench technology into 90/70nm generation for DRAM manufacturing.

PROCESS INTEGRATION AND RESULT

Process integration:

Fig. 1 shows the new cell design where the trench signal passes through BS (Buried Strap) interface, channel controlled by activated gate with V_{PP} voltage, through CP (Contact Plug) and then CB (Bit line contact) to bit line. Here the CP connects two AAs (Active Area) and CB is landing on CP (Fig. 2) with tungsten dual-damascene process. A folded bit line and SA structure are used for signal sensing and data write back. One CB is shared by two cells. Bit line (M0) here is just located at the top of CB and between two AAs.

CKB DT pattern gives a wider process window for dry etch process and DT-DT short becomes a minor concern as compare with current existing mirror cell technology. The depth of DT can achieve 7um without DT-DT's short as shown in Fig.3a. The profile provides an acceptable cell capacitance up to 25fF/cell (for 90nm generation) without any other trench area enhancement methods. HSG (Hemispherical Silicon Grain), WB (Wet Bottle) are needed for further capacitance enhancement to increase the value around 10%

(as shown in Fig. 4) and 25%, respectively. HSG was performed by using existing poly deposition tool with grain size from 15nm to 20nm, which save the extra cost of new equipment investment. Fig.3b shows the interface of HSG poly grain and Si substrate with very good interface without poly gap fill concern.

Fig. 5a shows the top view SEM image of AA pattern, which is cut and isolated from each DT by process of DC (DT Collar) photo layer to form a small STI isolation. DC process provides a pre-STI HDP filling and to enlarge the following STI HDP process without gap filled problem as shown in Fig. 5b. DED (Deposit-Etch-Deposit) HDP deposition is used to provides a good gap filling for STI aspect ratio up to 4 without voids formation as shown in Fig. 6.

SCE (Short Channel Effect) is getting worse as technology keep shrinking down and the transistor channel length getting shorter and shorter. From current study, the device I_{off} is a main concern (too large) to provide enough drive current (Id) at a certain low threshold voltage (Vt). And how to fulfill the requirement of array device and periphery devices at same time, bring real challenges to process design. Dual gate spacer is a candidate process to suppress the periphery device SCE and keep array device with acceptable process window. Dual means there is different gate spacer thickness between cell array and core/periphery. In process integration the oxide spacer in array will be removed except the core/periphery area and then covered by nitride spacer to finish the dual spacer process. In general the total spacer thickness at core/periphery is twice of that of the array. Fig. 7 shows the dual spacer structure for 90/70nm process integration, which indicates the profile, is very good for device requirements. Fig. 8 shows the short channel Vt roll-off is significantly improved by dual gate spacer process. From our study , the threshold voltage deviation on Si is better than single gate spacer process.

The new cell design as previous mentioned has a CP (Contact Plug) contact to serve a plug plate for CB landing. CP is designed with contact length around 3F due to two adjacent AA must be bridged. Poly plug of phosphorus doped poly with concentration higher than $1E20/cm^3$ reduces the resistance. In this process a silicon oxide layer will be deposited prior to plug poly deposition to serve isolation layer between GC and CP. In other word the gate oxide spacer is formed (replace portion of silicon nitride) and to reduce the parasitical capacitance between CP to GC as well as to improve BLC (Bit Line Coupling) margin (see Fig. 9).

Summary:

The new cell transistor had been characterized with wide Vpp working window as shown in Fig. 10. Fig. 11 shows excellent cell refresh performance with wide Vbb range. The testing result demonstrates the full retention performance of new cell and the success of process integration of deep trench with small isolation, dual spacer, as well as CP process.

REFERENCES

[1]   H. Akatsu et al., Symp VLSI Technol. Dig., 2002, pp. 52-53.

[2]   N.J. Son et al, *Electron Device,* IEEE Transaction, 2004, vol. 51, pp. 1644-1652.

[3]   J. Amon et al, IEDM Technical Dig. IEEE int., 2004, pp. 73-76.

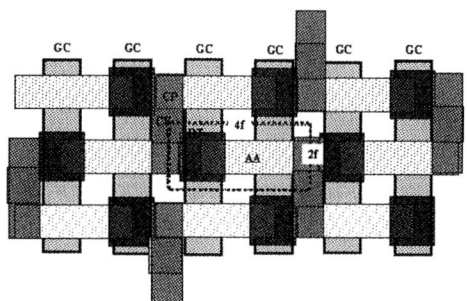

FIGURE 1. NEW CELL DESIGN FOR 70NM.

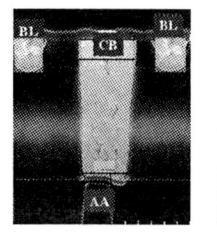

(a) OLD CELL   (b) NEW CELL

FIGURE 2. CROSS SECTIN VIEW OF CP/CB

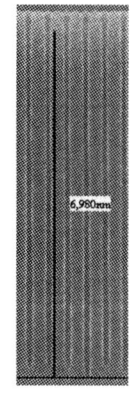

(a) DT DEPTH   (b) HSG SEM/TEM CROSS SECTION

FIGURE 3. CROSS SECTION OF DT PROFILE AND HSG

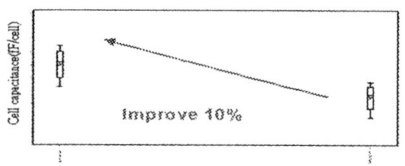

FIGURE 4. HSG ENHANCE CELL CAPACITANCE BY 10%.

(a)TOP VIEW   (b)CROSS SECTION.

FIGURE 5. NEW CELL TOP VIEW AND CROSS SECTION.

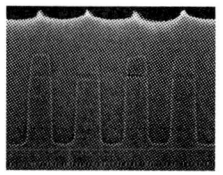   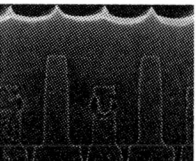

FIGURE 6. STI HDP GAP FILLED BY DED PROCESS

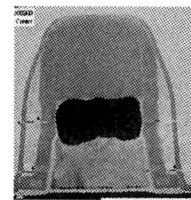   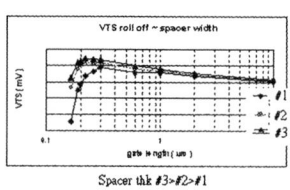

Spacer thk #3>#2>#1

FIGURE 7. TEM IMAGE OF   FIGURE 8. ROLL OFF CURVE
DUAL SPACER STRUCTURE V_S. SPACER THICKNESS

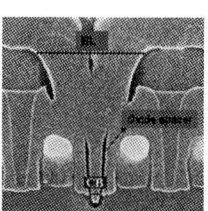

FIGURE 9. AN OXIDE SPACER BETWEEN CP AND GC

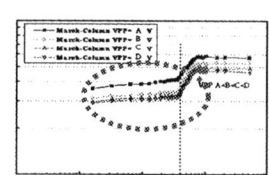

FIGURE 10. NEW CELL RETENTION V_S. V_{PP}

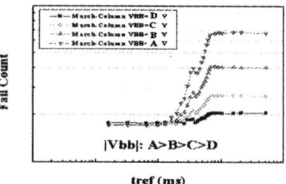

FIGURE 11. NEW CELL RETENTION V_S. V_{BB}

2006 International Symposium on VLSI Technology, Systems, and Applications

RELIABILITY ASSESSMENT OF THE EMBEDDED DRAM TECHNOLOGY WITH PMOSFET TRANSFER TRANSISTOR AND HIGH-K DIELECTRICS (Ta₂O₅) MIM CAPACITOR

R.F. Tsui, J.R. Shih, Kevin_Liu, Y.S. Tsai, H.W. Chin and Kenneth Wu

Reliability Assurance Division

Taiwan Semiconductor Manufacturing Company,

Hsin-Chu, Taitwan   E-mail: rftsui@tsmc.com

ABSTRACT

In this paper, wafer level and product level reliability characteristics of embedded DRAM technology with high-K dielectric Ta_2O_5 MIM capacitors have been analyzed. It is found although hot carrier injection can induce more apparent gate-induced-drain-leakage (GIDL) current than off-state bias temperature (BT) stress does, BT stress still dominate the failure bit count increase in real circuit operation. In addition, it is also found the competition between gate-induced-drain- leakage (GIDL) current and the MIM dielectric leakage dominate the failure bit count (FBC) evolution behavior after reliability stress. With transistor doping profile optimization, a new phenomenon of FBC reduction with burn-in time can be observed, and it is attributed to the leakage reduction of MIM capacitor with high-K dielectric Ta_2O_5 after stress.

INTRODUCTION

The same with the requirements of commodity DRAM, retention time performance and FBC are the most important issues for embedded DRAM technology High retention time performance implying low power consumption and low FBC implying low defect per million (DPM) can be achieved. It's well known that several leakage current mechanisms can degrade retention time performance, as shown in Fig. 1, which include 1) junction leakage current from the storage node; 2) subthreshold leakage current of the transfer transistor; 3) gate dielectric leakage current; 4) gate-induced drain leakage (GIDL) current at the storage node [1]; 5) isolation leakage current between capacitor to bit line or between the neighboring cells; 6) capacitor dielectric leakage current.

Conventionally, nMOSFET is widely used for transfer technology in DRAM technology to achieve smaller cell dimension. It has been confirmed that junction leakage and GIDL currents dominate the retention time performance. However, it has also been reported that pMOSFET has smaller GIDL current that nMOSFET's due to boron implant-induced more gradient LDD profile in pMOSFET. To have more robust retention time control in sub-0.1um embedded DRAM technology, pMOSFET as the transfer transistor can be a good candidate. In addition, high-K dielectric Ta_2O_5 MIM capacitor is also included to achieve high cell capacitance (Cs). In this study, the impacts of pMOSFET's GIDL current and dielectric leakage current on retention time performance are evaluated.

EXPERIMENTAL

In this study, transistors were processed with a standard sub-0.1μm technology. The key process parameters included gate oxide thickness 2.8nm and on mask minimum gate length around 0.18um for the transfer pMOSFET transistor formation. After the Co-salicide formation, MIM capacitors with high-K dielectrics Ta_2O_5 were processed, where the cross-sections of cell transistor and MIM capacitor are shown in Fig. 2. Tabel-1 shows the bias conditions of transfer transistor pMOSFET and MIM capacitor at different operation conditions. To evaluate the product reliability performance, a macro with 32Mbs was used for the high temperature operation life (HTOL) test at 125°C and 1.4xVcc. The characteristics of FBC and retention time were monitored at different read-out points. In addition, all the leakage current components in the cell array with 128Kbs were also monitored to explore the degradation mechanism of HTOL test.

RESULTS and DISCUSSION

Figure 3 shows the typical I-V characteristic of pMOSFET from off state to on-state. It can be observed that the GIDL current is sensitive to gate bias, which is due to thin oxide with 2.8nm used. As the design condition with V_{bb} equal to –0.4 volt, it is found there is the largest impact ionization rate, where substrate current peaks appear around –0.4 ~ -0.6V, as shown in Fig. 4. This operation condition implies there is the largest interface state generation due to hot carrier injection. For DRAM technology, saturation current degradation is not the most important parameter, instead of, GIDL current increase after stress may induce failure bit count increase. Figure 5 shows the GIDL current evolution during hot carrier stress. It can be observed there is apparent GIDL current increase at lower drain to gate voltage drop. In other words, when the cell is operated at on state for reading or writing any signal "1" or "0", it will enhance the GIDL increase after long-term operation.

Similarly, when the cell transistor is operated at off-state to store the signal in MIM capacitor, although there is no channel hot carrier, the high field across the gate to drain may also impose impact on the GIDL current after long-term operation. Figure 6 shows the GIDL current evolution during BT stress for the same stress time with HCI. Smaller GIDL current increase rate can be observed compared to HCI-induced increase. From the comparison of log [Id/(|Vdg|-1.2) vs. 1/(|Vdg|-1.2) before and after HCI and BT stress at the same Vds, as shown in Fig. 7, we can find most of the GIDL increments belong to band-to-defect tunneling, not band-to-band tunneling, which is the same with the conclusion in the reports [3]-[4]. In addition to the junction leakage, we also monitor the leakage currents through gate oxide, between capacitor to capacitor and through MIM capacitor, as shown in Fig. 8. We can observe there are little gate oxide leakage decrease and nearly no leakage current change in capacitor-to-capacitor isolation. However, there is apparent leakage reduction in MIM capacitor before the capacitance crashes during the stress and the reduction can be more than 20 times. Because retention time performance strongly depends on the leakage current and cell capacitance. The relationship between retention time and leakage currents can be expressed as $T_{ret} \propto C_s/I_{leak}$. Figure 9 shows the retention time performance comparison before and after HTOL test with different read-outs. We can observe the retention time performance can be improved with the BI time increase. This result contradicts to conventional concept, where product characteristics may be degraded or keep the same. This is the first time we observe such kind of behavior. Based on our above analysis on each component's leakage current, it is suspected that MIM leakage reduction during the stress leads to the retention time enhancement. To reconfirm this suspect, a special BI pattern with high bias was designed to stress MIM only and keep the pMOSFET transfer transistor at off state with low Vpp value. As shown in Fig. 10, it can be observed that the FBC reduces with BI time increases. It implies the retention can be improved and consistent with the proposed model.

CONCLUSION

Wafer and product level reliability assessments have been performed to the embedded DRAM technology with high-K dielectric Ta_2O_5 MIM capacitor. A new phenomenon of retention time increase or FBC reduction has been observed and well characterized. It is consider that although the GIDL current increase of transfer pMOSFETs may degrade the retention time performance after long-term stress, the leakage current reduction of MIM capacitor due to charge trapping after BI can cover GIDL current increase and dominates the product reliability performance enhancement.

REFERENCES

[1] S. Banejee, et. al, VLSI Symp. Digest., pp. 97, 1987.

[2] T. Hamamoto, et. al, IEEE ED. Vol.45, pp. 1300, 1998.

[3] Y.P. Kim, et. al, IRPS Proc., pp. 1, 2001.

[4] G.Q. Lo, et. al, IEEE EDL. Vol. 12, pp. 710, 1991.

Table-1

State	Vbb	Vpp	Vcp	Vnode
On "1"	-0.4	-		1.2
On "0"	-0.4	-	0.65	0.1
Off	-	1.6		1.2 or 0.1

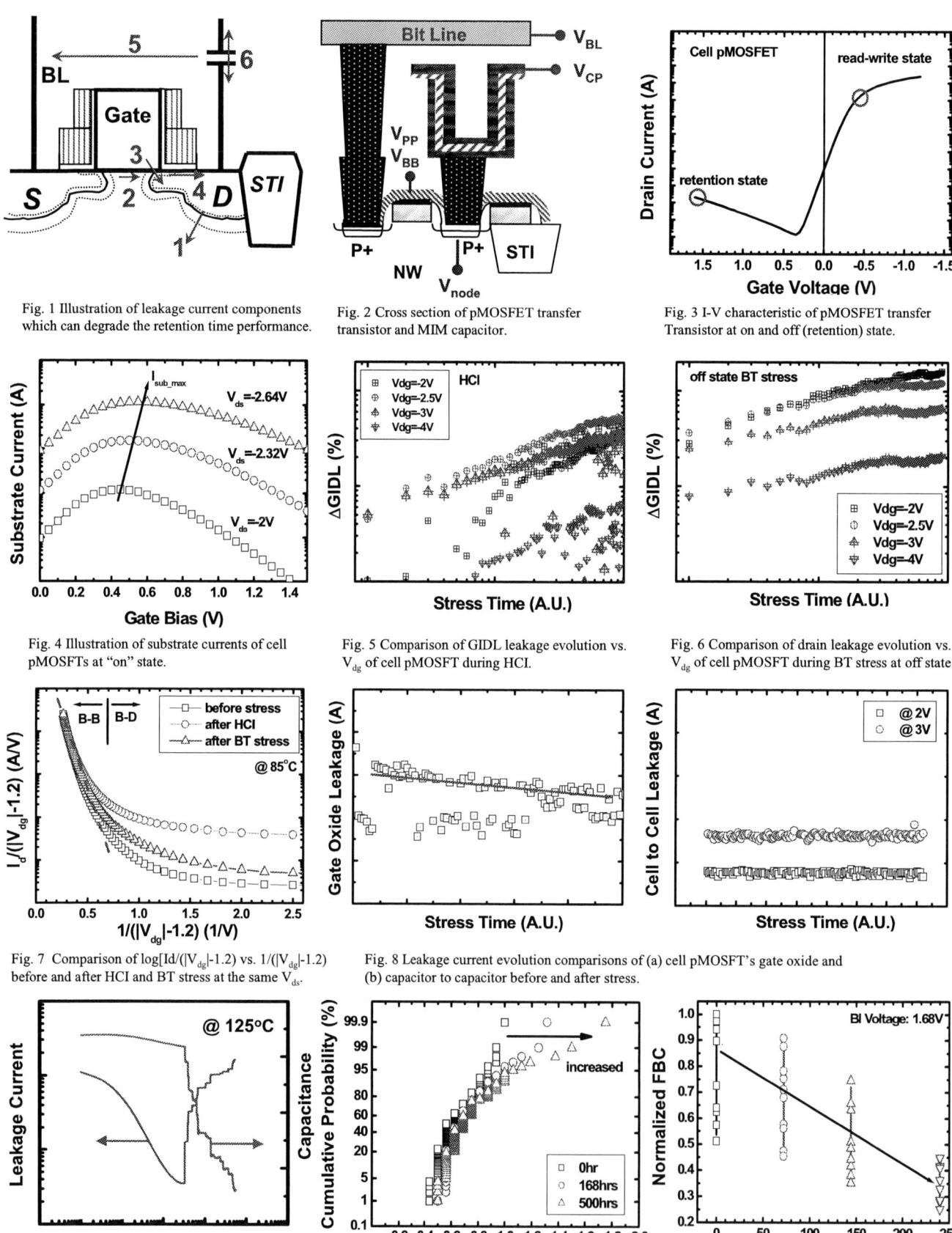

Fig. 1 Illustration of leakage current components which can degrade the retention time performance.

Fig. 2 Cross section of pMOSFET transfer transistor and MIM capacitor.

Fig. 3 I-V characteristic of pMOSFET transfer Transistor at on and off (retention) state.

Fig. 4 Illustration of substrate currents of cell pMOSFTs at "on" state.

Fig. 5 Comparison of GIDL leakage evolution vs. V_{dg} of cell pMOSFT during HCI.

Fig. 6 Comparison of drain leakage evolution vs. V_{dg} of cell pMOSFT during BT stress at off state.

Fig. 7 Comparison of $\log[I_d/(|V_{dg}|-1.2)]$ vs. $1/(|V_{dg}|-1.2)$ before and after HCI and BT stress at the same V_{ds}.

Fig. 8 Leakage current evolution comparisons of (a) cell pMOSFT's gate oxide and (b) capacitor to capacitor before and after stress.

Fig. 8(c) Leakage current and capacitance evolution of MIM capacitor during constant voltage stress.

Fig. 9 Cumulative plots of retention time performance of DUTs with 32Mb macro after HTOL test.

Fig. 10 FBC evolution during HTOL test with only MIM capacitors stressed at high bias.

2006 International Symposium on VLSI Technology, Systems, and Applications

Metal Gate Technology for 45nm and Beyond

[1,2]Kentaro Shibahara

[1]Research Center for Nanodevices and Systems, Hiroshima University
[2]Graduate School of Advanced Sciences of Matter, Hiroshima University
1-4-2, Kagamiyama, Higashihiroshima, 739-8527 Japan
Phone +81-82-424-6267, FAX: +81-82-424-3499, e-mail: ksshiba@hiroshima-u.ac.jp

Abstract

Metal gate is one of the most expected technologies for CMOS device performance improvement. However, its integration to CMOS devices is not so easy compared with poly-Si gate devices. In addition it needs a workfunction tuning method to adjust FET threshold voltage. In the presentation, these problems will be discussed based on author's research result of Mo and FUSI (Fully Silicided) gate referring state of the art of this field.

1. Introduction

Increase in EOT (Equivalent Oxide Thickness) due to depletion in poly-Si gate electrodes is serious obstacle for EOT scaling and device performance improvement. Metal gate is expected to solve this problem. In addition, recovery of reduced carrier mobility due to high-k gate dielectric introduction is also expected [1]. In this paper, status and issues of metal gate device development are described.

2. Processes and Structures of Metal Gate Devices

Figure 1 shows fabrication flow and cross section of various metal gate devices. Metal or metal nitride are stacked with poly-Si, which enables salicidation [2,3]. In this case, metal gate must be robust against high-temperature source-and-drain activation process. To avoid this difficulty, replacement gate process was proposed [4,5]. FUSI (Fully Silicided) process that convert an entire poly-Si gate to silicide is another candidate. A common issue for all structures is difficulty of dual gate formation for V_{TH} control of CMOS devices. In the case of poly-Si, high and low workfunctions are easily obtained by doping. A lot of methods to obtain dual workfunction with metal gates were proposed. Metal or metal nitride can utilize wet etching selectivity against HfO_2 [7,8], as shown in Fig. 2. In addition this dual metal integration process, selection of metal materials and composition control are usually demanded. To meet demand of precise workfunction control, metal carbide [9] and insertion of a very thin metal layer on gate insulator [10] were also investigated. Moreover, it should be noted that metal gate workfunction is affected by thermal treatment process [11].

3. Workfunction Tuning of Mo Gate

There are a lot of approaches to realize dual workfunction with a single metal gate. In this section, Mo workfunction tuning with nitrogen incorporation is described based on the authors' work [12-14]. As shown in Fig. 3, V_{TH} of Mo gate MOSFET was modified by incorporating nitrogen into Mo. Nitrogen was introduced by solid-phase diffusion (SPD) at $800°C$ for 1 min from TiN deposited on Mo [13,14]. Since original Mo workfunction was close to E_V of Si, it matches p-MOS application. Though Mo workfunction tunable range, -0.46 eV, was not sufficient for n-MOS application, it is a useful example of workfunction modification by pileup formation at the metal/gate-insulator interface. Figure 5 shows nitrogen depth profile in $Mo/SiO_2/Si$ MOS structures. Nitrogen pileup peak formed at the Mo/SiO_2 interface reduced by additional annealing without a TiN SPD source. Since this additional annealing caused reversible behavior in workfunction shift, it was clear that the nitrogen pileup was the origin of workfunction shift. Figure 4 shows a dipole model to explain the workfunction shift. Because of electronegativity difference between nitrogen atoms and SiO_2 the electric dipoles were formed. High density electric dipoles at the interface or in the insulator are observed as the workfunction shift. Thus, forming impurity pileup at the metal/insulator interface is a key for workfunction tuning.

4. FUSI Gate Technology

Workfunction of NiSi FUSI gate is tunable by pre-doping to poly-Si gate before full-silicidation [15-17]. The workfunction shift is caused by pileup of predoped impurity at the silicide/gate insulator interface. Figure 6 shows example of Sb pileup in NiSi MOS structures [18, 19] evaluated by the authors. Pileup peak height depended on silicidation temperature. Faster silicidation at $500°C$ resulted in smaller pileup and negligible workfunction shift in spite that silicidation at $450°C$ resulted in the shift of -0.35 eV. The precise location of Sb pileup was evaluated by XPS measurement [20]. Sb doped FUSI structure was cleavable at the gate oxide, as shown in Fig. 7(a). Though this feature would result in an reported adhesion issue [18,19,21], we have utilized this feature for sample preparation. The specimen were divided to upper and lower parts, as shown in Fig. 7(b). Sb-O related signals were observed for both the upper and the lower specimens, as shown in Fig. 8. Thus Sb penetration into SiO_2 was revealed. Recently, it was reported that location of the pileup was an important factor to decide workfunction shift direction [22].

The workfunction tunability is attractive feature for CMOS application, since the pre-doping is usually done by ion implantation. This means that workfunction shift direction and magnitude are controllable by selecting an ion specie and implantation dose. However, full silicidation process makes the control difficult. Figure 9 shows workfunction tunable range reported from three groups [18,19,23,24]. There are many discrepancies in this figure. The difference in silicidation condition is a considerable origin of the discrepancy. As shown in Fig. 6, pileup formation that was necessary for workfunction shift was affected by slight difference of silicidation process. Moreover, gate length dependence of silicide phase, explained in Fig. 10, is also known as a issue to be solved [25].

5. Conclusion

Metal gate CMOS can be fabricated by utilizing dual-metal integration or single metal whose workfunction is tunable with convenient process, as introduced in this paper. Though process issues concerning NiSi FUSI were described, other techniques have their own issues. To apply metal gate to mass production of leading edge CMOS, hard work is demanded.

Acknowledgements

This work was partly supported by STARC.

1-4244-0181-X/06/$20.00 ©2006 IEEE

References

[1] R. Chau et al., IEEE EDL **25**, 2004 p.408.
[2] S. Datta et al., IEDM 2003, p. 653.
[3] A. Vandooren et al., IEDM 2003, p.975.
[4] Y. Akasaka et al., SSDM 2004, p.196.
[5] A. Yagishita et al., IEEE TED **47**, 2000 p.1028.
[6] J. Kedzierski et al., IEDM 2003, p.441.
[7] S.B. Samavedam et al., IEDM 2002, p.433.
[8] Z.B. Zhang et al., 2005 VLSI Symp., p.50.
[9] J.K. Schaeffer et al., IEDM 2004, p. 287.
[10] I.S. Jeon, et al., IEDM 2004, p.303.
[11] C.Ren et al., IEEE EDL **25**, 2004 p.337.
[12] T. Amada et al., MRS Proc. **716**, 2002 p.299.
[13] M. Hino et al., SSDM 2003, p.494.
[14] T. Hosoi et al., Abst. MRS 2005 Spr. Meet., p.191.
[15] M. Qin et al., J.Electrochem. Soc. **148**, 2001 p.G271.
[16] J. Kedzierski et al., IEDM 2002, p.247.
[17] W.P. Maszara et al., IEDM 2002, p.367.
[18] K. Sano et al., SSDM 2004, p. 456.
[19] K. Sano et al., Jpn. J. Appl. Phys. **44**, 2005 p.3774.
[20] T. Hosoi et al., ISDRS 2005, WP4-05.
[21] J. Kedzierski et al., IEEE TED **52**, 2005 p.39.
[22] Y. Tsuchiya et al., IEDM 2005, p.637.
[23] C. Cabral et al., 2004 VLSI Symp., p.184.
[24] D. Aim et al., IEDM 2004, p.87.
[25] J.A. Kittl, 2005 VLSI Symp., p.72.

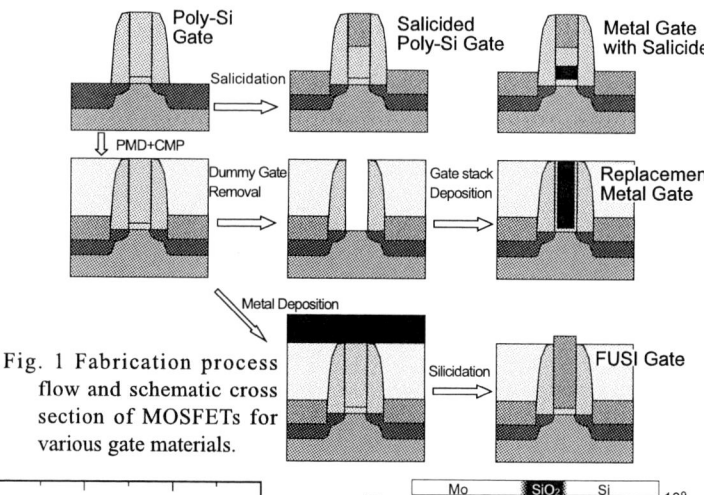

Fig. 1 Fabrication process flow and schematic cross section of MOSFETs for various gate materials.

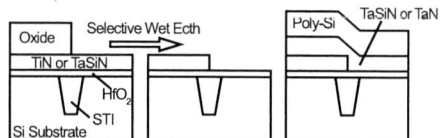

Fig. 2 An dual metal integration method utilizing selectivity between HfO_2 and metal nitride [7,8].

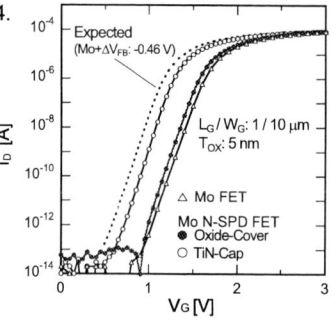

Fig. 3 I_D-V_G characteristics of Mo gate MOSFETs. V_{TH} depended on process flow after N-SPD for workfunction tuning.

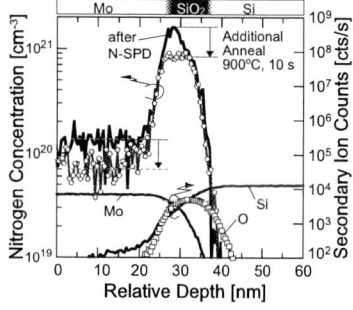

Fig. 4 SIMS depth profiles for Mo/SiO_2/Si MOS structures. Nitrogen was introduced by N-SPD. Additional annealing imitate S/D activation annealing.

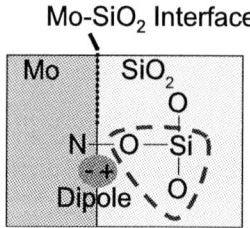

Fig. 5 A simple model to explain workfunction shift due to pileup formation at the Mo/SiO_2 interface.

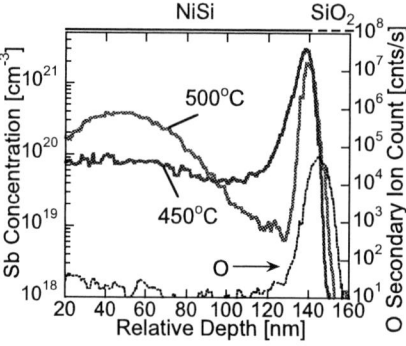

Fig. 6 Sb depth profiles in NiSi/SiO_2/Si MOS structures.

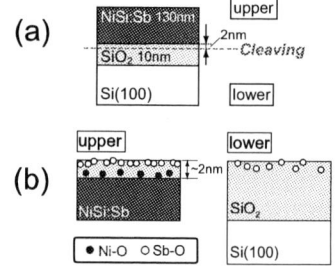

Fig. 7 (a) Sample structure for XPS measurement shown in Fig. 8. The specimen was cleaved and divided to upper and lower pieces. (b) A model to show the location of Sb pileup.

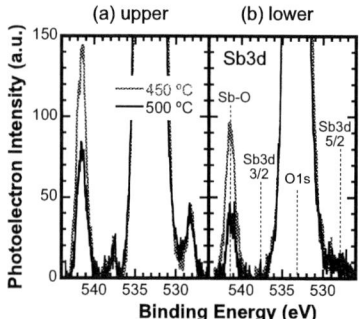

Fig. 8 Sb3d spectra taken by XPS with the specimen shown in Fig. 10.

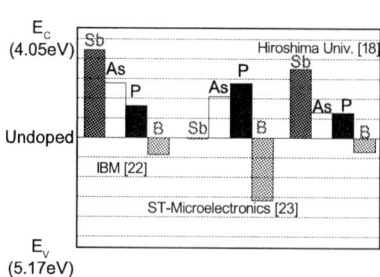

Fig. 9 Workfunction tunable range of NiSi FUSI gate reported from plural organizations.

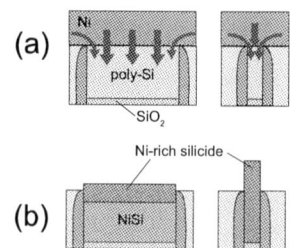

Fig. 10 Phase control difficulty of NiSi FUSI gate due to gate length. (a) before silicidation, (b) after silicidation.

106

2006 International Symposium on VLSI Technology, Systems, and Applications

Electron Trapping Processes in High-κ Gate Dielectrics and Nature of Traps

G. Bersuker, J. Gavartin*, J. Sim, C. S. Park, C. Young, S. Nadkarni, R. Choi, A. Shluger*, B. H. Lee[a]

SEMATECH, 2706 Montopolis Dr. Austin, TX 78741, USA; [a] IBM assignee
*Condensed Matter and Materials Physics, University College London, Gower Street, WC1E 6BT London, U.K.

Introduction. The high-k material property, which impacts electrical characteristics of the dielectrics, is a relatively high density of as-grown defects representing electron traps. Since electron trapping is reversible (Fig. 1), with the de-trapping kinetics being very similar to the trapping one [1], one may conclude that at a low voltage electrical stress of practical importance the electron trapping mostly occurs on the pre-existing ("as-processed") defects in the bulk of a high-k film. As can be seen in Fig. 1, initial fast increase of V_t during the first second of stress is followed by a slower V_t growth suggesting that two electron trapping processes having time scales different by about six orders of magnitude contribute to V_t change [2]. In this study, we propose a physical model of the fast and slow electron trapping processes, which adequately describes pulsed I_d-V_g and CVS low temperature measurements and consistent with the calculated characteristics of the oxygen vacancies in crystalline hafnia.

Fabrication and measurements. NMOS transistors were fabricated with ALD HfO_2 films deposited on the O_3 pre-treated (001) epi Si substrate and densified by the 700C N_2 anneal. A gate electrode was formed by a 10 nm TiN film, capped with the n-doped poly-Si layer. The fabrication process included a 1000C/10sec dopant activation anneal and 485C forming gas anneal.

Pulsed I_d-V_g measurements (100 ns rise/fall time, 100 μs pulse width) were performed on the 1μm/10μm transistors to monitor the kinetics of the fast electron trapping process [3]. The gate pulse voltage and temperature ranged from 1.65 V to 1.95 V, and 77K – 298K, respectively, Fig. 2. The transistors were also subjected to a substrate injection constant voltage stress at different gate biases V_g= 1.6V-2.1V in the same temperature range. The stress was interrupted for the I_d-V_g measurements to monitor variations of threshold voltage and transconductance values. An example of the threshold voltage change during the stress is shown in Fig.3. V_t shift values after the first second of the stress, which are mostly due to the fast charging process, were subtracted from these values. Stress conditions, which resulted in minimal G_m degradation, lower than 2.5%, were selected to avoid contribution to V_t from the interface damage. The analysis was done for the 3 nm HfO_2 gate stack, which shows electron trapping sufficient for reliable modeling. Comparison of the 3 nm and 2 nm gate stacks at CVS V_g=1.8 V are shown in Fig. 4.

Model and Extraction of Trap Characteristics. To model the fast charging process, we take into account that the electron injected into the dielectric conduction band can be effectively captured by a nearby defect when their energies are in resonance. To identify potential candidates for the electron traps, HfO_2 structure in a monoclinic phase was calculated using plane wave density functional theory with the spin polarized generalized gradient approximation. It was found that an appropriate candidate is a neutral oxygen vacancy, which creates small displacements of the surrounding Hf atoms and can trap an injected electron before it gains kinetic energy, leading to the formation of a negative O vacancy (V^-) with the estimated energy of about 0.4 eV below the conduction band, Fig.5. V^- can trap a second electron resulting in the V^{2-} defect of a slightly, by

about 0.1 eV, lower energy due to the electron repulsion. Electrons trapped at the O defects are delocalized over a rather wide area, in excess of 1 nm in diameter, Fig. 6. The time dependence of the density of the traps filled by the electrons injected from the substrate, n, which reflects that of V_t ($\Delta V_t = qnx$, where x is the trap position defined by the band bending, Fig. 7) and, hence, the drain current, $-\Delta I_d \propto \Delta V_t$, can be described as: $n = N_0(1-e^{-pt})$ Eq.(1). Here N_0 is the available total trap density and p is the probability of a single electron trapping event: $p \equiv \sigma J/q$ where σ is the defect capture cross-section, and J is the flux of the incoming electrons. By fitting Eq.(1) to the measured time dependence of the drain current during the pulse (Fig. 2), one can obtain an estimate for the characteristic trapping probability p and, subsequently, σ, Fig. 8. For all temperatures and pulse bias conditions, $\sigma \approx 2 \cdot 10^{-14}$ cm^2; this is consistent with the calculations of the effective dimension of the oxygen vacancy traps in Fig. 6.

Analysis similar to that discussed above shows that traps located in the dielectric further away from the substrate cannot be filled effectively by the electrons injected in the conduction band during CVS because their kinetic energy after even small acceleration exceeds the energy of the shallow traps associated with the neutral O vacancies (see Fig. 7). On the other hand, electrons trapped during the fast tunneling process are not subjected to the electrical field acceleration and, therefore, when emitted from the traps back to the conduction band (so-called secondary electrons) can be re-trapped by the nearby available traps, Fig. 7. This process of the secondary electrons re-trapping can be described by the equation similar to Eq.(1) where, however, the flux of the electrons emitted from the traps is given by $J = N \cdot 1/\tau \cdot exp(-E_i/kT)$, where N is the density of the filled fast traps, τ is the de-trapping time constant and E_i is the effective trap energy. The fitting for different stress voltages in the whole temperature range, Figs.9-10, required three terms in Eq.(1), $n_i = N_{0i}(1-e^{-p_i t})$ (i=1,2,3), suggesting that 3 types of processes contribute to slow charge trapping. At the lowest temperature T= 77K, ΔV_t time dependence is controlled by a single process with the lowest activation energy of ≤ 5 meV, which may be related to the polaronic behavior of an electron in the conduction band. At higher temperatures, two additional processes with activation energies of about 0.35 eV and 0.25eV (before the Poole-Frenkel correction for the barrier lowering of 0.1-0.15 eV) also contribute to the electron migration, Fig.11. These energies are consistent with the calculated energies of the V^- and V^{2-} defects, respectively, and in agreement with the earlier reports [4, 5].

Summary. The above results suggest that migration of the electrons, captured during the fast charging process, to other available traps represents the major process responsible for the intrinsic V_t instability in the high-k NMOS transistors. The extracted trap characteristics are consistent with those of the oxygen vacancies in the monoclinic hafnia.

References. [1] J.Sim et al., SSDM, 214, 2004; [2]G. Bersuker et al., NATO Symp. on Defects in High-k Dielect., St.Petersburg, 2005; [3]C. Young et al., IRPS Proc.,75, 2005; [4] G. Bersuker et al., Microel. Reliability, **44**, 1509, 2004; [5] G. Ribes et al., ESSDERC, 89, 2004.

1-4244-0181-X/06/$20.00 ©2006 IEEE   107

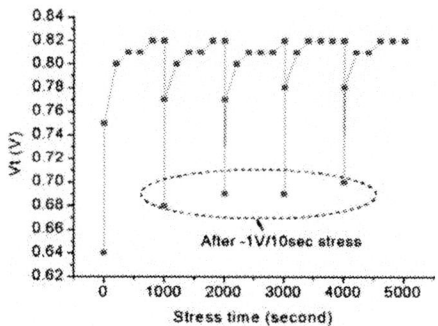

Fig. 1 Threshold voltage change in TiN/3nm HfO$_2$/1nm SiO$_2$ NMOS transistor during stress cycles, which include 1000 sec substrate injection stress at Vg=2.4V followed by 10 sec stress of the opposite bias (-1V).

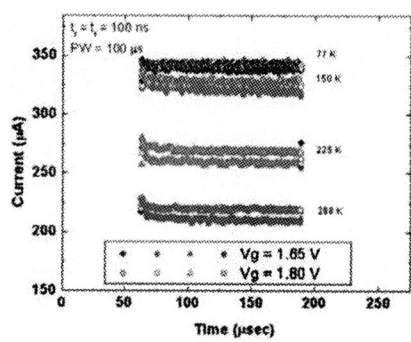

Fig. 2 Pulsed I$_d$-V$_g$ data with the pulse voltages V$_g$=1.65 V and V$_g$=1.80 V measured at different temperatures

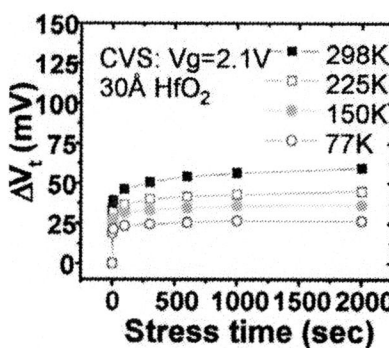

Fig. 3 Change of V$_t$ values during V$_g$=2.1 stress in the temperature range 298K~77K

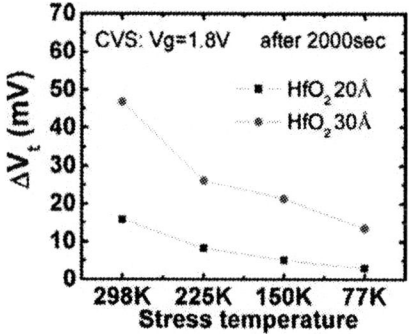

Fig. 4 V$_t$ change after the V$_g$=1.8V/2000 sec stress at different temperatures in 20A and 30A HfO$_2$ stacks

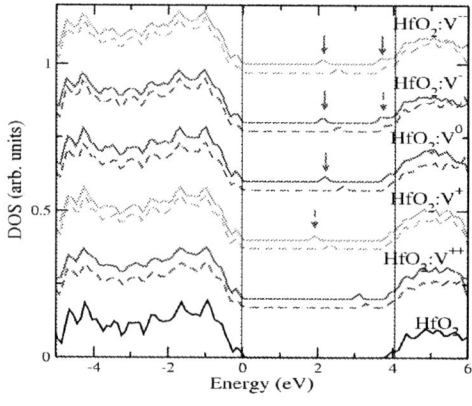

Fig. 5 Calculated electron density of states for the monoclinic HfO$_2$ with 4-fold (solid line) and 3-fold (dashed line) coordinated oxygen vacancy in different charge states. Arrows indicate gap levels.

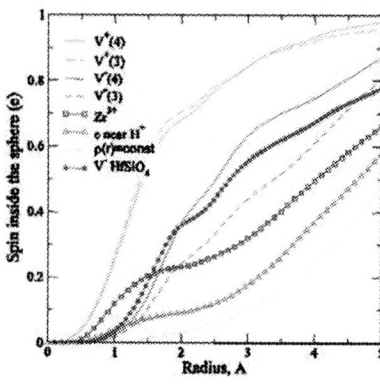

Fig. 6 The spin density integrated inside the sphere as a function of its radius R, for the electron captured by various defects in the monoclinic HfO$_2$.

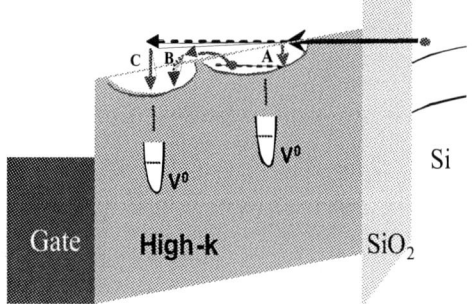

Fig. 7 Schematic of the proposed electron trapping processes: A- trapping near neutral O vacancy (V^0) via tunneling (fast process), B – electron de-trapping/re-trapping (slow process). Electron capture from the conduction band (C) is ineffective.

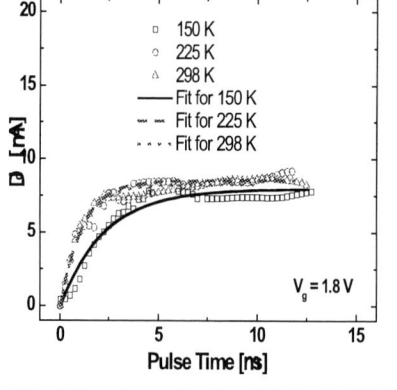

Fig. 8 Experimental and modeled pulsed I$_d$-V$_g$ data at different measurement temperatures.

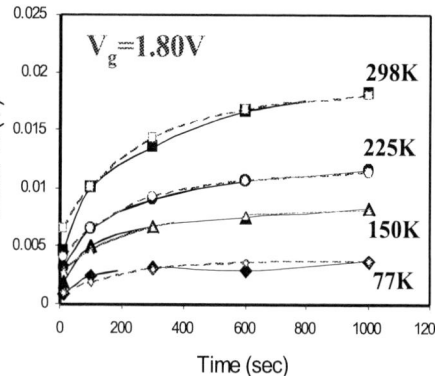

Fig. 9 Experiment and modeled (open symbols) V$_t$ shift during the V$_g$=1.80V stress at different temperatures

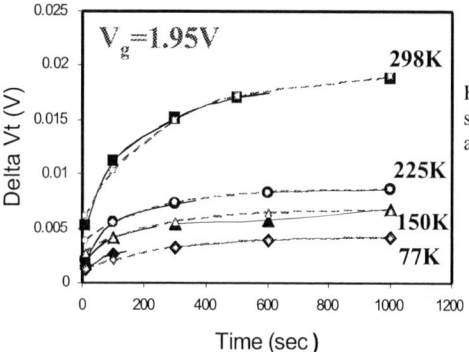

Fig. 10 Experiment and modeled (open symbols) V$_t$ shift during the V$_g$=1.95V stress at different temperatures

Fig. 11 Activation energies E$_i$ corresponding to p$_i$ values extracted from V$_g$=1.8V and 1.95V stress data in Figs. 9,10.

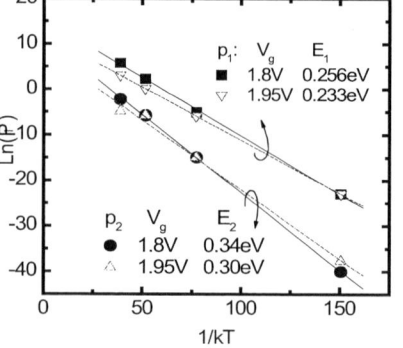

108

2006 International Symposium on VLSI Technology, Systems, and Applications

HfSiON Gate Dielectric for 45nm Node Low-Power Device

Tian-Choy Gan, Howard C.-H. Wang, Shang-Jr Chen, Ching-Wei Tsai, Peng-Soon Lim, Huan-Just Lin, Ying Jin, Hun-Jan Tao,
Shih-Chang Chen, Ying Keung Leung, Carlos H. Diaz, Mong-Song Liang, and Yuh-Jier Mii

Taiwan Semiconductor Manufacturing Company, No. 8, Li-Hsin Rd. 6, Hsinchu Science Park, Hsinchu, Taiwan, ROC
Phone: 886-3-6665157, Fax: 886-3-5637525, and Email: chwang@tsmc.com

Abstract

A 1.4 nm EOT stack film of HfSiON with interfacial oxide layer (IL) is demonstrated with excellent electrical characteristics and reliability for 45 nm node low-power technology. Mobility comparable to SiON is achieved along with adequate nMOS PBTI lifetime, TDDB lifetime, and breakdown voltage (V_{BD}). For the first time, we report lower V_{BD} for the HfSiON stack film despite of 3 orders gate leakage reduction compared to the same EOT SiON. It is attributed to IL breakdown in the proposed two-step breakdown mechanism. This possibly limits the scalability of such a stack film. On the other side, over-drivability of HfSiON with thick underlying oxide boosts input/output (I/O) device performance significantly.

Introduction

High-k gate dielectrics will be required for low-standby-power (LSTP) technology by 2006 according to 2003 ITRS. The requirement has stimulated considerable efforts toward the development of HfSiON gate dielectric due to good mobility of electrons/holes (as SiO_2) and significant reduction of gate leakage [1-2]. Promising transistor performance with a 1.6 nm EOT HfSiON stack film has been achieved [3]. However, electrical and reliability performance of HfSiON gate dielectric suitable for 45 nm node LSTP technology is not simultaneously demonstrated [4]. In this work, we report the electrical and reliability performance of a 1.4 nm EOT HfSiON stack film and investigate its scalability.

nMOS PBTI reliability

C-V simulation of the HfSiON gate dielectric reveals a 1.4 nm EOT after accounting for poly depletion and quantum mechanical effects (Fig. 1). Excellent nMOS PBTI lifetime of 25.6 years at 1.2 V is obtained (Fig. 2). Compared to 1.8 nm EOT HfSiON with superior thermal stability by nitrogen incorporation [5], this thinner film exhibits smaller current drive degradation under the same PBTI stressing electric field (Fig. 3). We attribute this to enhanced immunity to trap generation by PBTI stressing via further optimization of nitrogen incorporation. Cross-sectional TEM shows HfSiON remains in amorphous after ~ 1100°C annealing (Fig. 4).

TDDB, V_{BD}, and Film Scalability

TDDB lifetime of the 1.4 nm EOT HfSiON is plotted against V_G in Fig. 5, and operation voltage at 10-year lifetime is projected to be ~ 2 V. Breakdown voltage is also investigated to study intrinsic breakdown field for over-drive capability and/or burn-in test [6]. The tight distribution of V_{BD} indicates a uniform and high-quality HfSiON gate dielectric (Fig. 6). The averages of V_{BD} are 3.1 V and 3.9 V for nMOS and pMOS respectively.

Also presented in Fig. 6, nMOS V_{BD} of 1.4 nm EOT HfSiON is found to be lower than that of SiON, whereas gate leakage current (J_G) of HfSiON is ~ 3 orders of magnitude lower (Fig. 7). Two-step hard breakdown behavior is clearly observed in Fig. 7 for nMOS with HfSiON stack film — the first breakdown of IL followed by the second breakdown of HfSiON. This is because that at high voltage in inversion, electrons pass through the interfacial oxide band gap, but electrons pass above the conduction band edge of HfSiON [7]. Lower IL V_{BD} resulting from a thinner EOT is responsible for the lower V_{BD}

of the HfSiON stack film. This possibly limits the scalability of a HfSiON stack film.

To further understand the two-step breakdown mechanism, J_G-V_G characteristics of the HfSiON stack film are measured at different stages (Fig. 8). Before interfacial oxide breakdown, J_G remains unchanged (Fig. 8(a)). In post-oxide breakdown stage (Fig. 8(b)), gate leakage current increases significantly and is dominated by HfSiON. In post-HfSiON breakdown stage (Fig. 8(c)), HfSiON breaks down and the stack film behaves like a resistor as shown in the insert of Fig. 8(c). The J_G-V_G characteristics at different stages corroborate the two-step breakdown mechanism.

EOT, J_G, and Mobility

For 1.4 nm EOT HfSiON, J_G is reduced by ~ 3 orders of magnitude compared to SiO_2 and SiON. It is suitable for 45 nm node LSTP technology (Fig. 9). No hysteresis in C-V characteristics (Fig. 10) and superior sub-threshold slopes of ~ 70 mV/decade (Fig. 11) reveal excellent IL/silicon interface and HfSiON bulk film property. As a result, mobility comparable to SiON can be achieved for 1.4 nm EOT HfSiON gate dielectric (Fig. 12).

Device Performance

At a supply voltage of 1.2 V, I_{on} of 751 and 318 μA/μm have been obtained for low-power nMOS and pMOS devices respectively (Fig. 13). Fig. 14 shows simulated relative standby-power (P_{sb}) and relative inverter ring oscillator (F.O. = 3) speed for various generations of LSTP technologies. P_{sb} can be maintained the same by utilizing SiON in 65 nm node to replace SiO_2 in 90 nm node. However, speed enhancement in 45 nm node comes at the cost of 60% P_{sb} increase if no new material introduced. We show that by integrating HfSiON into 45 nm node, P_{sb} is reduced by 50% with 5% speed improvement. This clearly demonstrates the importance of HfSiON gate dielectric for 45 nm node LSTP technology.

We have studied 2.5V I/O devices with HfSiON and thick underlying SiO_2 (~ 4.5 nm). Because of thicker physical thickness of the stack film (EOT ~ 5.0 nm), TDDB lifetime is better than that of 5.0 nm SiO_2 and comparable with that of 6.7 nm SiO_2 (Fig. 15). This enables over-drivability (e.g. 3.3 V) to boost I/O device performance significantly (Fig. 16). Adequate hot carrier lifetime of I/O nMOS is confirmed at 3.3 V over-drive for a 5.0 nm EOT HfSiON stack film (Fig. 17).

Conclusion

Electrical characteristics and reliability of a 1.4 nm EOT HfSiON gate dielectric is demonstrated for 45 nm low-power technology. Lower V_{BD} possibly limits the scalability of a HfSiON stack film. In contrast, over-drive capability for HfSiON I/O devices is confirmed.

References

[1] H. C.-H. Wang et al., IEDM Tech. Dig., p.161, 2004.
[2] Y. Yasuda et al., Symp. on VLSI Tech., p. 40, 2004.
[3] N. Kimizuka et al., Symp. on VLSI Tech., p. 218, 2005.
[4] T. Watanabe et al., IEDM Tech. Dig., p. 507, 2004.
[5] H. C.-H. Wang et al., Symp. on VLSI Tech., p. 170, 2005.
[6] C. Hu, IEDM Tech. Dig., p. 319, 1996.
[7] M. Terai et al., Conf. on SSDM, p. 74, 2004.

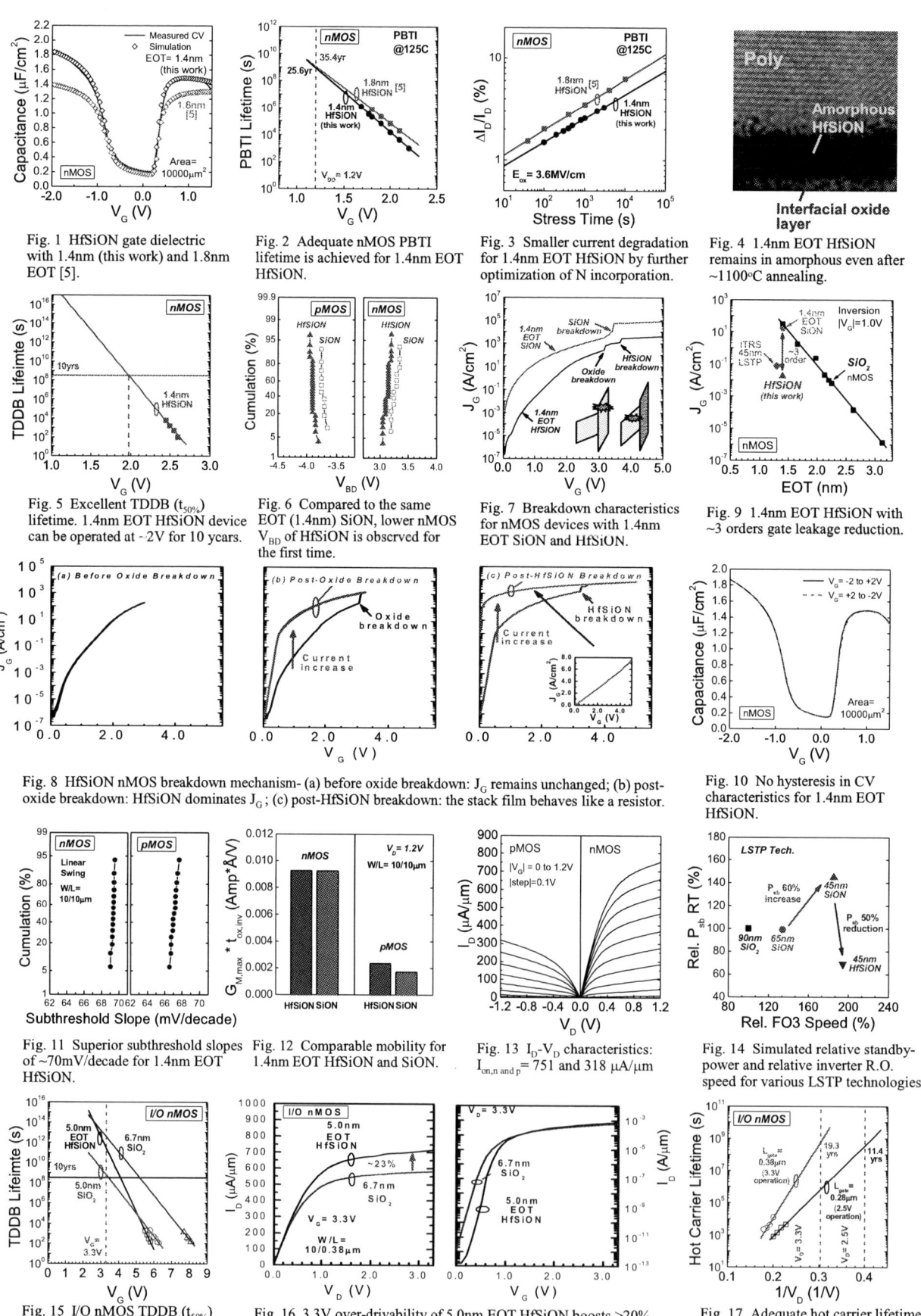

Fig. 1 HfSiON gate dielectric with 1.4nm (this work) and 1.8nm EOT [5].

Fig. 2 Adequate nMOS PBTI lifetime is achieved for 1.4nm EOT HfSiON.

Fig. 3 Smaller current degradation for 1.4nm EOT HfSiON by further optimization of N incorporation.

Fig. 4 1.4nm EOT HfSiON remains in amorphous even after ~1100°C annealing.

Fig. 5 Excellent TDDB ($t_{50\%}$) lifetime. 1.4nm EOT HfSiON device can be operated at -2V for 10 years.

Fig. 6 Compared to the same EOT (1.4nm) SiON, lower nMOS V_{BD} of HfSiON is observed for the first time.

Fig. 7 Breakdown characteristics for nMOS devices with 1.4nm EOT SiON and HfSiON.

Fig. 9 1.4nm EOT HfSiON with ~3 orders gate leakage reduction.

Fig. 8 HfSiON nMOS breakdown mechanism- (a) before oxide breakdown: J_G remains unchanged; (b) post-oxide breakdown: HfSiON dominates J_G; (c) post-HfSiON breakdown: the stack film behaves like a resistor.

Fig. 10 No hysteresis in CV characteristics for 1.4nm EOT HfSiON.

Fig. 11 Superior subthreshold slopes of ~70mV/decade for 1.4nm EOT HfSiON.

Fig. 12 Comparable mobility for 1.4nm EOT HfSiON and SiON.

Fig. 13 I_D-V_D characteristics: $I_{on,n\ and\ p}$= 751 and 318 µA/µm

Fig. 14 Simulated relative standby-power and relative inverter R.O. speed for various LSTP technologies.

Fig. 15 I/O nMOS TDDB ($t_{50\%}$) lifetime of 5.0nm EOT HfSiON is comparable with that of 6.7nm SiO$_2$.

Fig. 16 3.3V over-drivability of 5.0nm EOT HfSiON boosts >20% I/O nMOS current drive compared to 6.7nm SiO$_2$ gate dielectric.

Fig. 17 Adequate hot carrier lifetime confirmed at 3.3V over-drive for 5.0nm EOT HfSiON I/O nMOS device.

2006 International Symposium on VLSI Technology, Systems, and Applications

Detection of Trap Generation in High-κ Gate Stacks due to Constant Voltage Stress

C.D. Young, D. Heh, R. Choi, J.J. Peterson, J. Barnett, B.H. Lee[*], P. Zeitzoff, G.A. Brown, and G. Bersuker

chadwin.young@SEMATECH.org, 512-356-3612 (o), 512-356-7640 (f)
SEMATECH, 2706 Montopolis Drive, Austin, TX, 78741, U.S.A., [*]IBM assignee

Introduction

High-κ gate stack reliability has become one of the critical factors impacting the introduction of advanced gate stacks in future technology nodes. A high density of as-grown electron traps in the transition metal oxides [1] and the presence of the SiO_2 layer at the interface between the high-κ dielectric and the substrate, complicate evaluation of stress-induced defect generation in high-κ gate stacks [2,3]. Effectiveness of stress-induced trap generation in high-κ dielectrics remains a controversial issue, with the conclusions varying with the applied techniques and test conditions [4-10]. Thus, there is a strong need to de-convolute the contributions to trap generation from the interfacial SiO_2 layer (IL) and high-κ film [5-11]. In this study, by using a set of gate stacks with various combinations of the high-κ film and ILs thicknesses, and a developed analysis methodology for charge pumping (CP) data, we identified the IL in the stack as the potential reliability "weak link."

Sample Preparation and Characterization

High-κ/metal gate CMOS transistors were fabricated with atomic layer deposited (ALD) HfO_2 gate dielectrics on interfacial oxide layers (IL) of various thickness with the device parameters given in Table I. Constant voltage stress (CVS) with interspersed "sense" measurements were performed on W/L = 10/1 μm nMOS transistors (Fig. 1). The sense measurements included a stress-induced leakage current (SILC) and fixed amplitude, fixed base CP with a frequency sweep [5-10]; or DC I_d–V_g measurements [9-10]. A –1.5 V DC bias for 10 sec ("discharge") was applied to the gate before each SILC/CP measurement, and at the end of each 300 sec CVS/sense cycle to empty the electron traps prior to the next cycle [5-8].

Characterization Results and Discussion

The result of stress with I_d-V_g sense is shown in Fig. 2 where the V_t shift is mostly reversible, with a tendency to saturate (following a near-exponential dependence). Such V_t behavior is consistent with the process of injected electron trapping/de-trapping on the pre-existing bulk HfO_2 defects [2,7,8]. There is no significant stress-induced defect generation since there is no sizable increase of the V_t shift in the subsequent stress sequences. However, an unrecovered V_t shift of ~ 16 mV is detected. It has been shown that the unrecovered V_t shift can also be estimated from the CP measurements (Fig. 3) based on the trap density (N_t) increase after the stress and "imaging" these traps at the IL/high-κ interface [8]. It has been previously shown that as the CP frequency is reduced, electrically active traps farther away from the dielectric/substrate interface can be detected [9,10,11]. In an effort to correlate charge pumping frequency to measurement probing depth, we have modeled the CP process. To estimate the depth and energy profile of the traps contributing to CP at each given pulse frequency, rise/fall time (t_r/t_f), and amplitude, we have converted the N_t frequency dependence to a depth (with respect to the Si substrate) dependence, which is a function of the dielectric material properties such as the effective masses ($m_{e/h}$), barrier heights ($\Phi_{e/h}$), and capture cross-sections ($\sigma_{e/h}$) of electrons/holes, respectively [8,12]. The CP modeling results indicate that the probing depth may increase up to ~ 1.25 nm as the CP frequency is reduced to 2 kHz (Fig. 3). This depth roughly corresponds to the thickness of the interfacial SiO_2 layer with a starting thickness of 1.1 nm (the final effective thickness is higher due to significant roughness of the high-κ/IL interface [13]). After the stress, the density of the stress-generated traps is seen to increase with closer proximity to the high-κ film, while only a small increase is detected in the vicinity of the Si substrate (Fig. 4). These results point to more effective trap generation in the IL closer to the high-κ dielectric [7,8]. Data in Fig. 5 illustrate the trap generation rate [($N_{stress\ time}$ – $N_{initial}$)/stress time] as a

function of probing depth. Overall the rate increases with the probing depth, but decreases with the stress time. The stress-induced leakage current (Fig. 6) demonstrates a similar trend of a slowing generation rate with stress time (Fig. 5). This suggests that the trap generation occurs at the "precursor" defects, in which the density in the IL increases towards the interface with the HfO_2 (since the overall values increase as the CP measurement depth is increased). This conclusion is further supported by the observation that a majority of the irreversible V_t shift occurs after the first stress/detrapping CVS sequence, Figs. 2 and 4, reflecting on the limited initial supply of the precursor defects available for the conversion into the electron traps.

For further consideration, the 1.1nm SiO_2/3nm HfO_2 stack is compared to the 1.1nm SiO_2/5nm HfO_2 stack to evaluate how longer high-k deposition time may affect the IL. Indeed, one can see from the comparison in Fig. 7 that the stack with the thicker high-κ film shows a higher generation rate for the same stress time, which could be due to longer interaction with the IL associated with additional processing time required to deposit an additional 2 nm of HfO_2.

The CP measurements performed with a higher amplitude while keeping the base value constant at – 0.7 V shows increased N_t, Fig. 8. This can be understood by taking into account that an increased amplitude with fixed base increases the band bending, Fig. 9 (superimposed with the 1.8V amplitude band). This increased bending increases the energy region scanned in the upper half of the band gap during the CP measurement, where, as has been previously shown [14], the density of traps is higher. The charge pumping spatial and energy depth profile superimposed on the gate stack band diagram clearly shows that the stress-generated traps sensed by CP are relatively deep traps, mostly confined within the IL and IL/high-κ interface region, and not the pre-existing, bulk high-κ traps responsible for the reversible fast transient V_t instability.

With this understanding, the stress with CP measurement can be used to study the impact of gate stack processing on the IL. To correlate defect generation to the stress conditions in the IL and high-κ film, the stress voltages for the gate stacks with 3 nm and 5 nm ALD HfO_2 layers deposited on the 1 nm or 1.4 nm SiO_2 layer were selected to match electric fields across the IL. For the thicker IL, the depth profile of the N_t values is shifted towards the high-κ layer (Fig. 10): Both the initial and stress-induced N_t profiles in the 1.4nm SiO_2/3nm HfO_2 stack appears to be parallel shifted by about 0.3 – 0.4 nm with respect to the 1.1nm SiO_2/3nm HfO_2 data. This is consistent with the expectation that a similar HfO_2 deposition process would results in a similar effect on the underlying IL from the generated defect density profile standpoint. This is schematically illustrated in Fig. 11 where we took into account that the high-κ/IL interface is rather rough which significantly increases the effective thickness of the interfacial layer, in some instances by up to 0.5nm, with respect to its pre-high-κ deposition values [13]. Interaction with the high-κ film was suggested [4,15] to generate structural defects, in particular, Oxygen vacancies, which might represent the "precursor" defects exhibiting higher density in closer proximity to the high–κ layer. These precursor defects might be converted into electron traps by stress.

Summary

Using a set of HfO_2 gate stacks with different high-κ and IL thicknesses, we have determined that the stress-generated electron traps are located primarily within the IL. This study indicates that the traps are generated on the "precursor" defect sites, presumably formed due to an interaction of the IL with the high-κ film. Under the low voltage stress conditions of practical interest, these generated traps provide very limited contribution (compared to the pre-existing traps) to the threshold voltage instability.

References

[1] G. Bersuker, et al., Materials Today, p. 26, 01/2004
[2] B.H. Lee, et al., IEEE TDMR, Vol. 5, p. 20, 2005
[3] G. Ribes, et al., IEEE TDMR, Vol. 5, p. 5, 2005
[4] G. Bersuker, et al., JJAP, vol. 43, p. 7899, 2004
[5] R. Degraeve, et al., IEDM Proc., p. 935, 2003
[6] F. Crupi, et al., IRPS Proc., p. 181, 2004
[7] C.D. Young, et al., Micro. Rel., vol. 45, p. 806, 2005
[8] C.D. Young, et al., presented at IIRW, 2005
[9] A. Kerber, et al., IRPS Proc., p. 41, 2003
[10] T.H. Hou, et al., IRPS Proc., p. 581, 2004
[11] H.M. Bu, et al., as discussed at SISC 2004
[12] Y. Maneglia, et al., APL, Vol. 79, p. 4187, 1996
[13] M.P. Agustin, submitted to JAP, 2005
[14] J.-P. Han, VLSI Tech Digest, p. 161, 2003
[15] G. Bersuker et al., ECS PV2005-05, p.141, 2005

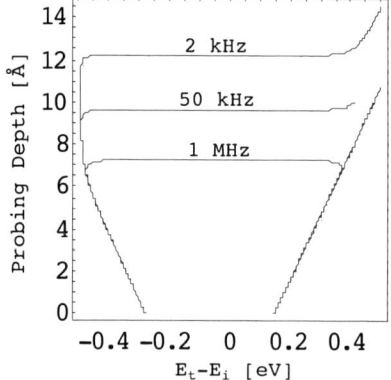

Fig. 3. 2-D spatial contours as a function of frequency for 1 MHz, 50 kHz, and 2 kHz.

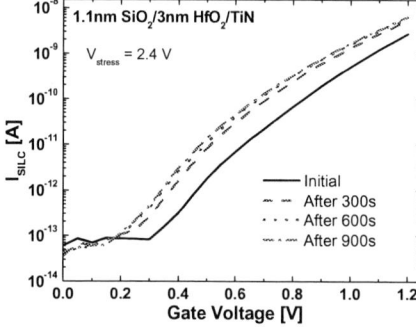

Fig. 6. SILC result correlates with Fig. 5 where the initial 300s shows the largest increase with only subtle increases thereafter since less traps are generated with increasing time.

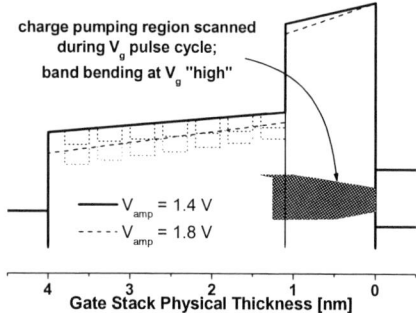

Fig. 9. The band diagram at V_g "high" for the two CP sense amplitudes. For a fixed base level, increased amplitude allows an increased CP scan range resulting in larger sensed N_t values.

Table I. Parameters extracted from NCSU CVC for the SiO_2/HfO_2 gate stacks used in this study.

Gate Stack IL $SiO_2/HfO_2/TiN$	EOT [nm]	V_{fb} [V]	V_t [V]	J_g (V_{fb}–1) [-A/cm²]
1.1 nm/3 nm	1.33	-0.67	0.60	2.85E-3
1.1 nm/5 nm	1.65	-0.69	0.68	1.38E-6
1.4 nm/3 nm	1.68	-0.64	0.61	1.86E-3

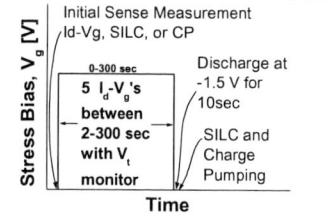

Fig. 1. Stress: CVS in inversion. Sense: interspersed I_d-V_g; or "discharge" for 10sec, with SILC and CP; repeat.

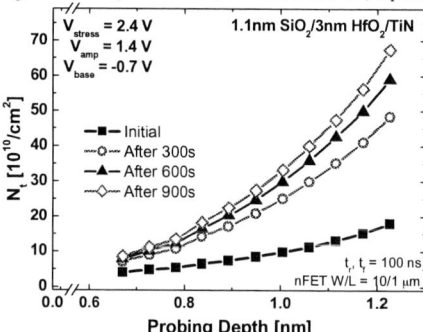

Fig. 4. Example of CP N_t data v. probing depth before and after a stress sequence with V_g = –1.5V discharge as in Fig. 1. As depth increases, trap generation is detected.

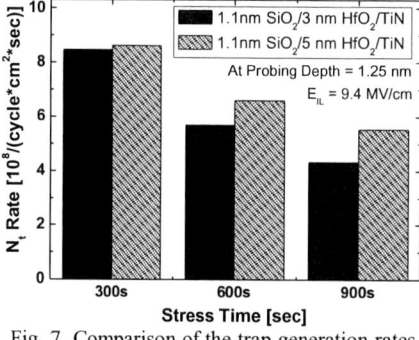

Fig. 7. Comparison of the trap generation rates for two high-κ thicknesses on a similar 1.1nm IL illustrating higher N_t values for the thicker high-κ film at a probing depth of 1.25 nm.

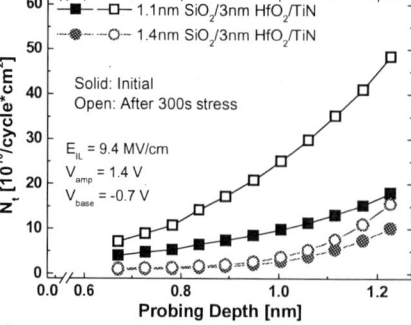

Fig. 10. N_t vs. probing depth before/after a 9.4 MV/cm stress. Trap distribution profiles in stacks with different ILs are shifted approximately by the difference in IL thickness.

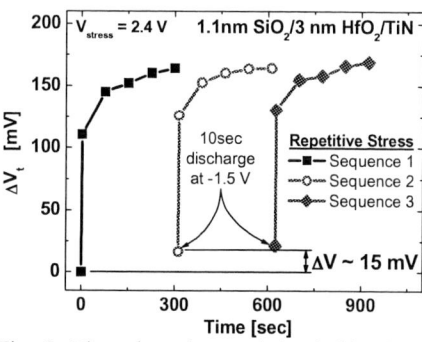

Fig. 2. Time dependence of threshold voltage during CVS. A discharge at V_g = –1.5V for 10 sec was performed before each stress sequence. Unrecovered V_t of ~15 mV is detected.

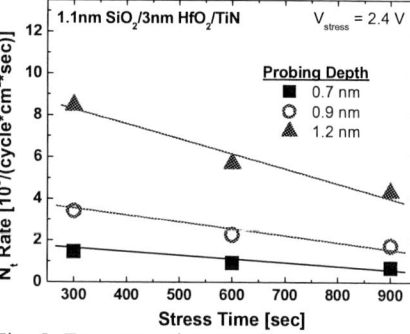

Fig. 5. Trap generation rate decrease with time suggests generation occurs at "precursor" defects unlike what typically happens for the SiO_2 case (not shown).

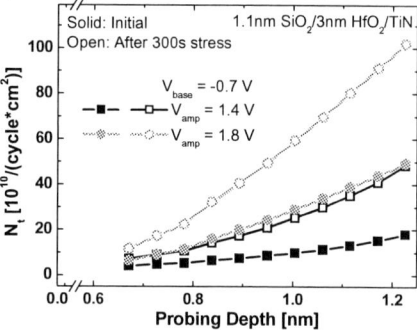

Fig. 8. CP N_t values v. probing depth for different CP sense amplitudes where a larger amplitude measures greater N_t suggesting larger trap densities in the upper half the band gap [19].

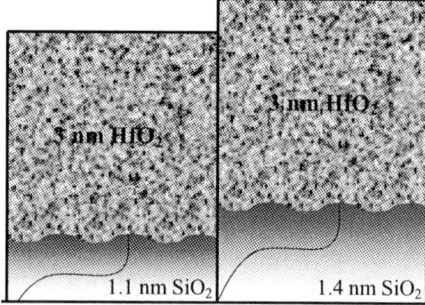

Fig. 11. Schematic representation of the gate stacks with different IL thicknesses. Shaded transition region corresponds to the portion of IL affected by interaction with the high-κ film [6].

112

2006 International Symposium on VLSI Technology, Systems, and Applications

Relationship of HfO_2 Material Properties and Transistor Performance

P. D. Kirsch[a], M. A. Quevedo[b], G. Pant[e], S. Krishnan, S. C. Song, H. J. Li[c], J. J. Peterson[d], B. H. Lee[a], R. W. Wallace[e], M. Kim[e] and B. E. Gnade[e]

SEMATECH, 2706 Montopolis Drive, Austin, Texas 78741, Assignments from: a) IBM, b) TI, c) Infineon, d) Intel

[e]Department of Electrical Engineering, University of Texas at Dallas, Richardson, Texas 75080

Ph: 512-356-7195; fax: 512-356-7640; e-mail: Paul.Kirsch@sematech.org

ABSTRACT

We report on the relationship between the materials science of a HfO_2/TiN stack and transistor performance. Atomic layer deposited (ALD) HfO_2 can be scaled to a physical thickness of 1.2 nm resulting in EOT=1.0 nm. In scaling HfO_2, the interfacial SiO_x layer (IL) is also scaled and the extent of HfO_2 crystallization is reduced. Reduced HfO_2 crystallinity is coincident with reduced threshold voltage instability (10 mV) and increased electron mobility (82% Univ. SiO_2). For these stable, high mobility devices, we find that HfO_2 can coordinate N as Hf-N without excessive nitridation of the IL.

INTRODUCTION

SiON gate dielectric scaling for complementary metal oxide semiconductor (CMOS) applications has slowed significantly. SiON has scaled to T_{phys}=1.2 nm for the 90 nm node [1], but is unlikely to scale further at the 65 nm node due to excessive gate leakage [2]. Despite the limitations of SiON, gate dielectric scaling is necessary, making high-permittivity (high-k) dielectrics an important area of research. Recently good device results have been reported for aggressively scaled HfO_2 (equivalent oxide thickness [EOT]=1.0 nm) [3, 4]. However little is known about the relationship between HfO_2 material properties and device results, especially as HfO_2 scales to T_{phys} < 2 nm. We report on the relationship between material properties and device performance in TiN/HfO_2 gate stacks.

EXPERIMENT

Atomic layer deposited (ALD) HfO_2 was grown with alternating pulses of $Hf[N(CH_3)C_2H_5]_4$ and O_3. Post-deposition annealing was done in N_2 or NH_3. The metal gate electrode was ALD TiN. The TiN was capped with a-Si followed by gate patterning, ion implantation, and activation anneal at 1000°C-5s. Blanket wafers were studied with Fourier transform infrared spectroscopy (FTIR), X-ray photoelectron spectroscopy (XPS), and 0.5° grazing incidence X-ray diffraction (GIXRD). Transmission electron microscopy (TEM) was done on device and blanket wafers. EOT and mobility values were extracted from measured capacitance-voltage (C-V) and current-voltage (I-V) data using the NCSU model. I-V characteristics were measured using 10×1 μm FETs with channel doping of ~2×10^17 B/cm^3. Mobilities were extracted using both DC and pulsed I-V (not shown) data.

RESULTS AND DISCUSSION

Fig. 1 shows FTIR results for ALD HfO_2 on blanket wafers prepared with HF last chemistry. Features indicative of SiO_2 (1000-1300 cm^-1) increase concurrently with HfO_2 features (600-800 cm^-1). Increase in the feature at 780 cm^-1 is expected with additional ALD cycles due to the thicker HfO_2 film. However, the additional SiO_2 intensity (1220 cm^-1) suggests that the interfacial oxide layer (IL) also grows due to the additional O_3 ALD half-cycle exposures. This result suggests that a thinner ALD HfO_2 film may minimize IL growth. Fig. 2 shows high resolution TEM images on fully processed transistor gate stacks (1000°C-5s) that are consistent with FTIR results. Fig 2(a) shows that the IL is 0.9 nm while Fig. 2(b) shows that the thicker HfO_2 film leads to a 1.4 nm IL. These results have also been observed for ALD HfSiO [5]. In addition to IL thickness changes observed with HfO_2 scaling, crystalline morphology changes. Fig. 3(a) shows a plan view TEM image of a 1.8 nm HfO_2 film. This image does not show clear evidence of large, 10 nm crystallites. Conversely, 5.2 nm HfO_2 in Fig. 3(b) shows clear lattice fringes and grain sizes near 10 nm. To further understand HfO_2 crystallinity, GIXRD was done on HfO_2 films capped with TiN/polySi, annealed to 1000°C-5s. The TiN/polySi cap was then removed with a wet etch and probed with a diffractometer. The top spectrum shows a 13 cycle (1 nm) HfO_2 reference film annealed to 1100°C-120s. This reference spectrum illustrates the ability of the diffractometer to detect crystallinity in ~1 nm films. The same film annealed to 1000°C-5s (bottom spectrum) does not show evidence of crystallinity in agreement with TEM. However, as the HfO_2

thickness exceeds 25 cycles (~2 nm), evidence of the monoclinic and tetragonal phases are seen.

The possible significance of these morphology results may be seen in Fig. 5 and Fig. 6. Threshold voltage (V_{TH}) instability and electron mobility improve for lesser HfO_2 physical thickness (T_{phys}) and may be related to lesser crystallinity and fewer grain boundaries. Because grain boundaries are known to be sites of incomplete bonding and therefore increased interfacial energy [7], reduced crystallinity may help address defect issues in HfO_2. The lesser degree of HfO_2 crystallinity may reduce charge trap-related defects such as O vacancies [8] or H-related defects [9]. Additional mobility data points are shown in Fig. 7 to further illustrate the improvement in peak and high field (1 MV/cm) mobility with reduced T_{phys}. As shown in Fig. 8, the improved long channel mobility and T_{inv} achieved with HfO_2 scaling results in improved linear drive current ($I_{d,lin}$) for the devices tested (0.1μm<L_g<1 μm). To separate the concurrent benefits of improved mobility and T_{inv} scaling, the inset in Fig. 8 shows $I_{d,lin}$ normalized to T_{inv}. As expected, 1.8 nm HfO_2 is superior.

To probe the IL quality, charge pumping (not shown) was done. Results show that HfO_2 devices had peak interface trap densities (N_{it}) in the mid 10^{10} cm^-2 range for 1.2 nm < T_{phys} < 3.3 nm. To further understand the IL, Fig. 9 shows N 1s XPS results for HfO_2 immediately after nitridation (a) and after 1000°C-5s anneal (b). Sample (b) was TiN/PolySi capped, annealed, wet etched, and probed as described earlier. From XPS, we estimate that 9% N is present in (a) and 8% N is present in (b). Trace (a) shows that the majority of the N is initially bound as Hf-N (396 eV). However, after capping with TiN/polySi, annealing to 1000°C-5s, and cap removal, the N is in both the Si-N and Hf-N chemical states [trace (b)]. The N chemical state redistribution may be attributed to the 1000°C-5s anneal. These XPS results for HfO_2 are significant because they show that N remains in the bulk of the HfO_2 and the IL is lightly nitrided SiO_x (~4% N). This light nitridation allows for EOT scaling without negatively impacting mobility.

Because mobility, EOT, and ultimately drive current have been shown to improve with reduced HfO_2 T_{phys}, an estimate of the T_{phys} scaling limit is important. Fig. 10 shows the C-V data for the HfO_2 scaling series to T_{phys}=1.2 nm. It is seen that the C-Vs are well behaved until T_{phys}=1.2 nm where the curve rolls off in accumulation and little EOT scaling is achieved. A TEM of the T_{phys}=1.2 nm is shown in the inset to Fig. 6. Fig. 11 shows the analogous leakage current-gate voltage (J_g-V_g) results for the scaling series. Note that as T_{phys} scales, the breakdown event occurs sooner and becomes softer as has been observed in the SiO_2 case [10]. The J_g (V_{fb}-1)-EOT relationship in Fig. 12 illustrates that at T_{phys}=1.2 nm, gate leakage current increases ~ 5× without significant EOT scaling, suggesting proximity to the practical scaling limit.

CONCLUSIONS

We have reported FTIR, TEM, GIXRD and XPS results alongside EOT scaling, threshold voltage stability and mobility data. ALD HfO_2 can be scaled to a physical thickness of 1.2 nm enabling EOT = 1 nm. In doing so, the interfacial SiO_x layer is also scaled and the extent of crystallinity is reduced. Furthermore, we find that HfO_2 can coordinate N as Hf-N without excessive nitridation of the IL. The positive attributes of scaling such as reduced crystallinity are concurrent with improved threshold voltage instability (10 mV) and mobility (82% Univ. SiO_2).

REFERENCES

[1]   Thompson et al., *IEDM*, p. 61, 2002.
[2]   Bai et al., *IEDM*, p. 657, 2004.
[3]   Doris et al., *VLSI Symp*, p. 214, 2005.
[4]   Kirsch et al., *ESSDERC*, p. 367, 2005.
[5]   Quevedo et al., *to be presented at* IEDM, Dec. 2005
[6]   Datta et al., *IEDM*, p. 653, 2003.
[7]   Callister, *Material Science and Engineering*, 3rd Ed. p. 77, 1994.
[8]   Takeuchi et al., *APL*, Vol 87, 062105, 2005.
[9]   Kang et al., *APL*, Vol 84, 3894, 2004.
[10] Timp et al., *IEDM*, p. 615, 1998.

1-4244-0181-X/06/$20.00 ©2006 IEEE   113

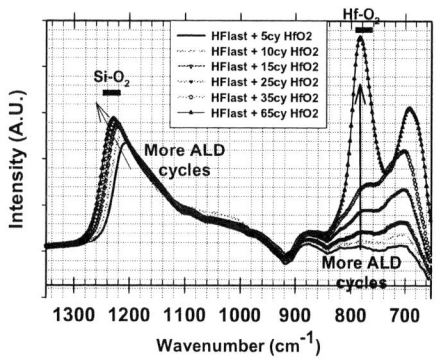

Fig. 1: Increased HfO_2 thickness (illustrated by arrow) also results in thicker interfacial SiO_2.

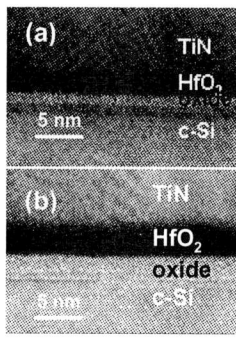

Fig. 2: HRTEM of transistor gate stacks (1000°C-5s activation) showing (a) 0.9 nm IL for 1.8 nm HfO_2 and 1.4 nm IL for 3.3nm HfO_2

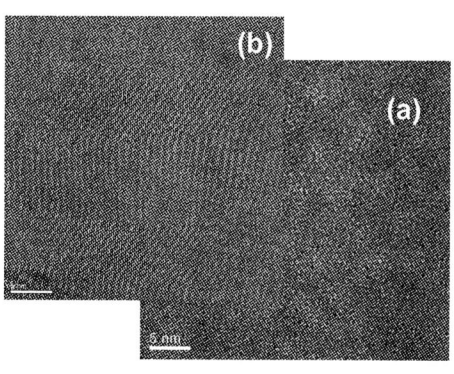

Fig. 3: Plan view TEM for (a) 1.8 nm HfO_2 and (b) 5.2 nm. Only the thick film (b) shows clear lattice fringes and 10 nm grains.

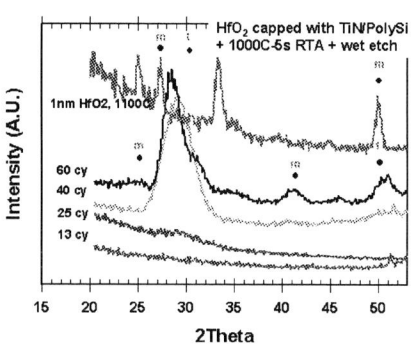

Fig. 4: GIXRD shows HfO_2 crystallinity increases with T_{phys}. T_{phys} range is 1 nm (13 cy) to 5 nm (60 cy).

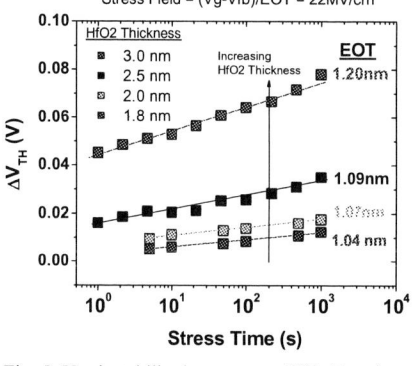

Fig. 5: V_{TH} instability improves as HfO_2 T_{phys} is scaled thinner possibly related to reduced crystallinity.

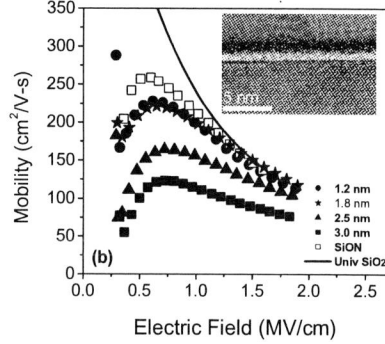

Fig. 6: Peak / high field mobility improve as HfO_2 T_{phys} is scaled thinner possibly related to reduced crystallinity. HRTEM of 1.2 nm HfO_2 (inset).

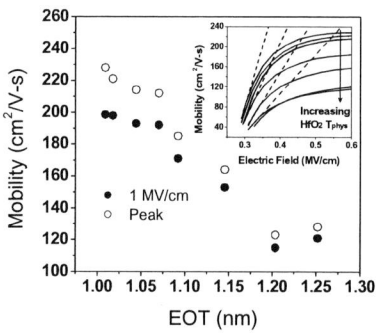

Fig. 7: Peak / high field mobility for HfO_2/TiN stacks. EOT scaling is achieved by reducing $HfO_2 T_{phys}$. Inset shows effective mobility improvement for coulomb scattering limited region.

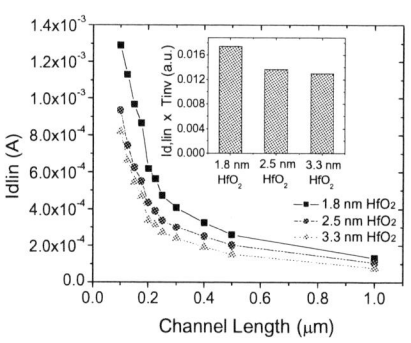

Fig. 8: Id,lin improvement with 1.8 nm HfO_2 vs. thicker films. Normalization of Id,lin to Tinv also shows scaling benefit (inset).

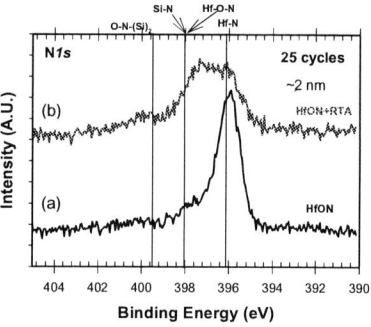

Fig. 9: N 1s XP spectra for HfO_2 (a) after nitridation and (b) after 1000°C-5s. Hf-N intensity (BE=396 eV) shows that N remains in HfO_2 after 1000°C-5s.

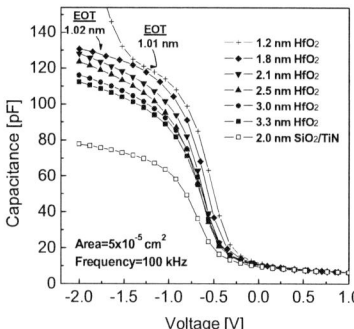

Fig. 10: C-V results for HfO_2 ranging from 1.2 nm < T_{phys}<3.3 nm.

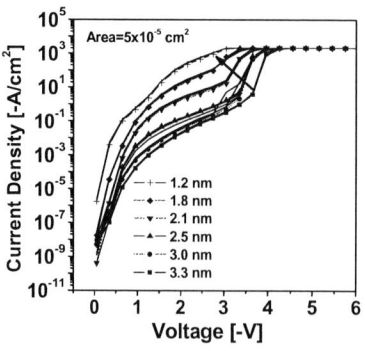

Fig. 11: Gate current density for the HfO_2 thickness series in accumulation.

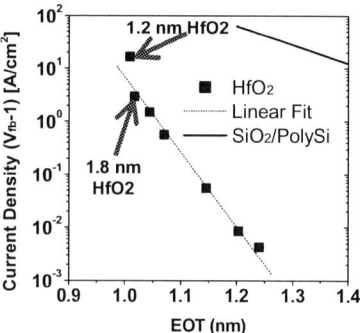

Fig. 12: Relationship between gate current density and EOT. Deviation from trendline at T_{phys}= 1.2 nm suggests proximity to material limit.

2006 International Symposium on VLSI Technology, Systems, and Applications

$Hf_xTa_yN_z$ Metal Gate Electrodes for Advanced MOS Devices Applications

Chin-Lung Cheng[1], Kuei-Shu Chang-Liao[2*], Tzu-Chen Wang[2], Tien-Ko Wang[2],and Howard Chih-Hao Wang[3]

[1]Institute of Mechanical and Electro-Mechanical Engineering, National Formosa University, Huwei, Taiwan, R.O.C.
[2]Department of Engineering and System Science, National Tsing Hua University, Hsinchu, Taiwan, R.O.C.
[*]TEL: (886)-(3)-5742674, FAX: (886)-(3)-5720724, E-mail: Lkschang@ess.nthu.edu.tw
[3]Taiwan Semiconductor Manufacturing Company, Hsinchu, Taiwan, R.O.C

ABSTRACT

With a small amount of hafnium (Hf) incorporated into tantalum nitride (TaN) metal gate, excellent thermal stability of electrical properties can be achieved up to post-metal anneal at 950 °C for 45 s. Besides, a work function tuning range from 4.2 to 4.6 eV can be obtained by suitably adjusting nitrogen (N) concentration in $Hf_xTa_yN_z$ metal gate.

INTRODUCTION

The thermally stable metal gate electrodes are desirable to supersede polycrystalline-silicon-gate electrode for reducing the gate depletion and dopant penetration problems. However, the work function of TaN/SiO_2 and HfN/SiO_2 structures are increased with increasing temperature and finally merged to 4.7 ~ 4.8 eV [1], [2]. In this work, the composition effects of Hf, Ta and N in $Hf_xTa_yN_z$ metal gate on the thermal stability and work function tuning of MOS devices were investigated.

EXPERIMENT

MOS capacitors with $TaN/Hf_xTa_yN_z/SiO_2/Si$ structure were fabricated. A SiO_2 layer with 3 nm in physical thickness was thermally grown on the cleaned wafers. Next, $Hf_xTa_yN_z$ (~50 nm) films were deposited using reactive dc magnetron cosputtering of 99.9% pure Hf and Ta targets in an Ar/N_2 ambient with various powers. The fabrication details for all samples are summarized in Table 1. A 100 nm TaN film was then deposited via sputtering to serve as capping layer and to prevent the top surface of $Hf_xTa_yN_z$ from being oxidized during post-metal anneal (PMA). Subsequently, PMA using rapid thermal annealing was carried out in N_2 gas for 45 s at 750, 850 and 950 °C, respectively. A TaN/Hf_xTa_yN gate stack was then patterned by helicon-wave plasma etching. A 500 nm thick Al film was deposited on the backsides of all samples.

RESULTS AND DISCUSSION

A. The composition effects of Hf and Ta in Hf_xTa_yN metal gate

Theory tells that ΔJ_T vs. voltage (V) and ΔJ_v vs. voltage (V) curves would each exhibit a peak when the applied voltage is equivalent to the barrier at the anode [3]. By using the dependence of ΔJ_T and ΔJ_v on gate voltage, the work function of $Hf_{0.40}Ta_{0.46}N_{0.14}$ gate electrode at PMA=850 °C was estimated to be 4.53 eV as shown in Fig. 1. Figure 2 (a) shows that by suitably adjusting the Hf and Ta contents in the Hf_xTa_yN films, thermally stable work function and a midgap work function value are achievable, making it a suitable candidate for FD-SOI transistors. Figure 2 (a) also indicates that a decrease of Hf content in Hf_xTa_yN gate electrode leads to an increase of work function. Figure 2 (b) shows the increase of Ta content in Hf_xTa_yN gate electrode leads to the decrease of EOT due to the combination of physical thinning of SiO_2 and formation of a high-k layer. The relatively low EOT of the $Hf_{0.27}Ta_{0.58}N_{0.15}/SiO_2$ gate stack could be attributed to the fact that a Ta-based gate dielectric possesses a higher dielectric constant than a Hf-based one. To investigate the PMA induced electrical-property variation, the conductance (G_p) as a function of frequency (ω) for all samples after various PMA temperatures were measured and plotted as $G_p/\omega A$ versus ω by biasing the Si surface in depletion condition. The results shown in Fig. 3 suggest that the Interface trap density (Dit) for $Hf_{0.27}Ta_{0.58}N_{0.15}$ sample is the lowest and the thermal stability can be maintained up to 950 °C for 45 s. Figure 4 reveals the defect generation rate (P_g) as a function of PMA temperature and a MOS device with Ta-rich Hf_xTa_yN gate electrode (e.g., the $Hf_{0.27}Ta_{0.58}N_{0.15}$ sample) possesses significantly lower P_g. A high-temperature annealing could lead to the creation of extrinsic states, which is usually associated with the bonding defects at the metal/dielectric interface. Figure 5 indicates that the diffusion of Hf and Ta for $Hf_{0.27}Ta_{0.58}N_{0.15}$ gate electrode is insensitive to PMA temperature and Hf has higher diffusion capability than Ta. This may provide the physical explanation of the best electrical properties and the lowest electrical thickness of the $Hf_{0.27}Ta_{0.58}N_{0.15}$ sample.

B. The composition effects of N in $Hf_xTa_yN_z$ metal gate

Figure 6 (a) shows that a work function tuning range from 4.2 to 4.6 eV can be achieved by suitably adjusting N concentration in $Hf_xTa_yN_z$ metal gate. The shifts on the values of EOT and J_g for $Hf_xTa_yN_z$ metal gate with various nitrogen compositions are slight as shown in Fig. 6 (b). Figure 7 (a) shows that the depth of Hf and Ta diffusion is decreased with increasing nitrogen concentration in $Hf_xTa_yN_z$ metal gate. The suppression of Hf and Ta diffusion in $Hf_xTa_yN_z$ metal gate was attributed to the presence of N in the gate electrode, which can act as a diffusion barrier. Negligible variation on D_{it} for $Hf_xTa_yN_z$ metal gate with various nitrogen compositions has been observed upon PMA treatment up to 950 °C for 45 s as shown in Fig. 7 (b).

CONCLUSIONS

This work demonstrated a thermally robust high-quality $Hf_{0.27}Ta_{0.58}N_{0.15}$ gate electrode for advanced MOS device applications. With a proper amount of Hf , Ta and N in $Hf_xTa_yN_z$ metal gate, the work function can be adjusted from 4.2 to 4.6 eV. Negligible variation of the electrical characteristics including work function, EOT, interface trap density, and defect generation rate for $Hf_{0.27}Ta_{0.58}N_{0.15}$ gate electrode has been observed after a PMA up to 950 °C for 45 s.

REFERENCES

[1] M. S. Joo et al., "Thermal instability of effective work function in Metal/high-k stack and its material dependence," IEEE Electron Device Lett., vol. 25, (2004) 716.

[2] C. Ren et al., "Fermi-level pinning induced thermal instability in the effective work function of TaN in TaN/SiO_2 gate stack," IEEE Electron Device Lett., vol. 25, (2004) 123.

[3] S. Zafar, et al., "A method for measuring barrier heights, metal work functions and fixed charge densities in $metal/SiO_2/Si$ capacitors," *Appl. Phys. Lett.*, vol. 80, pp. 4858-4860, 2002.

1-4244-0181-X/06/$20.00 ©2006 IEEE

Table 1 The detailed process conditions of $Hf_xTa_yN_z$ metal gate.

Samples (XPS analysis)	Power of Ta	Power of Hf	Ar/N_2 (sccm)	PMA (°C)
$Hf_{0.48}Ta_{0.26}N_{0.16}$	150 W	200 W	24/4.3	750
	150 W	200 W	24/4.3	850
	150 W	200 W	24/4.3	950
$Hf_{0.40}Ta_{0.46}N_{0.14}$	150 W	150 W	24/4.3	750
	150 W	150 W	24/4.3	850
	150 W	150 W	24/4.3	950
$Hf_{0.27}Ta_{0.58}N_{0.15}$	150 W	100 W	24/4.3	750
	150 W	100 W	24/4.3	850
	150 W	100 W	24/4.3	950
$Hf_{0.21}Ta_{0.40}N_{0.39}$	150 W	100 W	24/8	750
	150 W	100 W	24/8	850
	150 W	100 W	24/8	950
$Hf_{0.15}Ta_{0.34}N_{0.51}$	150 W	100 W	24/16	750
	150 W	100 W	24/16	850
	150 W	100 W	24/16	950

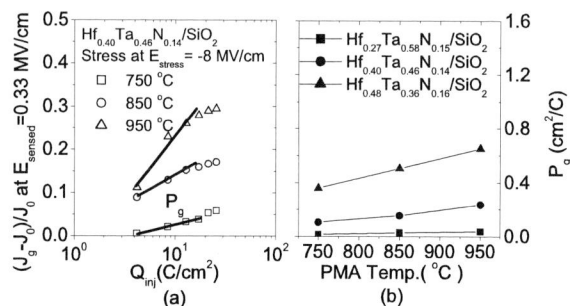

FIGURE 4. (a) Stress-induced leakage current (SILC) $[(J_g-J_0)/J_0]$ as a function of injected charge (Q_{inj}) for various samples under constant-voltage condition at room temperature. (b) The defect generation rate (P_g) as a function of PMA temperature..

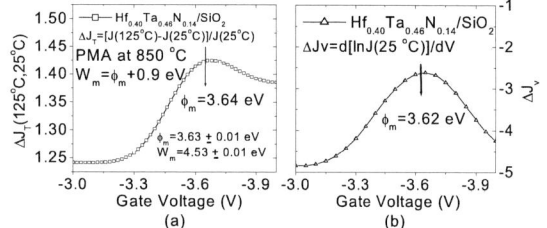

FIGURE 1. The dependence of (a) ΔJ_T and (b) ΔJ_v on gate voltage. The ΔJ_T and ΔJ_v are defined as: $\Delta J_T = [J(125\ °C)-J(25\ °C)]/J(25\ °C)$ and $\Delta J_v = d[lnJ(25\ °C)]/dV$, respectively.

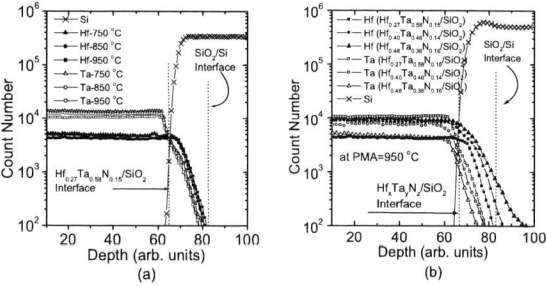

FIGURE 5. (a) The profiles of $Hf_{0.27}Ta_{0.58}N_{0.15}$ metal gates at different PMA treatments (b) of Hf_xTa_yN metal gate with various Hf and Ta compositions at PMA= 950 °C, analyzed by time-of-flight secondary ion mass spectroscopy (TOF SIMS).

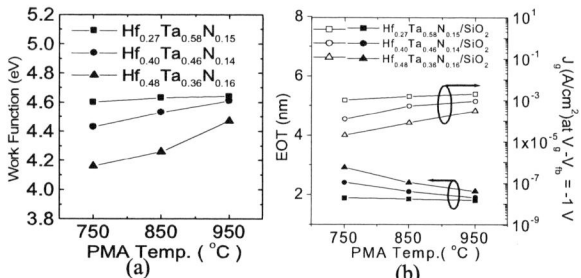

FIGURE 2. (a) Work functions of Hf_xTa_yN metal gate (b) The EOT and J_g at V_g-V_{fb}=-1 V of MOS devices with various Hf and Ta compositions in $Hf_xTa_yN_z$ metal gate as a function of PMA temperature.

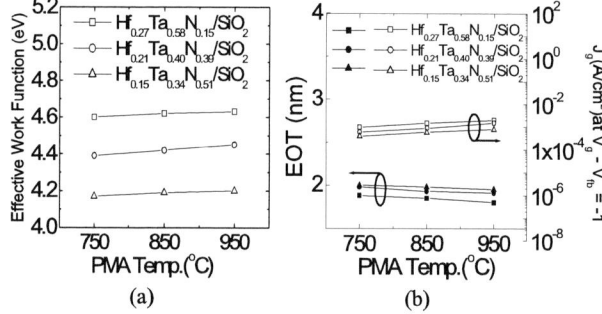

FIGURE 6. (a) Work functions of $Hf_xTa_yN_z$ metal gate (b) The EOT and J_g at V_g-V_{fb}=-1 V of MOS devices with various nitrogen compositions in $Hf_xTa_yN_z$ metal gate as a function of PMA temperature.

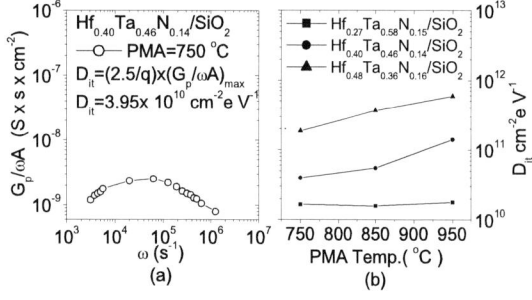

FIGURE 3. (a) The conductance of $Hf_{0.40}Ta_{0.46}N_{0.14}/SiO_2/Si$ capacitor after PMA=750 °C is measured as a function of frequency and plotted as $G_p/\omega A$ versus ω by biasing the Si surface in depletion condition. (b) Interface traps density (D_{it}) of MOS capacitors with various Hf and Ta compositions in $Hf_xTa_yN_z$ metal gate after various PMA treatments.

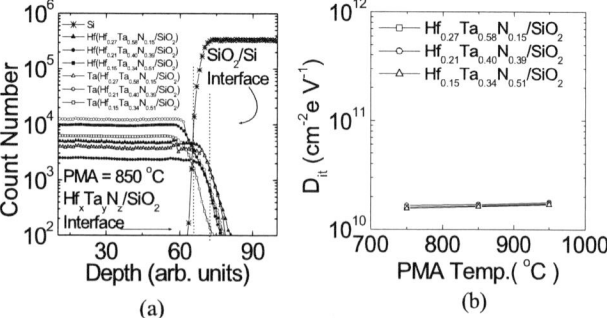

FIGURE 7. (a) Various elements profile of $Hf_xTa_yN_z$ metal gate with various nitrogen compositions at PMA= 850 °C analyzed by TOF SIMS. (b) D_{it} of MOS capacitors with various nitrogen compositions in $Hf_xTa_yN_z$ metal gate after various PMA treatments.

2006 International Symposium on VLSI Technology, Systems, and Applications

Impact of WSi$_x$ Metal Gate Stoichiometry on Fully Depleted SOI MOSFETs Electrical Properties

J. Widiez*, M. Vinet, B. Guillaumot*, X. Garros, S. Minoret, T. Poiroux, O. Weber, L. Thevenod, P. Holliger, B. Previtali, V. Barral, K. Sidi Ali Cherif, P. Grosgeorges, A. Toffoli, S. Maîtrejean, M. Cassé, F. Martin, D. Lafond, O. Faynot, M. Mouis** and S. Deleonibus.

CEA-DRT-LETI - 17 rue des Martyrs 38054 Grenoble Cedex 9, France, email: julie.widiez@cea.fr
*also with STMicroelectronics, 850 rue J. Monnet 38926 Crolles, France **IMEP, ENSERG, 23 rue des Martyrs, 38016 Grenoble, France.

ABSTRACT

For the first time, we report fully depleted SOI MOS transistors with WSi$_x$ gate on HfO$_2$. Gate work function, dielectric properties and channel mobility are presented in terms of Si/W ratio and compared to TiN gate devices. A 35% electron mobility gain was obtained with a WSi$_x$ gate device as compared to a TiN gate transistor. It was found that both mobility and dielectric characteristics were drastically improved by decreasing the Si/W ratio of WSi$_x$ films.

INTRODUCTION

Metal gates become necessary to adjust the threshold voltage of FD devices. So far, metal gates with work functions close to mid-gap, such as TiN [1], TaSiN [2], W [3], NiSi [4] have been reported. In the past, WSi$_x$ material has been widely studied for tungsten polycide gates (WSi$_x$/poly-Si) [5-7] but only few papers deal with WSi$_x$ directly deposited on the gate dielectric [8-11]. In this paper, WSi$_x$ is used as gate material in a MOSFET. We compare the characteristics of FD SOI transistors fabricated with Si-rich CVD-WSi$_{2.7}$, W-rich CVD-WSi$_2$ and CVD-TiN metal gates on HfO$_2$ and SiO$_2$. Gate work function, dielectrics properties and long-channel carrier mobility are studied.

EXPERIMENT

Devices were fabricated on SOI wafers. The transistor channel was left undoped (Na=10^{15}cm^{-3}). A damascene replacement gate process [12] was used. Source and drain epitaxy was performed to minimize the external resistance. Thin films of HfO$_2$ were deposited using an Atomic Layer Deposition (ALD) process with a (HfCl$_4$,H$_2$O) chemistry. Post deposition annealing was performed at 600°C in N$_2$ followed by the metal gate deposition. A 40nm CVD WSi$_x$ layer was deposited using WF$_6$ and Cl$_2$H$_2$Si precursors followed by an Ar anneal at 800°C necessary to thermally stabilize the WSi$_x$ layer [13]. The deposition temperature was used to modulate the Si/W ratio from 2 at 450°C to 2.7 at 550°C (Fig.1&2). SEM and AFM images (Fig. 1&3) show that WSi$_2$ has smaller grain size compared to WSi$_{2.7}$. XRD analyses show that the tetragonal WSi$_2$ crystalline phase exists in both materials. However, lowering the deposition temperature induces the formation of the W$_5$Si$_3$ crystalline phase (Fig. 4). Reference TiN gate devices were made with a 10nm CVD TiN deposited at 680°C using TiCl$_4$ and NH$_3$ precursors and encapsulated with a 50nm highly doped polysilicon layer. SiO$_2$ devices were used as controls.

RESULTS AND DISCUSSION

A. Gate work function: All long-channel devices show perfect sub-threshold characteristics (Fig. 5). WSi$_2$ and WSi$_{2.7}$ gate devices exhibit the same threshold voltage. This is expected since the work function is determined by the interface properties and both materials have the same 2.7 Si/W ratio at the interface between the metal and the oxide (Fig. 2). A positive 200mV shift in the threshold voltage is observed between TiN and WSi$_x$ metal gate devices. The effective WF of WSi$_x$ and TiN on HfO$_2$ were determined to be 4.55eV and 4.75eV (Table 1). On SiO$_2$, the TiN and WSi$_x$ work function were measured to be both near the mid-gap (Table 1), a different value than on HfO$_2$.

B. Dielectric characteristics: Capacitive extractions (Fig. 6) indicate that the silicon film thickness (t$_{Si}$) is 10nm +/- 2nm. Equivalent Oxide Thickness (EOT) measurements are summarized in Table 1. A 4Å

EOT increase was observed with WSi$_x$ on SiO$_2$ gate dielectric compared to the TiN gate. This may be due to the fluorine (F) and/or chlorine (Cl) diffusion into SiO$_2$ (Fig. 7). SiO$_2$ dielectric constant can decrease or free oxygen atom can diffuse to the SiO$_2$/Si interface inducing gate oxide re-growth [10]. Compared to SiO$_2$, HfO$_2$ blocks F and Cl atoms diffusion at the interface (Fig. 7), limiting the EOT increase. The slight EOT enhancement observed on WSi$_x$/HfO$_2$ gate stack was induced by the WSi$_x$ post deposition annealing at 800°C [10]. EOT as thin as 10Å were obtained on TiN/HfO$_2$ devices. The leakage current of the metal gates with HfO$_2$ was 4 orders of magnitude smaller than with SiO$_2$ for the same EOT (Fig. 8) [12]. Due to F and/or Cl diffusion in SiO$_2$, leakage current in WSi$_x$/SiO$_2$ gate stack is 2 decades higher than in TiN/SiO$_2$ dielectric. Besides, leakage current in WSi$_2$/HfO$_2$ stacks was reduced compared to Si-rich WSi$_{2.7}$. Fig. 9 shows that the leakage current yield is strongly affected with WSi$_{2.7}$. On the other hand, for WSi$_2$, the leakage current was well-controlled and lower than with TiN/HfO$_2$ (due to higher EOT).

C. Long channel mobility: Electron and hole effective mobility values were extracted using the split-CV technique (Fig. 10, only electron mobility are shown here). With SiO$_2$, electron mobility with WSi$_x$ gate is clearly improved (+34%) as compared to TiN gate. Low-temperature mobility measurements (Fig. 11 & 12) prove that Coulomb scattering is more pronounced with TiN gates than with WSi$_x$ gates. This trend is confirmed with HfO$_2$ despite the slight difference in EOT and t$_{Si}$. Moreover, the electron and hole effective mobilities are 20% higher with WSi$_2$ than with WSi$_{2.7}$ gate transistors. The degradation is identical at high and low effective fields. This indicates that the degradation is due to a surface roughness-like scattering mechanism. Temperature studies (Fig. 13) confirm this assumption; the additional mobility term (Fig. 14) decreases with the effective field and is nearly independent of the temperature. The extraction of the roughness parameters Δ and L$_C$ (Fig. 13) demonstrates that Δ is larger with WSi$_{2.7}$ than with WSi$_2$ gate. We can speculate that the higher roughness of WSi$_{2.7}$ (Fig.1&3) induces a higher roughness scattering-like mechanism at the interface between the channel and the dielectric.

CONCLUSION

WSi$_x$ is integrated for the first time as gate material on FD SOI MOSFETs with HfO$_2$ and appears to be a promising metal gate material. An effective work function of 4.55eV has been measured. 35% mobility enhancement is obtained with WSi$_x$ devices compared to TiN transistors though a reduction of Coulomb scattering. Better dielectric characteristics and higher mobilities have been demonstrated with WSi$_2$ compared to Si-rich WSi$_{2.7}$ gate transistors. Low temperature studies evidence that the poorer characteristics of Si-rich WSi$_{2.7}$ gate transistor is due to higher surface roughness-like scattering mechanism.

REFERENCES

[1] M. Vinet *et al.*, *SSDM*, p.768, 2004. [2] A. Vandooren *et al.*, *SOI conf.*, p.205, 2002. [3] B. Doris *et al.*, *IEDM*, p.631, 2003. [4] J. Kedzierski *et al.*, *IEDM*, p.247, 2002.[5] S.L. Hsu, *VLSITSA*, p.90, 1993 [6] L.C. Chen *et al.*, *IEEE EDL*, 15(9), p. 351, 1994. [7] H. Kujirai *et al.*, *IEDM*, p.395, 2001. [8] L. Krusin-Elbaum *et al.*, *IBM J.*, 31(6), p.634, 1987. [9] B. Sell *et al.*, *Micro. Eng.* 55, p.197, 2001. [10] K.C. Saraswat *et al.*, *IEEE EDL*, 1(2), 1980. [11] K. Nakajima *et al.*, *IEEE TED*, 52(10), p 2215, 2004. [12] B. Guillaumot *et al.*, *IEDM*, p.355, 2002. [13] S. Maîtrejean *et al.*, *MRS*, 2003. [14] L. Thevenod *et al.*, *Microelec. Eng.*, 80, p. 20, 2005.

The authors would like to thank CEA-LETI and Crolles pilot lines. This study was partly funded by the Advanced Devices Program of the Alliance (STM, Philips and Freescale Semiconductors).

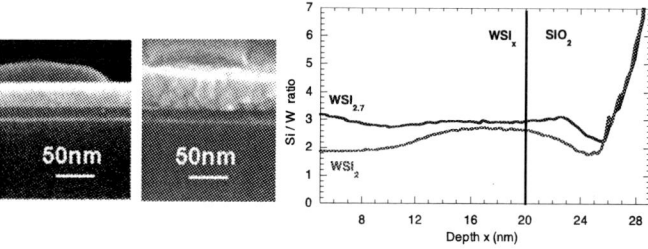
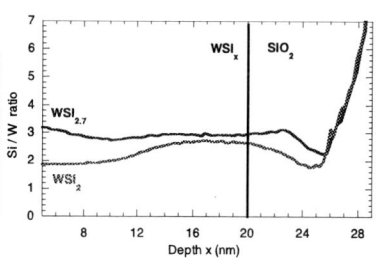
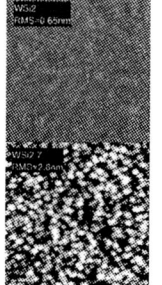
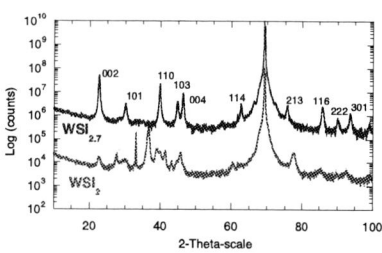

Fig. 1: SEM images of WSi$_2$ (*left*) and WSi$_{2.7}$ (*right*) metal gate on HfO$_2$ gate dielectric FD SOI MOSFETs. Smaller grain size is observed in WSi$_2$ material.

Fig. 2: SIMS analysis of WSi$_x$ on SiO$_2$ fabricated at different temperatures: the bulk average Si/W ratio is 2.7 at 550°C and 2 at 450°C. The same stoichiometry is reached at the interface after the anneal at 800 °C.

Fig. 3: AFM on 1x1µm² WSi$_2$ (up) and WSi$_{2.7}$ (down) samples with the same Z-range (20nm).

Fig. 4: Diffraction diagrams of WSi$_{2.7}$ and WSi$_2$ gate materials. WSi$_{2.7}$ material is composed of only the tetragonal phase WSi$_2$ whereas, WSi$_2$ material has two tetragonal cristalline phases: WSi$_2$ and W$_5$Si$_3$.

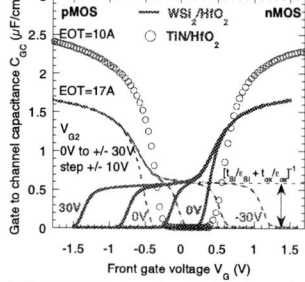
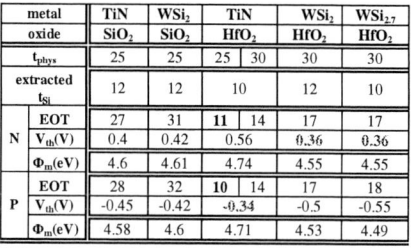

metal		TiN	WSi$_2$	TiN		WSi$_2$	WSi$_{2.7}$
oxide		SiO$_2$	SiO$_2$	HfO$_2$		HfO$_2$	HfO$_2$
t$_{phys}$		25	25	25	30	30	30
extracted t$_{Si}$		12	12	10		12	10
N	EOT	27	31	**11**	14	17	17
	V$_{th}$(V)	0.4	0.42	0.56		0.36	0.36
	Φ$_m$(eV)	4.6	4.61	4.74		4.55	4.55
P	EOT	28	32	**10**	14	17	18
	V$_{th}$(V)	-0.45	-0.42	-0.34		-0.5	-0.55
	Φ$_m$(eV)	4.58	4.6	4.71		4.53	4.49

Fig. 5: I$_{DS}$ versus V$_G$ at V$_{DS}$=+/- 1.2V in n- and p-channel FDSOI MOSFETs with a metal gate in TiN (w/o symbols) and in WSi$_x$ made at 450°C (WSi$_2$) and at 550°C (WSi$_{2.7}$).

Fig. 6: Capacitance versus front gate voltage at various back gate biases in p- and n-channel FDSOI devices with a WSi$_x$ metal gate on HfO$_2$ dielectric. t$_{Si}$=12nm. TiN/HfO$_2$ CV curves are shown (circles).

Table 1: Extracted values of EOT (Å), t$_{Si}$ (nm), threshold voltage (V$_{th}$) and effective gate work function (Φ$_m$) for the different nMOS and pMOS gate stacks.

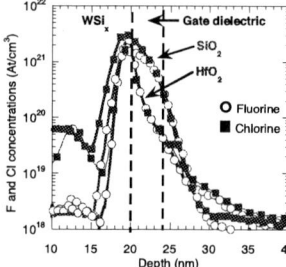
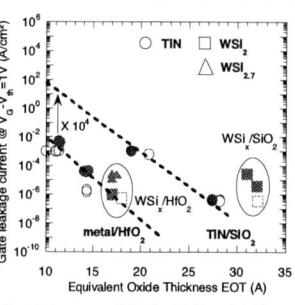
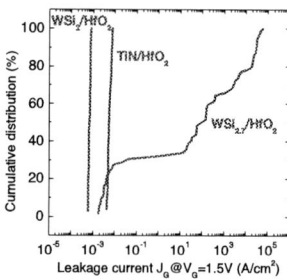
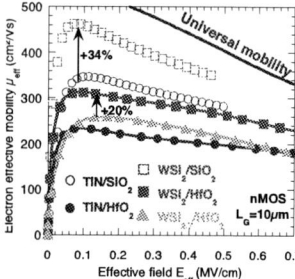

Fig. 7: SIMS analysis of fluorine (F) and chlorine (Cl) concentrations in WSi$_x$ films on SiO$_2$ and on thick HfO$_2$ dielectrics.

Fig. 8: J$_G$ versus EOT for WSi$_x$ and TiN on HfO$_2$ and on SiO$_2$ for nMOS (full symbols) and pMOS (empty symbols). Best WSi$_{2.7}$ points are shown.

Fig. 9: Cumulative distribution of the gate leakage current at V$_G$=1.5V for WSi$_x$ and TiN metal gate nMOSFETs.

Fig. 10: Long channel electron effective mobility versus the effective field in FDSOI devices with metal gate on SiO$_2$ and HfO$_2$.

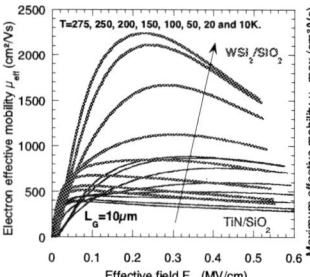
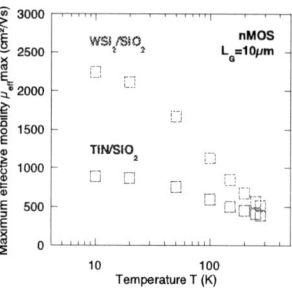
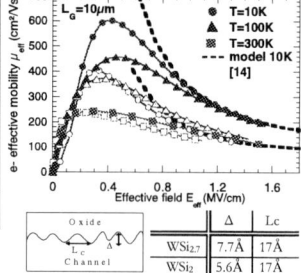
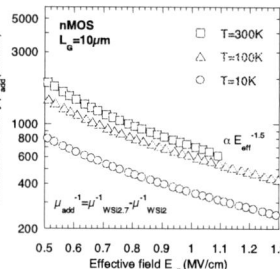

Fig. 11: Evolution of the electron effective mobility versus the effective field with the temperature (T from 10K to 275K) for WSi$_2$ and TiN gate on SiO$_2$.

Fig. 12: Maximum electron effective mobility evolution in function of temperature (300K to 10K) for TiN and WSi$_2$ on SiO$_2$.

Fig. 13: *top:* Electron µ$_{eff}$ versus E$_{eff}$ for WSi$_2$ (closed symbols) and WSi$_{2.7}$ (open symbols). *Bottom:* Obtained values of Δ and Lc for the WSi$_{2.7}$ and WSi$_2$ from model of [14], in dashed lines.

Fig. 14: Extracted additional electron mobility between WSi$_2$ and Si-rich WSi$_{2.7}$ on HfO$_2$ at 300, 100 and 10K.

2006 International Symposium on VLSI Technology, Systems, and Applications

NMOS and PMOS Metal Gate Transistors with Junctions Activated by Laser Annealing

S. Severi [a,b], E. Augendre [a], A. Falepin [a], C. Kerner [a], J. Ramos [a], P. Eyben [a], ,W. Vandervost [a], C. Curatola [c], S. Felch [d], F. Nouri [d],
P. Kraus [d], V. Parihar [d], T. Noda [e], R. Schreutelkamp [d], T. Y. Hoffmann [a], P. Absil [a], K. De Meyer [a,b], M. Jurczak [a], S. Biesemans [a]

a) IMEC Interuniversity Microelectronics Center, Kapeldreef 75, 3001 Leuven Belgium, b) K.U.Leuven, ESAT-INSYS, Kasteelpark Arenberg 10,
c) Philips Research Leuven, Kapeldreef 75, B-3001 Heverlee, Belgium, d) Applied Materials, Sunnyvale, CA, USA, e) Matsushita Electric Industrial Co.,
Ltd. 19 Nishikujo-kasugacho, Minami-ku, Kyoto, 601-8413, Japan
Severis@imec.be, Tel. Phone: +32 16 288579, Tel. Fax: +32 16 281844

Abstract

We demonstrate for the first time the integration of metal gate electrode and non-melt laser annealed junctions in both NMOS and PMOS transistors. We report the highest drive current so far in laser annealed devices with good Short Channel Effects (SCE) control down to 40nm gate length. Overlap length is quantified by CV and SSRM, values of 2 nm for both NMOS and PMOS laser-annealed transistors are reported for the first time.

Introduction

Formation of ultra shallow junctions suitable for scaled CMOS planar devices requires advanced annealing alternatives to conventional spike RTA. High doping activation and shallow junction profiles have been demonstrated with low temperature annealing [1] as well as high temperature millisecond Flash [2] or sub-millisecond Laser [3] annealing. However the integration of diffusion-less junctions creates challenges related to contact and overlap resistance [4], [5]. Typically, junction implant parameters optimized for spike RTA need to be redesigned to solve these issues. In addition, special care has to be taken in the case of millisecond anneals to minimize the damage in both the gate dielectric and junction depletion region (usually not an issue in the case of spike RTA). In this paper we investigate the compatibility of a new sub-millisecond laser annealing technique with an advanced planar CMOS process. By means of characterizing the main transistor parameters we contribute to the understanding of the various aspects related to this new annealing technique, like implant re-optimization and adequate implant defect annealing.

Experimental

The process flow description for both NMOS and PMOS transistors is shown in Fig. 1. After conventional Vt adjust implants, 1.4 nm SiON gate dielectric is formed. The NMOS and PMOS transistors have been processed on separate wafers with PVD TaN gate electrode, [6], and Ni Fully Silicided (FUSI) gate, [7], respectively. After halo implantation, several splits on Ge and B implantations for PMOS and As implantations for NMOS have been performed. A low temperature 60nm nitride spacer is formed prior to the deep junction implants. The implantation energy is optimized in order to obtain similar junction depth after spike or laser annealing. After formation of NiSi, BEOL completes the flow.

Results and Discussion

By using a metal gate electrode we decouple the influence of the anneal on gate and SD activation while avoiding poly-depletion effects. Fig. 2 confirms previously reported results that laser anneal can form diffusion-less and highly activated junctions. The doping activation can be further improved by increasing the a-Si layer depth prior to anneal, as shown in Fig. 3. For shallower Ge PAI the end-of-range defects are located closer to the dopant and could interact faster with the boron during the Laser pulse compared to deeper PAI, leading to a lower activation level. Nevertheless the transistor performance is degraded, as shown in Fig.4, due to the formation of regrowth defects in the junction to gate overlap region and to mobility degradation, as shown later. A shallower Ge pre-amorphization alleviates the issue of residual defects formation and a higher B energy improves the mobility and device performance. Capacitance measurements on multi-finger structures on short channel devices (with CD-SEM measured gate lengths) allow the extraction of the main transistor main parameters, namely overlap capacitance, Leff, $R_{S/D}$ and mobility. We used a similar method as presented in [9]. We extract the Leff from CV measurement subtracting the parasitic capacitance Cov at 1V below V_{FB}, as shown in Fig. 5. The related junction overlap length is defined at a doping active concentration of $1e19cm^3$, as deduced from overlap capacitance simulation (here not shown). We show instead a lateral junction profile extracted from SSRM measurements, Fig. 6. The extracted overlap length in case of an extension implantation of B 1keV $1e15cm^{-2}$ (see Fig. 7) is in agreement with the 3 nm shown by the SSRM profile at an active doping concentration of $1e19cm^{-3}$. The overlap length shrinks from 7nm to 2nm moving from spike RTA to laser annealed transistors. We found that the S/D resistance can be reduced by 35% for B 1keV laser anneaed junctions compared to spike RTA. If the extension energy is decreased, the resistance becomes equal to the spike reference value and in case of deep Ge pre-amorphization is even higher due to the defects in the overlap junction region. The peak hole mobility versus L_{EFF} and the linear Vt roll-off are shown in Figs.8 and 9. The long channel mobility is degraded for all the different laser implant conditions compared to the spike annealed transistors. This could partially counteract the gain obtained in S/D resistance. In addition deep Ge pre-amorphization results in further mobility degradation. Laser annealing leads to improved SCE control compared to spike due to abrupt non-diffused junctions and halo profiles.

It is known that laser annealing preserves the as-implanted profile for n+ junctions [3]. Nevertheless the junction overlap is longer for Laser than spike RTA on poly-Si gate transistors, Fig.10, when targeting the same final junction depth. This relates to As implantation through the thin gate oxide for the considered laser anneal cases (either 5keV As or 3keV P). Using a metal gate electrode the overlap length is much reduced (cf. 5keV As case) due to the higher implant stopping power of TaN compared to poly-Si. The Ion-Ioff trade-off of PVD TaN NMOS transistors annealed with Laser is shown in Fig. 11. We have used the same halo dose for spike RTA and laser-annealed devices. During laser annealing the B halo does not diffuse and the dopant activation is much higher than in the case of spike RTA. Despite the high Vt we still achieve transistor performance that is good compared to that reported earlier, Fig.12. Lowering the halo dose will allow a further improvement in performance without significant SCE degradation.

Conclusion

We have presented the highest performance for laser-annealed NMOS and PMOS transistors reported so far. We discussed how the different implantation conditions affect the main transistor parameters (such as potential for reduction of junction overlap length and S/D resistance) and proposed some possible solutions for further performance improvement.

Reference

[1]R. Lindsay et al., IWJT, 2004, [2] T. Ito et al., VLSI 2003, [3]K. Adachi et. al., VLSI 2005, [4]S. Severi et al., IEDM 2004, [5]A. Shima et al., VLSI 2005, [6]K. Henson et al., IEDM 2004, [7]K. G. Anil et al., VLSI 2004, [8]H. Y. Wong et al., EDL April 2005, [9]K. Romanjek et al., EDL August 2004, [10]A. Shima et al., IEDM 2003, [11]A. Shima et al., VLSI 2004

1-4244-0181-X/06/$20.00 ©2006 IEEE   119

	NFET		PFET	
	Reference	Laser	Reference	Laser
Gate stack:			Gate stack:	
TaN / 1.4nm SiON			NiSi FUSI / 1.4nm SiON	
Pockets and Extensions:			Pockets and Extensions:	
	B 10keV	B 10keV	As 40keV	As 55keV
	–	Ge PAI	–	Ge PAI
	As 1keV	As 5keV	BF₂ 1keV	B 0.5/1keV
HDD:			HDD:	
	–	–	–	Ge PAI
	As 25keV	As 30keV	B 3keV	B 5keV
	1050°C spike	1300°C Laser	1050°C spike	1300°C Laser

Fig.1: Process flow conditions for spike RTA and laser annealed transistors.

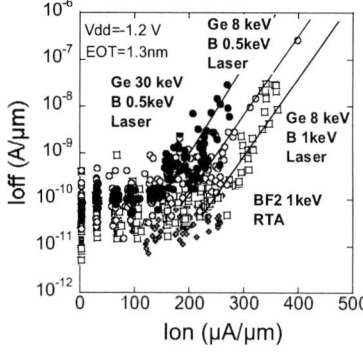

Fig. 4: Ion-Ioff for spike RTA devices vs. implant splits with laser.

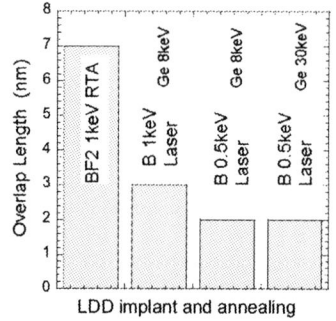

Fig.7: Overlap length for the laser annealing implant splits compared to the overlap obtained with spike anneal

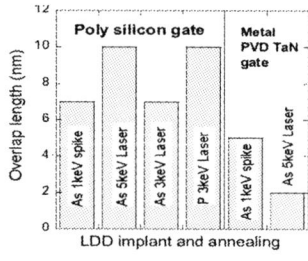

Fig.10: Overlap length values of junction implanted on poly-silicon or TaN gate transistors.

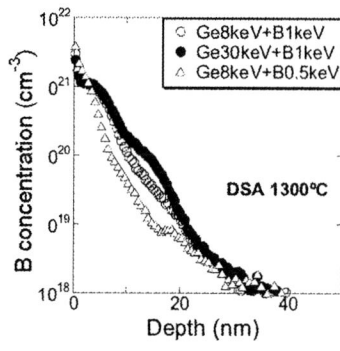

Fig.2: SIMs profile of p+ junctions after laser annealing

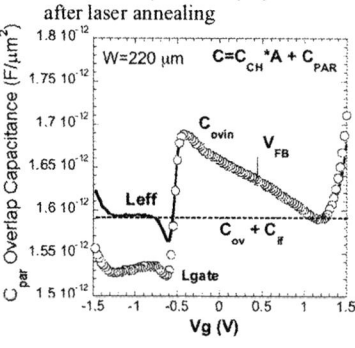

Fig.5: Overlap capacitance extracted from linear interpolation of the capacitance as a function of the area.

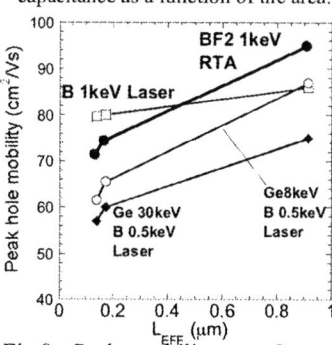

Fig.8: Peak mobility vs. L_{EFF} extracted for PMOS devices.

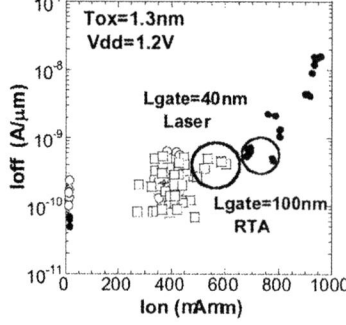

Fig.11: Ion-Ioff comparing spike RTA with Laser annealed transistors with the same implanted B halo dose.

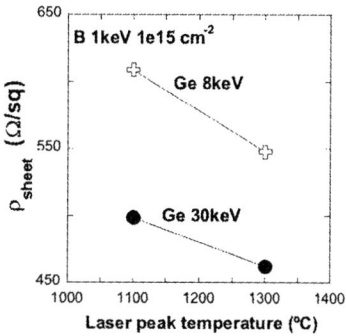

Fig.3: Influence of Ge pre-amorphization on junction sheet resistance

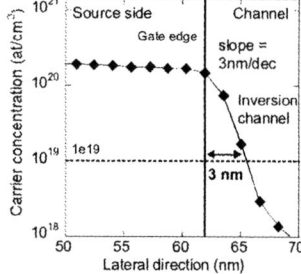

Fig.6: SSRM showing the overlap under the gate of B 1keV diffusion-less junction.

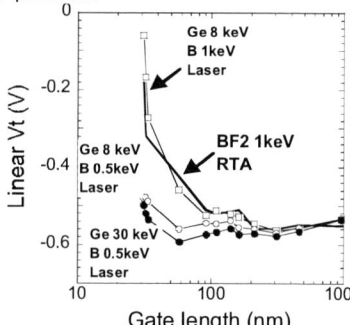

Fig.9: Vt roll-off for PMOS transistors shows improved SCE for Laser annealed devices.

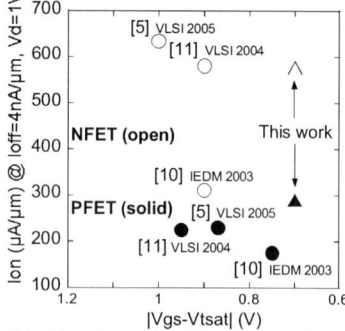

Fig.12: Benchmark chart comparing the most recent published data on laser annealed transistors.

2006 International Symposium on VLSI Technology, Systems, and Applications

Tunable Workfunction for Fully Silicied Gates (FUSI) and Proposed Mechanisms

Y. H. Kim, C. Cabral. Jr. E. P. Gusev*, L. Gignac, M. Gribelyuk, and M. Ieong

IBM T.J. Watson Research Center,Yorktown Heights, NY 10598, USA,

Phone +1-914-945-1439; Fax +1-914-945-2141; Email: yhkim@us.ibm.com

* Qualcomm Research Center, San Jose, CA

ABSTRACT

We present fully silicided gate (FUSI) work function modulation and mechanisms using different NiSi alloys as well as different phases ($Ni_{31}Si_{12}$ and Ni_{rich}-Pt-Si). It is shown that the interface layer between gate silicide and dielectric is the key to modulate the workfunction due to Fermi level pining on HfSiO and HfO_2. Ge, Yb, Pt, and Al were fully explored to identify good candidates for N and PFET. A~400meV workfunction shift was achieved toward the conduction band edge using NiAlSi demonstrating a mobility of 300 cm^2/Vs at peak, matching NiSi control devices on Hf_xSiO_y. $Ni_{rich}PtSi$ has shown 4.85eV workfunction and good mobility. Different gate materials were fully explored to understand the mechanism of workfunction modulation depending on the dielectrics. It was found that the workfunction of metal rich silicides are strongly dependent on the pure metal workfunction and it can be changed by controlling the number of metal atoms which are available at the interface.

INTRODUCTION

Gate oxide scaling has been one of driving forces for high performance device technology. Since physical thickness scaling reached the fundamental limit for SiO_2, high-k dielectrics and metal gates have been attractive options to eliminate poly Si depletion as well as to achieve higher capacitance. The main motivation is to find thermally stable and suitable workfunction metals. FUSI metal gates have recently received much attention as an alternative metal gate solution because of the relative process simplicity. NiSi has been one of best candidates for gate electrode applications because it has good stability with gate oxides and high performance characteristics without poly Si depletion [1,2]. NiSi has mid gap workfuntion (4.6 eV) with SiO_2, and slightly lower workfuntion (4.45eV) with Hf-based high-k dielectrics due to Fermi-Level pining [5]. Workfunction modulation can be achieved with respect to different silicide materials, alloys, ion implant, and phases [1,2,3,4,6]. However, ion implantation could cause lots of issues including delamination, dopant fluctuation control in small geometries, and diffusion into gate stacks. Therefore, the investigation and optimization of thermally stable and controllable FUSI materials is very important. In this study, systematic alloy and phase examination was carried out for understanding the mechanism of NiSi workfunction modulation.

EXPERIMENTAL

Transistors and capacitors were fabricated using various dielectrics (Hf_xSiO_y, HfO_2, and SiON) with 50 ~70nm of poly Si for FUSI study and 70 ~100nm of CVD SiGe was deposited for FUGE gate. Souce/Drain activation was carried out at 980ºC 5sec. NiSi FUSI and NiGe FUGE gates were formed at 450°C~550°C 30sec followed by forming gas anneal at 450°C 30min.

RESULTS AND DISCUSSION

Fig.1 shows NiAlSi process flow. From EDX, TEM, and EELS analysis, we found that Al was segregated toward top and bottom of the stack, while negligible amounts were seen in the middle of the stack. TEM images (Fig.2) indicate the monolayer formation of Al between HfO_2 and NiSi, while no visible layer was found at the NiSi and NiPtSi interface. Pt exists through out the NiSi stack. EDX confirms that a higher number of Al atoms were present at the interface between dielectric and gate electrode. Transistor characteristics are shown in Fig.4. NiSi gate with Hf-based high-k dielectrics showed Fermi-Level pinning resulting in high Vth. However, the addition of Al and the change of silicide phase help to achieve low Vth by modifying the interface between gate electrode and dielectric. Furthermore the addition of Pt to Ni_3Si shifts workfunction toward valence band edge. NiAlSi and $Ni_{rich}PtSi$ shows similar mobility values compared to NiSi control. Fig.7 explains possible mechanisms for NiSi, Ni_3Si, $Ni_{rich}PtSi$, Pt_3Si, and NiAlSi workfunction modulation. It is speculated that Hf-Si bonds are dominant in NiSi interface system, workfunction of NiSi is pinned to certain energy level (4.45eV) leading to high Vth. However, metal rich phase silicides (Ni_3Si, $Ni_{rich}PtSi$, and Pt_3Si) show workfunction modulation because they have fewer Hf-Si bonds compared to metal atoms. Metal segregation (NiAlSi) and phase change (Ni_3Si) are the main techniques to modulate workfunction. It is worth mentioning that the workfunction of metal rich phase FUSI (Ni_3Si and Pt_3Si) are closer to the pure metal workfunction (Fig.8). In this manner, Pt addition to Ni_3Si shows an additive effect to shift workfunction toward valence band resulting from the higher workfunction of Pt compared to that of Ni. Additionally, Yb, Ge addition to NiSi and fully germanized gate (FUGE) were also explored. 20% Ge did not change Vfb (flat band) with SiON and HfO_2 (Fig.9 and 10). FUGE devices showed Vfb shift (700mV) compared to NiSi control with 3nm SiON. However, capacitance was decreased significantly (Fig.11). Also, Yb addition to NiSi showed Vfb shift toward n+poly Si device but large EOT (equivalent oxide thickness) loss was found (Fig.12). Since Vfb can be easily biased by fixed charge or any reaction with a thick oxide, one should carefully scrutinize impurity diffusion with low doped substrate including appropriate physical analysis, capacitance, interface state density, and transistor characteristics (mobility and swing,) with thin oxides. Ge tends to form Ge_xO_y and Yb is aggressively oxidized when it is exposed O and high temperatures. EOT loss and performance degradation can not be tolerated for high performance technology.

CONCLUSION

Many of Ni silicide materials (Yb, Ge, Pt, and Al) were fully explored to identify good candidates. N+poly Si compatible FUSI gate material NiAlSi (4.25eV) and P+poly Si compatible quarter gap $Ni_{rich}PtSi$ (4.85eV) FUSI gate were identified and showed good transistor characteristics. Ge and Yb showed large degradation. The possible mechanism of FUSI workfunction modulation was presented and has a strong correlation with pure metal workfunctions. It was shown that metal segregation and silicide phase change are the two important techniques to modulate workfuntion.

REFERENCES

[1] E. Gusev, et al.Tech. Dig. IEDM. , p 89 (2004)

[2] S. Severi, et al.Tech. Dig. IEDM. , p 99 (2004)

[3]M. Terai, et al. VLSI Symp. Proc. , p 68 (2005)

[4] K. Takahashi, et al Tech. Dig. IEDM. , p 91 (2004)

1-4244-0181-X/06/$20.00 ©2006 IEEE 121

[5] T. Nabatame, et al Tech. Dig. IEDM. , p 83 (2004)
[6] J. A. Kittl et al. VLSI Symp. Proc. , p 72 (2005)
[7] M.J.H. van. Dal , et al Tech. Dig. IEDM. (2005)

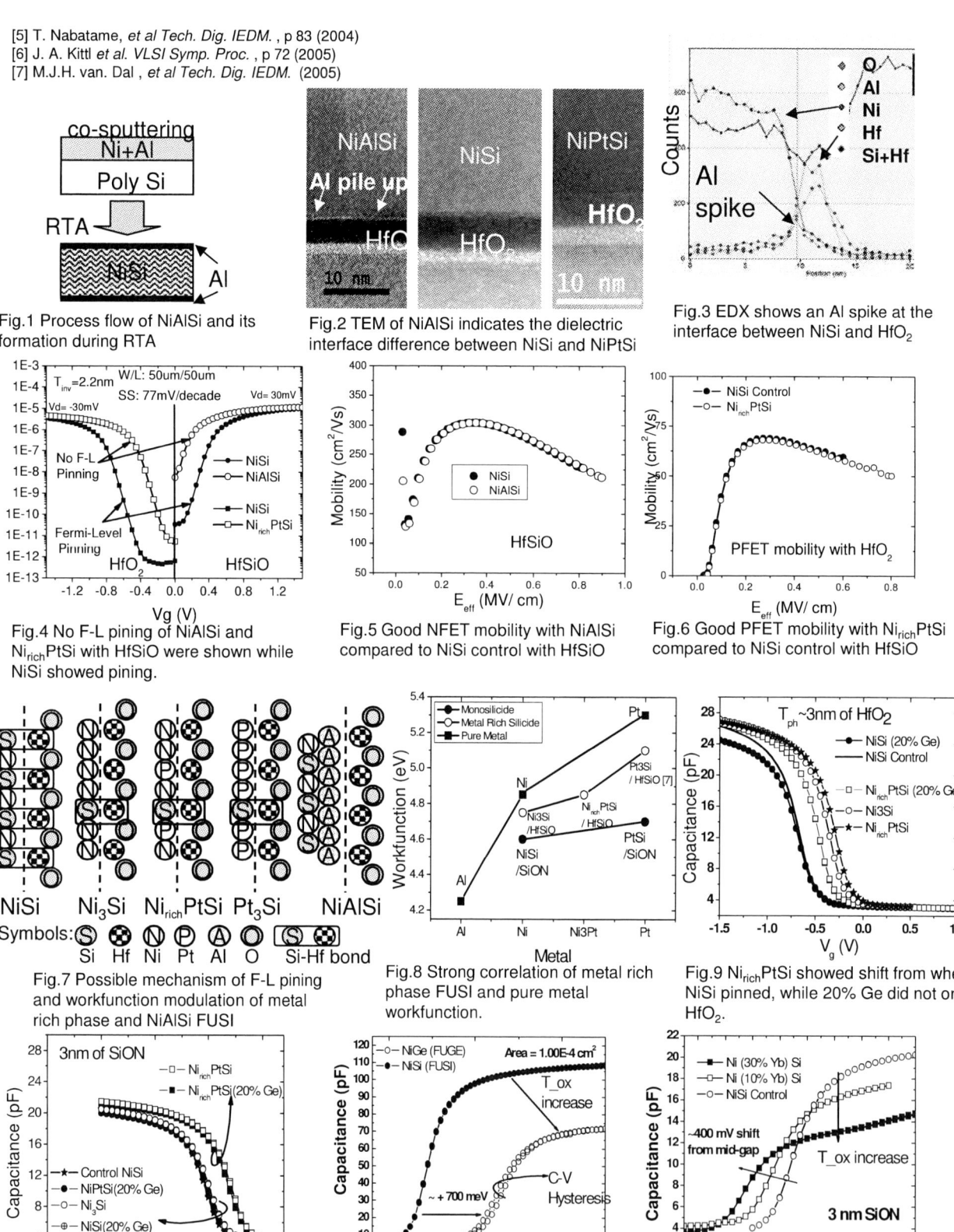

Fig.1 Process flow of NiAlSi and its formation during RTA

Fig.2 TEM of NiAlSi indicates the dielectric interface difference between NiSi and NiPtSi

Fig.3 EDX shows an Al spike at the interface between NiSi and HfO$_2$

Fig.4 No F-L pining of NiAlSi and Ni$_{rich}$PtSi with HfSiO were shown while NiSi showed pining.

Fig.5 Good NFET mobility with NiAlSi compared to NiSi control with HfSiO

Fig.6 Good PFET mobility with Ni$_{rich}$PtSi compared to NiSi control with HfSiO

Fig.7 Possible mechanism of F-L pining and workfunction modulation of metal rich phase and NiAlSi FUSI

Fig.8 Strong correlation of metal rich phase FUSI and pure metal workfunction.

Fig.9 Ni$_{rich}$PtSi showed shift from where NiSi pinned, while 20% Ge did not on HfO$_2$.

Fig.10 Ni$_{rich}$PtSi showed shift from where NiSi pinned, while 20% Ge did now show any shift with SiO$_2$. Also Ni$_3$Si did not show any shift on SiON

Fig.11 FUGE shows Vfb shift but with large EOT loss

Fig.12 Yb addition to NiSi shows Vfb shift but large EOT loss

2006 International Symposium on VLSI Technology, Systems, and Applications

Roadblocks and Critical Aspects of Cleaning for Sub-65nm Technologies

Paul W. Mertens, G. Vereecke, R. Vos, S. Arnauts, F. Barbagini, T. Bearda, S. Degendt, C. Demaco, A. Eitoku, M. Frank,
W. Fyen, L. Hall, D. Hellin, F. Holsteyns, E. Kesters, M. Claes, K. Kim, K. Kenis, H. Kraus, R. Hoyer, T.Q. Le, M. Lux,
K-T. Lee, M. Kocsis, T. Kotani, S. Malhouitre, A. Muscat, B. Onsia, S. Garaud, J. Rip, K. Sano, S. Sioncke, J. Snow,
J. Van Hoeymissen, K. Wostyn, K. Xu, V. Parachiev, M. Heyns

Imec
Kapeldreef 75
B-3001 Leuven, Belgium

ABSTRACT

This study will review some of the critical aspects of cleaning for sub-65 nm technologies. These issues include: surface preparation for high k dielectrics on Si and on Ge, metal gate cleaning and removal of small particles without creating damage to structures.

HIGH-DIELECTRIC CONSTANT GATE INSULATOR

One of the key issues in surface preparation for ALCVD of high-k dielectrics is to obtain a surface that is most suited to obtain a strong 2-dimensional growth of the dielectric film, avoiding roughness and pin holes in the high k layer. For ALCVD it is required to start from a uniform coverage of the surface with silanol groups, in order to obtain uniform film growth. The biggest challenge is to obtain such a uniform coverage while limiting the thickness of the (chemical) SiO_2 layer. The quality of the nucleation of the film is readily observed on growth characteristics. Fig. 1 shows the growth characteristics for an ALCVD HfO_2 layer [1]. Linear growth without incubation delay is obtained on chemical oxide. HF-last surfaces show a major incubation delay due to the nucleation inhibition on H-passivated surface.

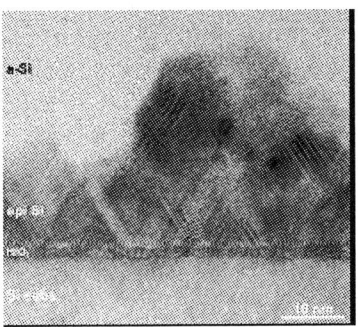

Fig. 2: X-TEM viewgraph of poly-silicon grown on thin (20 cycles) ALCVD Hf Ox on very thin chemical oxide. Some localized pinholes in the layer are revealed by the epitaxial alignment to the substrate.

ALTERNATIVE SEMICONDUCTOR: Ge

Due to its high carrier mobility, the interest in germanium as a transistor semiconductor material has strongly increased. In order to build transistors surfaces need to be cleaned and prepared properly by wet chemical treatments. It was found that the chemical behaviour of Ge is quite different from Si. As an example, a standard cleaning 1 mixture (SC1) at 20C used for 60 sec typically removes far less than 1 nm of Si, while it removes several hundreds of nm of Ge. The major difference is that GeO_2 readily dissolves in water in contrast to SiO_2. The GeO, however, seems to be a lot more resistant than the GeO_2.

In order to obtain good surface preparation for critical applications such as high k dielectric deposition or epitaxial growth the surface chemistry needs to be characterized. It was found that water, HCl and HF remove all native GeO_2 but leave GeO, HI treatment removes oxide and a lot of the GeO while HBr effectively removes all Ge-oxide from the surface (see Fig. 3) [2].

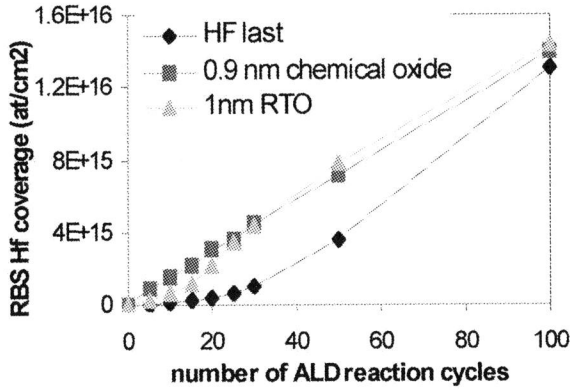

Fig. 1: Growth kinetics of ALCVD HfO2 layer on three different surfaces measured with a "large" spot size [1].

Fig. 2 shows a case where a very thin (0.4nm) chemical oxide was grown prior to a 20-cycle HfO_2 deposition. Then a poly-silicon layer was deposited at 550 C on top. Some of the deposited silicon is epitaxially aligned to the bulk substrate. This reveals the presence of pinholes in such thin chemical oxide, making it unsuited for growth of good quality gate dielectric films. Such pinholes in the chemical oxides are believed to result in the incubation delay in large spot size (more macroscopic) growth characteristics. Therefore chemical oxides result in a trade-off between being thinner at the expense of having more defects.

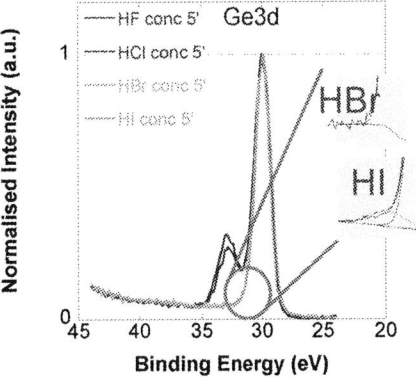

Fig. 3: XPS spectra of Ge surfaces that were treated with an oxidizing wet chemical, followed by a 5 min dip in four different hydrohalogenic acids [2].

1-4244-0181-X/06/$20.00 ©2006 IEEE   123

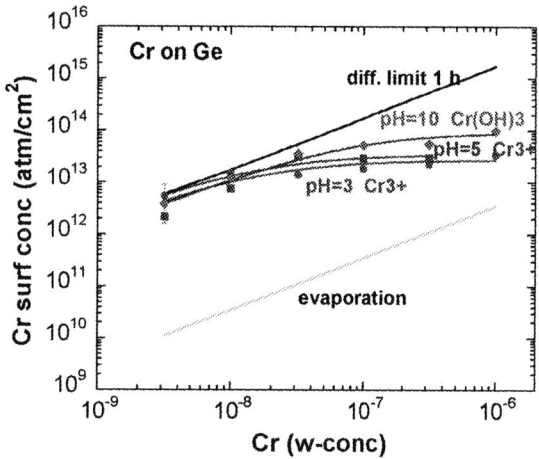

Fig. 4: Deposition of Cr in 1 h on Ge surfaces from three different Cr spiked solutions: pH=3 (HNO3) (E.R.=0.15nm/min), pH=5.7, pH=10 (NH4OH) (E.R.=0.01 nm/min) [3].

The tendency for metallic contamination to deposit on Ge surfaces from aqueous solutions is an important aspect for setting up appropriate specifications for wet chemical treatments. The deposition of Cr from three different solutions during 1 h immersion was studied and also compared to the behaviour on Si surfaces [3]. Prior to immersion into a metal spiked solution all wafers received a dilute APM treatment resulting in GeOx. The solutions under study are: pH=3 (HNO3) (E.R.=0.15nm/min), pH=5.7, and pH=10 (NH4OH) (E.R.=0.01 nm/min). At high pH in general a strong tendency for metals to deposit has been observed ("transport limited deposition") just like on silicon (Fig. 4). At low pH, however, on silicon the deposition of most metals is know to be suppressed. This was not observed on Ge. The suppressed deposition on Si wafers at low pH is attributed to the competition with adsorption of H+ on the surface. A similar phenomenon does not seem to take place on Ge surfaces.

METAL GATE STACK

Advanced metal gate stacks have shown to be very vulnerable with respect to chemical attack during wet cleaning. In gate stacks composed of a 100nm amorphous silicon on top of a 5nm metal layer were found to be heavily attacked after cleaning in HF based cleaning mixtures as shown in Fig. 5 [4].

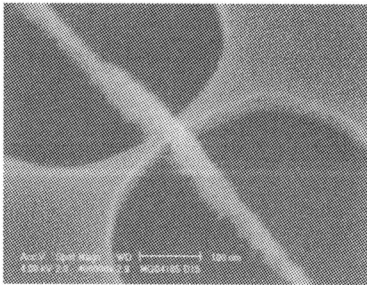

Fig. 5: Severe attack observed with SEM after post etch cleaning of the gate electrode stack composed of 100nm amorphous Si on top of 5nm metal [4].

SMALL PARTICLE REMOVAL

In semiconductor manufacturing, as feature sizes are scaling down below 65 nm, particles with a diameter of a few tens of nanometer need to be considered as killer defects. In a lot of established cleaning processes particles are chemically removed relying on etching and undercutting. In future technologies the specifications on allowable substrate loss per wafer clean no longer allow for pure chemical particle removal. Additional physical mechanisms, such as megasonic agitation, or high velocity liquid aerosol jets, to remove contaminant particles are therefore required. At the same time, as particle size decreases, the ratio of adhesion force over cleaning force increases, thereby compromising the particle removal efficiency (PRE). In addition, wafer surfaces contain fine structures with fairly high aspect ratio, such as gate electrodes or low k isolation patterns, which become vulnerable to sideward impact by physical forces. The combination of all these trends results in a strong reduction of the process window to the extent that removal of nano-particles is becoming a major challenge in future IC manufacturing. Different aspects of this challenge are treated below.

The particle removal in megasonic cleaning is known to increase with increasing concentration of dissolved gas as is illustrated in Fig. 6 [5]. More specifically the particle removal is attributed to the occurrence of gaseous bubble collapse, commonly referred to as transient gaseous cavitation [6].

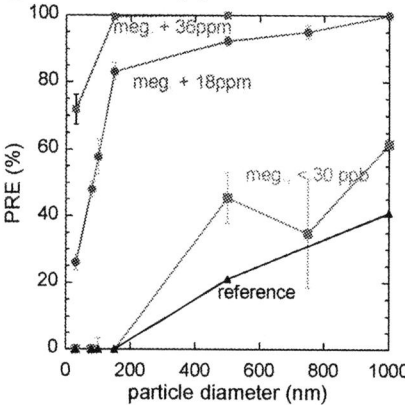

Fig. 6: Effect of dissolved O2 gas on the removal efficiency of SiO2 particles by megasonic in UPW at 20C for 5 min. Reference is for rinse and Marangoni dry only [6].

Fig.7 shows that, even though cleaning of nano-particles is becoming increasingly more difficult as particle size decreases, it is actually possible to remove particles with a diameter as small as about 30 nm by megasonic cleaning. In general the decrease in PRE at smaller particle size is at least partially associated with the outspoken non-uniformity in cleaning performance, shown in the left part of Fig. 8.

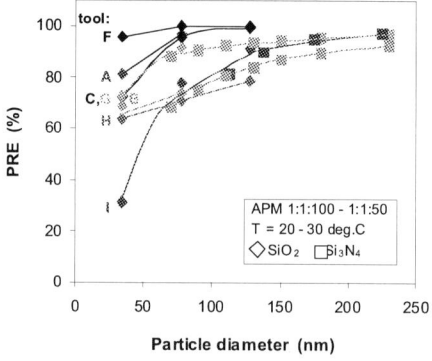

Fig. 7: Removal efficiency of SiO2 or Si3N4 - particles for different megasonic cleaning systems, as a function of the particle diameter[7, 8].

As the use of physical forces to assist the cleaning turns out to be beneficial, at the same time also damage can be caused to fine poly lines. The damage typically shows up as the removal or bending of small pieces of line with a length of about 1 to 2 µm. Because of the localized character of the damage it has also been attributed to cavitation. Full wafer inspection with a KLA-Tencor AIT allowed to determine defects statistics and the spatial distribution of defects on wafers, as shown in the right part of Fig. 8. Also very strong non-uniformity of the damage was observed. From Fig. 8 it can be seen that the regions with highest PRE tend to correspond with the regions of most damage. Fig. 9 shows a typical frequency distribution for damage of these poly lines as a function of line-width and aspect ratio. The distribution is characterized by a strong increase in damage frequency for line-widths below 100 nm. This observation is valid for all tools and process conditions tested in this study. It indicates that present physically assisted cleaning tools will suffer a dramatic reduction in process window with new technology generations. The trade-off between defects created by physically assisted cleaning versus the particle cleaning efficiency is shown in Fig. 10 for many different type of tools and techniques.

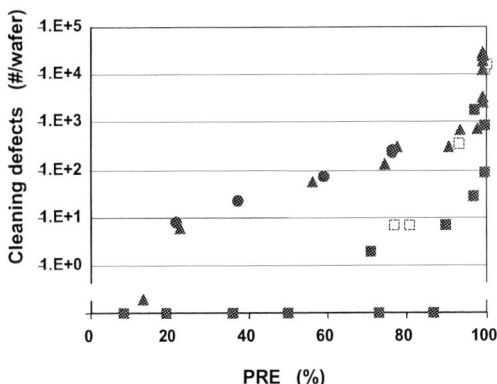

Fig. 10: Trade-off of structural damage to 70nm poly-si lines versus particle (SiO2 78nm diameter) removal efficiency for different physical assisted cleaning methods[7,10].

The challenge for future cleaning with physical assistance is represented in Fig. 11. Suppressing the damage, requires the elimination of the high-energy tail of the distribution of the cleaning events. In state of the art cleaning tools several factors contribute to the spread of the distribution of the cleaning events of megasonic systems: Non-uniformity over the wafer, often related to non-uniformity in the piezo-electric tranducers, power transients in time [11] A lot of effort will be needed to eliminate these factors.

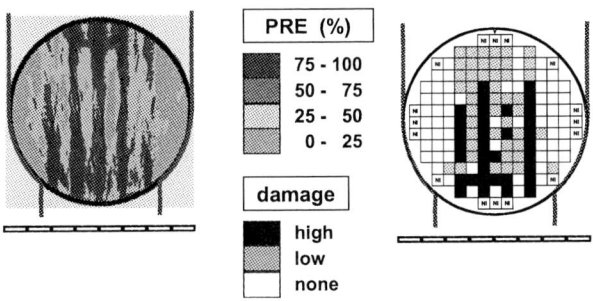

Fig. 8: Wafer maps for particle removal efficiency (left, 34-nm SiO2 particles) and megasonic damage (right, 70-nm poly-gate-stack lines, NI = not inspected) for tool C, using a standard Teflon carrier and aerated APM 1:1:50 at 30 °C. The wafer carrier and the transducer array are shown to scale on the maps. Low PRE and low damage on the left and right sides of the wafers resulted from shadowing of the acoustical waves by the sidewalls of the carrier [8, 9].

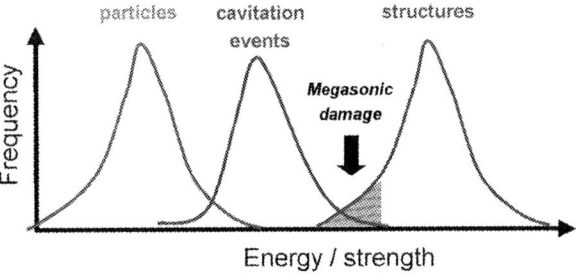

Fig. 11: Schematic diagram illustrating the narrowing process window for particle removal, in this case for physcial (e.g. megasonic) assisted cleaning.

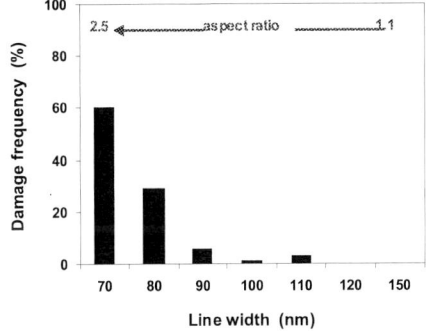

Fig. 9: Poly-gate lines damage frequency distribution as a function of line-width and aspect ratio (APM 1:1:80 at 30°C, 5 min, 100 % power)[9].

REFERENCES

1. Annelies Delabie et al,- AVS ICMI Conference Santa Clara 2003.
2. Onsia et al, Solid State Phenom. **103-104**, p. 27, 2005.
3. S. Sioncke et al, ECS transactions Vol. 1 No3, p. 220.
4. S. Garaud et al, submitted to UCPSS2006.
5. S. Cohen, et al: United States Patent 5,800,626 (1998).
6. F. Holsteyns, et al, Proc. 8th Int. Symp. Cleaning Technology in Semiconductor Device Manufacturing, Electrochemical Society Proc. Vol. PV 2003-26 (2004), p. 161.
7. G. Vereecke et al, Solid State Phenom. **103-104**, p. 141, 2005.

8. G. Vereecke et al, Proceedings of the SCP 9th International Symposium on Wafer Cleaning and Surface Preparation, June 8 – 9 2005.
9. G. Vereecke, et al, 8th Int. Symp. Cleaning Technol. in Semicond. Dev. Manufacturing, 204th Meeting of The Electrochemical Society, Orlando, FL, October 12-16, 2003 (Proc. 8th Intern. Symp. Cleaning Technol. in Semicond. Dev. Manufacturing, Electrochemical Society PV 2003-26, 2004, pp. 145-152).
10. G. Vereecke, et al, submitted to UCPSS2006.
11. Y. Shimada, H. Itou, Y. Ishii, T. Shirasu, and A. Tomozawa: Proc. UCPSS 2000, Solid State Phenomena **76-77** (2001) p. 241.

Effect of H_2 Addition during Cu Thin Film Sputtering

Masahiro Ooka and Shin Yokoyama

Research Center for Nanodevices and Systems, Hiroshima University,

1-4-2 Kagamiyama, Higashi-Hiroshima, Hiroshima 739-8527, Japan

Introduction

Recently, Cu is widely used in ULSI for metal interconnection. The merits of Cu are (1) resistivity of Cu (1.7 $\mu\Omega$·cm) is lower than that of Al or Al alloys and (2) higher resistance to electromigration [1]. There are many deposition techniques for Cu film on Si wafer. The popular techniques for filling trenches or vias are electroplating, electroless plating and metal organic chemical vapor deposition. Today, Cu electroplating is most popular technique. A conformal and conductive seed layers are necessary for Cu electroplating. The seed layer is generally deposited by sputtering.

Oh et al.[2] reported that the plasma H_2 pretreatment of Cu seed layer can improve the surface condition for electroplating of Cu. We think that H_2 adding is expected to improve coverage characteristic for sub-micron hole during Cu sputtering.

In this paper, we study the effect of H_2 adding to Cu sputtering and also investigate the influence on the subsequent Cu electroplating.

Experiment

We used DC magnetron sputtering system. The source DC voltage was set to 500 V. Pressure is 5 mTorr. Plasma source gas was Ar or mixture of Ar and H_2.

Cu surface was measured by atomic force microscope (AFM) and X-ray diffraction (XRD). AFM image shows roughness of Cu surface. And XRD spectra show Cu crystal orientation and grain size.

And next, we tried Cu electroplating. The electroplated Cu films were grown on the sputtered Cu layers (250nm). The electrolyte was composed of Cu sulfate (200 g/l), sulfuric acid (27 cm^3/l), polyethylene glycol (1.3 g/l), and chloride ions (1000 ppm). The solution temperature was 25°C. And we changed the plating current (1–20 mA/cm^2).

Results and Discussions

Effect of H_2 adding on the morphology of sputtered Cu surface

A 300 nm thick thermal SiO_2 layer was grown on the Si wafers and then a TiN film was sputtered with the thickness of 30 nm. Finally, Cu was sputtered with the thickness of 650 nm on it. The plasma source gas is the mixture of H_2 (0 - 60%) and Ar.

Cu surface was observed using AFM. The AFM images are shown in Fig. 1. Average roughness tends to decrease by H_2 addition. Power spectrum calculated from AFM line profiles using fast Fourier transform (FFT) is shown in Fig. 2. The ratio of the high wavenumber component (4-8 μm^{-1}) with respective to low wavenumber component (1–3 μm^{-1}) increases by the H_2 addition. It is suggested that Cu surface becomes much smoother by H_2 adding (see Ra in Fig. 1).

Next, crystalinity of the sputtered Cu film is investigated by XRD (Fig. 3). All the measured Cu films are almost (111) oriented. However, small amount of (200) peaks are observed. Figure 4 shows (111) and (200) orientation peaks. All peaks decrease with increasing H_2 adding. However, (200) peaks more rapidly decreases than (111) peak. This means that the more preferred orientation [to (111)] Cu film can be obtained by H_2 adding. In case of H_2 adding, hydrogen-termination may occur on the surface of deposited Cu. And also O atoms on Cu surface may be scavenged by H radicals (see Fig. 5). H_2 adding causes the Cu surface smooth and suppresses oxidation.

Effect of H_2 adding to resistivity of Cu

Resistivity of sputtered Cu films was measured using van der Pauw method [3]. Table I shows the Cu resistivity. The resistivity tends to decrease with increasing added H_2. And Cu film which was grown and stored in atmospheric ambient for 3 months has high resistivity due to oxidation in the atmosphere. Suppression of Cu oxidation by scavenging O by H_2 addition may contributes to suppression of the increase of resistivity.

Characteristics of electroplated Cu on the sputtered Cu

Cu film (300nm) was electroplated on the sputtered Cu (250 nm) seed layer. The deposition rate tends to increase with plating current [see Fig. 6(a)]. About 300 nm thick electroplated Cu films were deposited and their surface roughness was measured by AFM. Figure 6(b) shows a relation between the roughness and film thickness. The roughness seems to depend on Cu film thickness and not depend on plating current.

The electroplated (5mA/cm^2) Cu on sputtered Cu (using Ar or Ar/H_2 mixture plasma) layer is compared. It is found that the electroplated Cu surface roughness does not depend on Cu seed layer, i.e., H_2 adding during sputtering deposition of seed layer does not affect the final surface roughness.

Next, crystalinity of sputtered and electroplated Cu film was compared by using XRD peaks (Fig. 8). After electroplating, (111) and (200) peaks increase. Figure 9 shows an effect of H_2 addition at (111)/(200) peak ratio. The ratio of electroplated Cu film is similar to that of sputtered Cu film. Therefore, H_2 addition sputtering can control the orientation of electroplated Cu.

Conclusions

By addition of H_2 to Ar plasma in Cu sputtering, it is found that the surface becomes smooth and the resistivity tends to decrease with increasing the amount of H_2. However, we found that Cu surface roughness becomes same after electroplating of 300 nm. Orientation of electroplated Cu can be controlled by H_2 addition during Cu sputtering for the seed layer.

References

[1] J. Musil et. al.: Czch. J. Phys. 45 (1995) 249.

[2] J. Oh et al.: Jpn. J. Appl. Phys. 40 (2001) 5294.

[3] L. J. van der Pauw: Philips Research Reports 13 (1958) 1.

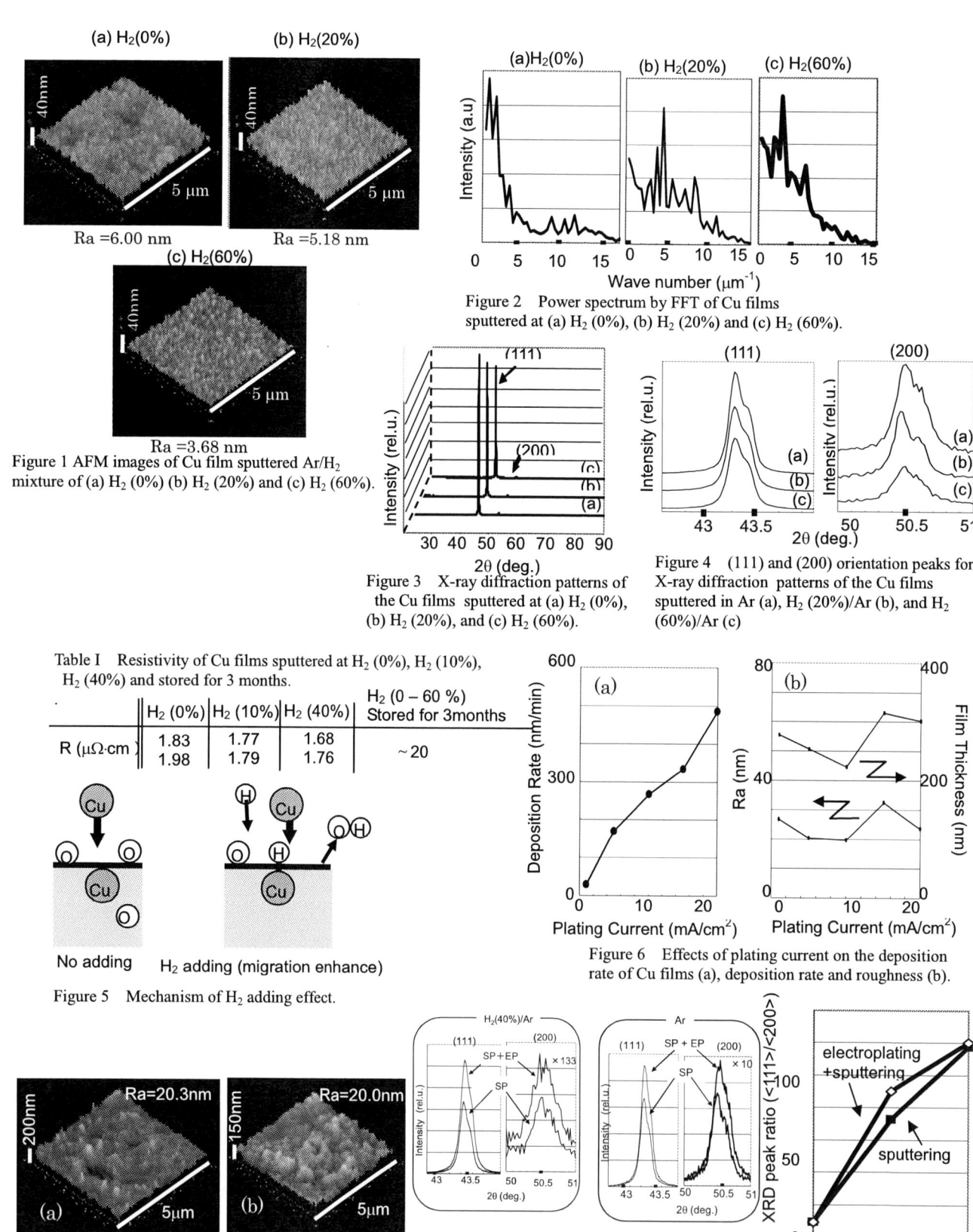

(a) H₂(0%)   (b) H₂(20%)

Ra =6.00 nm   Ra =5.18 nm

(c) H₂(60%)

Ra =3.68 nm

Figure 1 AFM images of Cu film sputtered Ar/H₂ mixture of (a) H₂ (0%) (b) H₂ (20%) and (c) H₂ (60%).

Figure 2 Power spectrum by FFT of Cu films sputtered at (a) H₂ (0%), (b) H₂ (20%) and (c) H₂ (60%).

Figure 3 X-ray diffraction patterns of the Cu films sputtered at (a) H₂ (0%), (b) H₂ (20%), and (c) H₂ (60%).

Figure 4 (111) and (200) orientation peaks for X-ray diffraction patterns of the Cu films sputtered in Ar (a), H₂ (20%)/Ar (b), and H₂ (60%)/Ar (c)

Table I Resistivity of Cu films sputtered at H₂ (0%), H₂ (10%), H₂ (40%) and stored for 3 months.

	H₂ (0%)	H₂ (10%)	H₂ (40%)	H₂ (0 – 60 %) Stored for 3months
R (μΩ·cm)	1.83	1.77	1.68	~ 20
	1.98	1.79	1.76	

Figure 5 Mechanism of H₂ adding effect.

Figure 6 Effects of plating current on the deposition rate of Cu films (a), deposition rate and roughness (b).

Figure 7 AFM images of electroplated Cu film. Cu seed layer is sputtered at (a) H₂ (0%) and (b) H₂ (40%).

EP: electroplating SP: sputtering

Figure 8 (111) and (200) peaks for XRD patterns of sputtered and electroplated Cu films. Cu seed layer is sputtered Ar and H₂ (40%)/Ar.

Figure 9 Effect of H₂ addition ratio on (111)/(200) peak ratio for XRD patterns.

Resistance Increase in Metal Nano-wires

Hsueh-Chung Chen, Hsien-Wei Chen, Shin-Puu Jeng, Chii-Ming M. Wu and Jack Y.-C. Sun

Taiwan Semiconductor Manufacturing Company, Ltd. (TSMC), Hsinchu, Taiwan, R.O.C.

Abstract

As the dimension of copper interconnect scales into the nano-meter regime, the resistivity of copper rapidly increases, primarily due to an electron scattering effect and other dimensional dependent factors, such as film quality. In this paper, we attempt to use a simplified parameter, dimension impact factor (DIF), which includes both surface and grain boundary scattering, to characterize the dimensional dependency of metal resistivity. Among the metal studied, silver has the largest DIF while aluminum has the lowest value. The chief reason is that aluminum has a short electron mean free path (MFP), meaning that it tends to be less affected by dimensional scaling, and has a higher electron specular ratio. In addition to the factor of MFP, resistivity can be affected by other dimensional dependent factors, such as film quality.

Introduction

In order to meet the demands for higher performance and lower cost, the feature size of device is scaled approximately every two years. Unfortunately, metal resistivity increases as metal linewidth scales into the nano-meter regime, and high wire resistance degrades circuit performance. In this paper, we attempt to use a simplified parameter, dimension impact factor (DIF), which includes both surface and grain boundary scattering, to characterize the dimensional dependency of metal interconnect. We clearly demonstrate that resistivity can be affected by other dimensional dependent factors, such as film quality. The effect of temperature on metal resistivity is also studied.

Experimental

Both Cu and W wires are fabricated by standard single damascene process. In order to overcome the lithography limitation, a back-fill process by depositing a thin TEOS liner inside trenches is employed to form wires with a 70 nm linewidth. In order to alleviate the poor deposition quality, caused in the higher aspect ratio of the trench, wide trenches with a thin dielectric thickness are fabricated. Together with CMP overpolish, various aspect ratio wires are obtained (Fig. 1 and Fig. 2). Each resistivity data point shown in this paper is obtained from a minimum of ten sets of TEM measurement and 4-point probe. For Al-Cu film, standard physical vapor deposition (PVD) method is used.

Two-Dimensional Resistivity Model Equation (1) is a 2-D model, which combines both surface scattering and grain boundary effects by assuming grain size is the same dimension of linewidth.

$$r = r_{bulk} + \frac{3}{2}(1-P) \cdot \frac{r_{bulk} \cdot l}{\sqrt{Xarea}} + \frac{3}{2}\left(\frac{R}{1-R}\right) \cdot \frac{r_{bulk} \cdot l}{\sqrt{Xarea}} \quad ....(1)$$

where λ is the mean free path of electrons, P is the fraction of electrons scattered specularly at surface (i.e., no momentum loss),

R is fraction of electrons scattered away by grain boundary, $Xarea$ is the cross section of wire, and $\sqrt{Xarea}$ is defined as the effective width. In here,

$$\left[(1-P)+\left(\frac{R}{1-R}\right)\right] \cdot r_{bulk} \cdot l \quad\text{..}(2)$$

is defined as the dimension impact factor (DIF). Larger DIF means metal resistivity increases faster as dimension shrinks.

Results and Discussions

Resistivity of Metallic Thin Films Fig. 3 shows resistivity as a function of blanket film thickness for thermal evaporation (TE) silver [1], TE aluminum [2], PVD tungsten [3], PVD aluminum-copper, and ECP copper. Each metal exhibits different degree of size dependence. The resistivity of copper and silver rises rapidly, while tungsten, aluminum, and aluminum-copper films show relatively small changes. It is notice that the behavior of aluminum and aluminum-copper is similar. The fittings are done with the 1-D version of Eq. (1).

Resistivity of Metal Nano-wires Fig. 4 plots the resistivity of copper as a function of wire cross-section and filling aspect ratio. Resistivity increases as the cross section of copper wires decreases. However, the copper wires with a low-aspect filling ratio exhibit lower resistivity values as compared to those with high ratios. Due to the nature of metal filling process, the properties of both PVD copper seed layer and electro-plated copper are strongly dependent upon the aspect ratio of metal trenches. The copper films inside high aspect trenches likely have more grain or twin boundaries, compositional variation, and micro voids. [4] As a result, their resistivity tends to be higher. Likewise, there are seams in the fine tungsten wires, as shown in Fig. 2(e). The seam creates additional surface and would further increase resistivity. In this study, we attempt to minimize these deposition related variations, and only those wires with low-aspect-ratio filling are used. Fig. 5 shows the resistivity of tungsten, silver and copper nano wires. The data of silver wires is re-plotted from reference 7. For comparison, simulation curves of alpha tungsten [3] and aluminum wires are plotted in Fig. 5. As expected, both copper and silver wires show a significant size effect.

Model Fitting Using the electron mean free path listed in Table 1, the resistivity of each metal wire can be fitted by Eq. (1). The parameters used for ECP copper and CVD tungsten wires are $\rho_{bulk,Cu}$=2.2 μΩ-cm, P=0.1, R=0.2 and $\rho_{bulk,W}$=13.8 μΩ·cm, P=0.45, R=0.1, respectively. These scattering parameters are close to those extracted from thin films in Fig. 3, where metal wires show more influence by size effect. In order to realize the impact of dimension dependence, the resistivity curves of copper and tungsten are then extrapolated to 10 nm. Meanwhile, simulation curve for aluminum and alpha-tungsten is done utilizing the scattering parameters obtained from the fitting of metallic thin films.

1-4244-0181-X/06/$20.00 ©2006 IEEE

Dimension Impact Factor The dimension dependence found in Fig. 3 and Fig. 5 can be explained by the dimension impact factor. Table 1 lists the electrical resistivity, electron mean free path and scattering parameters. Tungsten and aluminum have the lowest DIF, and they are about 64% and 48% of copper, respectively. Moreover, tungsten and aluminum also has the highest electron specular ratio of 0.45 and 0.5, respectively, meaning that there are less scattered electrons. The resistivity of tungsten and aluminum are less affected by the effect of surface and grain boundary scattering.

Unfortunately, the bulk resistivity of both tungsten and aluminum is relatively high. As a result, their application could be limited to non-performance driven products, such as memories, or devices with short interconnect length, where RC delay is not a major concern. Nevertheless, it is worth pointing here that although the impurity and beta phase in CVD W increase the bulk resistivity, our simulation data suggests that the alpha phase tungsten could have a lower resistivity than copper beyond the 22 nm generation, where the resistivity of aluminum wire is expected to be lower than copper wires starting from the 45 nm generation.

Conclusion

The changes in resistivity of metal nano-wires are studied. The mean free path of electron scattering has a determining role in resistivity increase. The resistivity of aluminum and tungsten lines remains relatively stable as compared to that of silver and copper during scaling. Poor film quality increases metal resistivity, as demonstrated in our trench filling experiment. High aspect ratio filling probably is not preferred for future generations. Good control of surface and large grain size are also important to achieve stable wire resistance.

References

[1] J. W. C. De Vries, Thin Solid Films, 167, 1988, pp. 25-32.
[2] Juan M. Camacho and A. I. Oliva, Microelectronics Journal, 36, 2005, pp. 555-558.
[3] S. M. Rossnagel, I. C. Noyan, and C. Cabral, Jr., J. Vac. Sci. Technol., B 20(5), 2002, pp. 2047-2051.
[4] K. Mirpuri and J. Szpunar, Micron 35, 2004, pp. 575-587.
[5] D.R. Lide, (ed.) in Chemical Rubber Company handbook of chemistry and physics, CRC Press, Boca Raton, Florida, USA, 79th edition, 1998.
[6] K. Fuchs, Proc. Cambridge Philos. Soc., 34, 1938, pp. 100-108.

Table 1 Physical and Scattering Properties of Selected Metal Wires

Metal Wire	Resistivity[5], ρ ($\mu\Omega \cdot cm$)	Mean Free Path[6], λ (nm)	P	R	Dimension Impact Factor (DIF)
Ag	1.6	52.7	0.05	0.4	1,363
Cu	1.67	39.3	0.1	0.2	759
Al	2.65	14.9	0.5	0.3	367
W	5.2	14.2	0.45	0.1	488

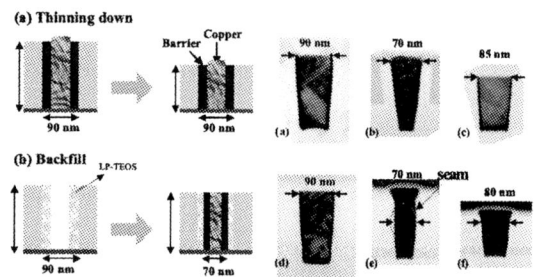

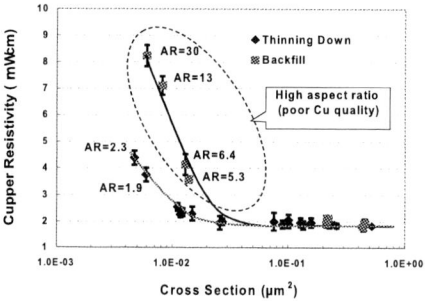

Figure 4. Copper resistivity as a function of metal line cross section. High filling aspect ratio results in poor metal filling quality hence leads to high metal resistivity.

Figure 1. Schematics of nano-wire fabrication (a) "thinning-down" and (b) backfill processes.

Figure 2. Cross sections of copper and tungsten nano-wires (a) copper wire by the standard damascene process; (b) copper wire by the backfill process; (c) copper wire by the "thinning-down" process; (d) tungsten wire by the standard damascene process; (e) tungsten wire by the backfill process; (f) tungsten wire by the "thinning down" process

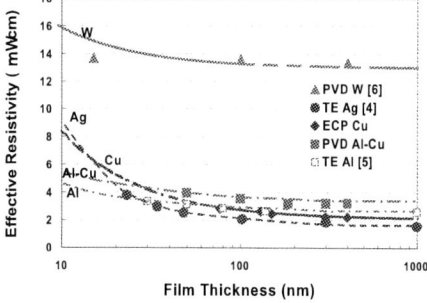

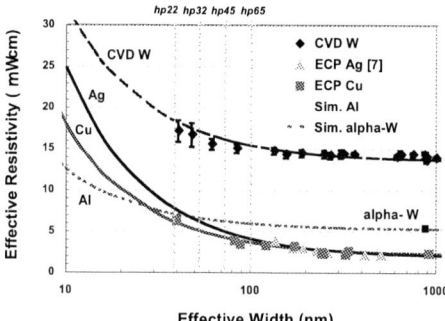

Figure 3. Resistivity of Al, Al-Cu, W, Cu, and Ag thin film as a function of film thickness. The data of Ag, Al and W are adopted from reference 1, 2 and 3, respectively.

Figure 5. Measured resistivity of tungsten and copper nano-wires as a function of effective line width. The data of silver wires are adopted from reference 7. The data of Al-Cu and alpha-phase tungsten are from simulation based on equation (1).

2006 International Symposium on VLSI Technology, Systems, and Applications

Die-Based Electromigration Characterization For
Copper / Low-K Dual Damascene Interconnects

Shou - Chung Lee, Anthony S. Oates
Taiwan Semiconductor Manufacturing Company
121, Park Ave. 3, Hsinchu Science-Based Industrial Park,
Hsinchu, Taiwan, R.O.C.

ABSTRACT

We demonstrate that electromigration (EM) failure distributions can be determined from multi – link via – chain test structures by the analysis of successive failures of in the chain. Failure distributions from one multi – link structure are consistent with that of conventional single link structure. This approach to testing results in improved defect sensitivity and allows the variation of EM failure characteristics across wafers to be determined.

INTRODUCTION

Cu / low-k dual – damascene interconnects EM failure modes has been widely discussed recently. Generally, intrinsic failure modes are associated with voids in the line [1], and extrinsic modes are associated with voids associated with vias [2]. Of particular interest is the observation the Cu / low- k failure modes tend to be dominated by early failure distributions from vias. The statistical multi – link structure has been proposed by Ogawa etc [3] to detect the Cu via reliability early failures. Here we point out a more application for the multi – link structures. In this study, the analysis of resistance versus time trace of the multi – link structure has been investigated to characterize the behavior of the Cu / low-k interconnects. We show that the successive EM failure time in a single multi – link structure can form a complete lifetime distribution. This implies that EM reliability can be evaluated in one die rather than across a wafer, as is the current industry practice. This die-based EM characterization is capable of providing more accurate die-to-die variation information and is especially important to understand process and reliability issues for 300mm wafers.

EXPERIMENTAL

Cu / low-k EM reliability are usually evaluated using single link structures designed to investigate either "down-stream" electron flow, in which the electrons flow from the upper metal level to the lower level (V1M1) or "up-stream" electron flow, in which the electron flow is in the opposite direction (V1M2). Here we characterize dual damascene Cu / low-k EM failure distributions using multi – link test structure consisting of chains of identical V1M1 or V1M2 elements with 250µm long Cu lines to extract the failure time of each elements. The test structure schematic diagrams are shown in Fig.1.

RESULTS AND DISCUSSIONS

Failure modes of single link EM structure

The failure modes of single link V1M1 and V1M2 structures were discussed first as a basis for the further discussions of the analysis of the multi – link structures. For V1M1 structures, two types of failure modes were observed from the resistance trace shown in Fig.2. The first failure mode is large resistance jumps with the magnitude of 500 ~ 1000 Ω, which is a catastrophic failure for the circuit operation. The second failure mode is small resistance jump with the magnitude ~ 100 Ω and following by a progressive

resistance increase. Failure analysis indicated that the large resistance jumps ~ 1000 Ω were associated with the small slit – like voids under vias (Fig.3 (A)), while small resistance increase ~ 100 Ω were associated with voids in the lower metal level (Metal-1) which tend to span the Cu line thickness (Fig.3 (B)). The major reliability concern of V1M1 structure is the slit – like voids directly under vias because of its low failure time and high resistance increase. For V1M2 structures, the failure mode is the voids in the upper metal line (Metal-2) away, or above vias. This failure mode of V1M2 structure is identical with the failure mode with small resistance increase of V1M1 structure. Fig. 4 shows the resistance trace of the single link V1M2 structure. The small resistance jump around 50 ~ 100 Ω is due to the beginning of void formation process in which the void just span to the barrier layer so that the barrier layer resistance shows up. The progressive resistance increase following by the small resistance jump is due to the void growth process that the Cu atoms were continuously displaced along the line length direction, and more barrier layer will shunt the current so that the resistance increases gradually. Fig.5 confirms that the voids occurred in the metal line away from vias.

Analysis of multi – link EM structure

Fig.6 plots the resistance race of the V1M1 50-link structure. The obvious stair - like resistance jumps with consistent magnitude of ~ 1000Ω in single link structure were observed. Successive EM failures of different Cu line elements can be clearly seen according to the large resistance jumps in the 50-link structure. The failure time distribution of the successive EM failures of V1M1 50 -link structure were plot in Fig.7, in which this die is from wafer center area. Results of single link V1M1 structure are also included in Fig.7 for comparisons. The failure time distribution of one 50-link structure is in excellent agreement with the single link distribution, implying no other effect influence the failure time of individual Cu test line in multi – link structure. The failure analysis confirmed the same failure mode as the single link structure i.e. slit – like voids under vias, as shown in Fig.8. It is noted that there are less than 50 failures were observed in this 50-link structure since we just consider the failures due to small slit – like voids under vias.

Fig.9 shows the typical resistance trace of V1M2 50-link structure. It is difficult to find the small resistance jump due to the void formation in the plot since both void formation and growth processes from different lines occur simultaneously and result in resistance increase with magnitudes close to each other. To improve the sensitivity to detecting the EM failure signal due to the void formation of each Cu line, the resistance derivative with time plot, i.e. dR / dt , were also generated in Fig.9. Strong signals from the main peaks clearly represent the failure time due to void formation and correlate well with the small jumps in the test structure resistance. Peaks with smaller magnitude than the main peaks are also evident; their cause may be due to voids at locations other than vias. Since these smaller peaks tend to occur at later times they do not represent the most important reliability limiting situation, and so we focus on the larger peaks. Fig.10 shows the failure time distribution of one 50-link structure analyzed by the resistance derivative technique. The

---

1-4244-0181-X/06/$20.00 ©2006 IEEE   131

multi-link distribution is in good agreement with the distribution derived from single link structures. Fig.11 shows the failure analysis confirms that the V1M2 50-link structure failure is due to the voids in the upper level metal line above the via and is consistent with the results of single link structures.

Application of multi – link EM structures: Die-based EM characterization

The analysis of successive failures of Cu lines in multi – link structures is useful for investigating the die-based EM reliability. Fig. 12 demonstrated the failure time distribution comparisons from different die locations using the V1M1 50-link structures in a diameter of 300 mm wafer. The median time to fail (MTF) of edge dies are obviously 2 times lower than that of center die, which provide a more clear picture to the variations of different dies. The multi – link structures is not only useful for the die-based EM reliability characterization, but is a very cost effective structure. It is possible to use only one, or few, multi – link sample to characterize the EM reliability that usually to be done with the sample size of 10 to 30 for of the single link structures. So the EM parameters such as activation energy (*Ea*), current exponent (*n-value*), and critical current density (*jc*) etc. can be measured using the multi – link structures.

CONCLUSIONS

Cu / low-k EM reliability can be characterized by using only one sample of multi – link structure as opposed to the current industry practice of using multiple test structures from across a wafer. Based on our results from multi – link structures, the failure modes and failure time distributions of both V1M1 and V1M2 structures are consistent with the results from single link structures. This technique enables die-to-die variations of EM reliability to be accurately addressed with the benefit of a saving of test resources.

REFERENCES

[1] Baozhen Li, TDMR, 2004, vol.4, No.1 pp. 80-85
[2] Jason Gill, IRPS, 2002, pp. 298-304
[3] E. T. Ogawa, IRPS, 2001, pp.341-349

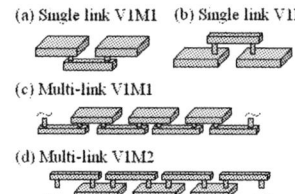

FIGURE 1. SINGLE LINK AND MULTI-LINK TEST STRUCTURE DIAGRAMS

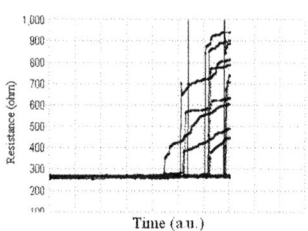

FIGURE 2. RESISTANCE-TIME TRACE OF THE SINGLE LINK V1M1 STRUCTURE

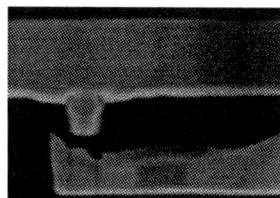

FIGURE 3 (A). FAILURE ANALYSIS OF SINGLE LINK V1M1 STRUCTURE WITH LARGE RESISTANCE JUMP ~1000 Ω

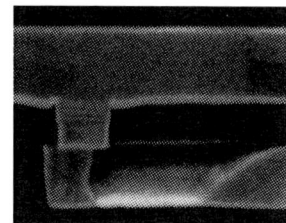

FIGURE 3 (B). FAILURE ANALYSIS OF SINGLE LINK V1M1 STRUCTURE WITH SMALL RESISTANCE JUMP ~ 100Ω

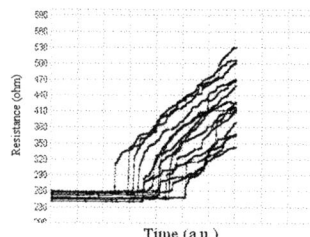

FIGURE 4. RESISTANCE-TIME TRACE OF THE SINGLE LINK V1M2 STRUCTURE

FIGURE 5 . FAILURE ANALYSIS OF THE SINGLE LINK V1M2 STRUCTURE

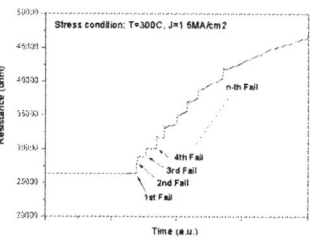

FIGURE 6. RESISTANCE-TIME TRACE OF THE 50-LINK V1M1 STRUCTURE

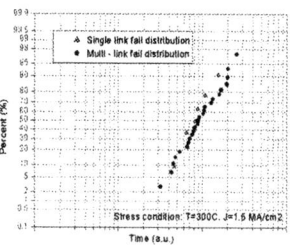

FIGURE 7. FAILURE DISTRIBUTIONS OF THE SINGLE LINK AND 50-LINK V1M1 STRUCTURES

FIGURE 8. FAILURE ANALYSIS OF THE 50-LINK LINK V1M1 STRUCTURE

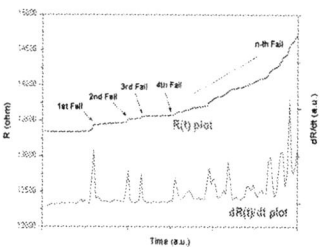

FIGURE 9. RESISTANCE AND RESISTANCE DERIVATIVE WITH TIME PLOTS OF THE 50-LINK V1M2 STRUCTURE

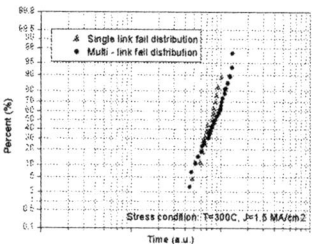

FIGURE 10. FAILURE DISTRIBUTIONS OF THE SINGLE LINK AND 50-LINK V1M2 STRUCTURES

FIGURE 11. FAILURE ANALYSIS OF THE 50-LINK V1M2 STRUCTURE

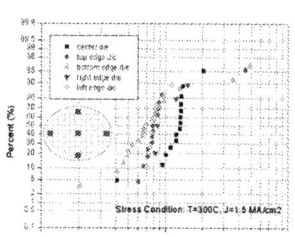

FIGURE 12. FAILURE DISTRIBUTIONS OF THE V1M1 50-LINK STRUCTURE FROM DIFFERENT DIE LOCATIONS

2006 International Symposium on VLSI Technology, Systems, and Applications

Integration of Cu Damascene with Pore-sealed PECVD Porogen Low-k (k=2.5) Dielectrics for 65nm Generation

M.L. Yeh, C.C. Chou, T.I. Bao, K.C. Lin, I.I. Chen, K.P. Huang, Z.C. Wu, S.M. Jeng C.H. Yu and M.S. Liang

Advanced Module Technology Division, Research & Development
Taiwan Semiconductor Manufacturing Company, Science Park, Hsin-Chu, Taiwan, R.O.C.
Tel: (886)-3-6665307, Fax: (886)-3-5773671, E-mail: ccchoug@tsmc.com

Abstract

Owing to the k extendability of porogen LK formed with the incorporation and removal of organic porogen precursors, the porogen LK is the competitive candidate for inter-metal dielectrics (IMDs) of 65nm generation and beyond. However, its porosity raises major challenges in the Cu/LK integration. Chemical and metal penetrability of the porogen LK film revealed the necessity of a protective pore sealing layer in dual damascene. Pore sealing materials were evaluated and SiCxHy film demonstrated exceptional barrier property against metal diffusion and good step coverage over the trench profile. By introduction of this SiCxHy layer, 10% capacitance reduction was achieved despite the higher k of the material. With the well-controlled thickness, SiCxHy pore sealing also demonstrated no via-Rc shift compared to the scheme without pore sealing, therefore excellent protection on the trench structure without via performance degradation was accomplished.

Introduction

As the size of IC device shrinks, the adoption of low-k material as the IMD is essential. To meet the more challenging RC delay requirement in 65nm generation and beyond, porogen LK formed with the incorporation and removal of organic porogen precursors is the potential candidate for extreme low-k material with k<2.5 and appropriate mechanical properties. However, the porosity of the porogen LK film raises many integration challenges such as moisture uptake, chemical absorption and metal diffusion. In this paper, a SiCxHy film served as the protective pore sealing layer was investigated. Excellent electrical performance including Rs, line-line leakage and capacitance, and via-Rc will be presented.

Low-k Film Formation

The low-k material used in this study is a PECVD deposited organosilicate glass (OSG) with the incorporation of organic porogen precursors. The porogen is removed in the subsequent electron beam (EB) curing process and forms pores in the average size of 1.5nm. The EB condition was optimized to ensure the complete porogen removal and appropriate mechanical properties with k<2.5. The hardness and the Young's modulus of this film were measured 1.3 and 8.5 GPa, respectively, using a nanoindentor. The leakage current density at 1 MV/cm and the breakdown field of this LK film are 1.5E-11 A/cm^2 and 4.3 MV/cm, respectively, from mercury probe measurements.

Cu/ELK Integration

The Cu/ELK interconnects was integrated using damage-minimized low-k etching and ashing [1] for patterning and optimized Cu-CMP process [2] for damascene completion. No capping layer or middle etch stop layer was used in the IMD film stacking. Nitrogen-free IMD processes were introduced to resolve the amine induced PR poisoning commonly associated with Cu/OSG integration using via-first approaches [3].

The Necessity of Pore Sealing

The major challenge for integrating Cu/porogen LK process is the porosity of the porogen LK film. The vulnerable porogen LK film revealed obvious surface porosity after etching process, as shown in the top-view SEM of the etched trench profile in Fig. 1. The pits on the etched surface provided penetration channels for moisture, chemicals and metals in the subsequent processes. Fig. 2 shows the Auger electron spectrum of the LK film before and after etching. The increased post-etching carbon and fluorine depth profiles represesed the etching chemical penetration into the porous film. The penetrability of the porous film was further identified from the trench-bottom metal penetration shown in the cross-sectional TEM of Fig. 3. TaN for Cu barrier penetrated through the etched surface and formed Ta clouding. These facts all indicated that the requirement of a protective pore sealing layer is inevitable in the via-first dual damascene interconnects.

Result and Discussion

(1) *Pore Sealing Material Selection*: Several materials were evaluated, and SiCxHy and CxHy were selected for their better interface adhesion to Cu and ELK. Fig. 4(a) and 4(b) shows the cross-sectional TEM of Cu/ELK interconnect with SiCxHy and CxHy pore sealing respectively. While Ta clouding was only alleviated with CxHy pore sealing, SiCxHy served as good barrier to metal penetration and no Ta clouding was observed. The barrier property is also demonstrated in the thickness loss of the pre-metal clean step: Thickness loss of SiCxHy is only 1% to that of CxHy.

(2) *Step Coverage*: To meet the necessary step coverage for fully protection of porogen LK film, a low-deposition rate SiCxHy film was introduced as this pore sealing layer. Fig. 5 reveals the etched trench sidewall and bottom covered by the SiCx film with good step coverage.

(3) *Line-Line Leakage Current and Cu Line Resistance*: Fig. 6, 7 and 8 shows that tightly distributed 0.11/0.10 um line-line leakage current, 0.11um wide Cu line Rs and 0.10um via-chain via_Rc were achieved with no degradation in yield.

(4) *Capacitance Reduction*: The capacitance reduction performance of pore sealing was demonstrated in the C-1/R plot where C and R are the line resistance and capacitance in the 0.6um-pitched lines. From the results with various pore sealing layer thickness shown in Fig. 9, the line capacitance was reduced as the pore sealing layer thickness increased. Though the insertion of higher-k pore sealing layer provided the greater capacitance, the integrated 10% capacitance reduction obtained with 2nm pore sealing layer showed evident damage resistance on the porogen LK film. Furthermore, no via-Rc shift was observed from the result of 2nm pore sealing layer, shown in Fig. 10, therefore the SiCx pore sealing achieved excellent protection on the trench structure without via performance degradation.

Conclusion

A 65nm generation Cu/porogen LK interconnect technology with protective pore sealing layer was developed successfully. Though the higher-k material is introduced, the overall line capacitance was reduced by 10% without via-Rc degradation. The excellent damage resist property of pore sealing layer provided a feasible integration of the vulnerable porogen LK film, which makes the Cu/ELK (k<2.5) BEOL interconnection applicable for 65nm generation and beyond technologies.

Reference

[1] Y.L. Yang et al., IITC, pp12 (2003)
[2] S.M. Jang et al., Symp. on VLSI Tech., pp 18 (2002)
[3] K. Mosig et al., IITC, pp292 (2001).
[4] L.P. Li et al., Symp. on VLSI Tech., pp105 (2003)

1-4244-0181-X/06/$20.00 ©2006 IEEE

[5] K.C. Lin et al., Symp. on VLSI Tech., pp 66 (2004)

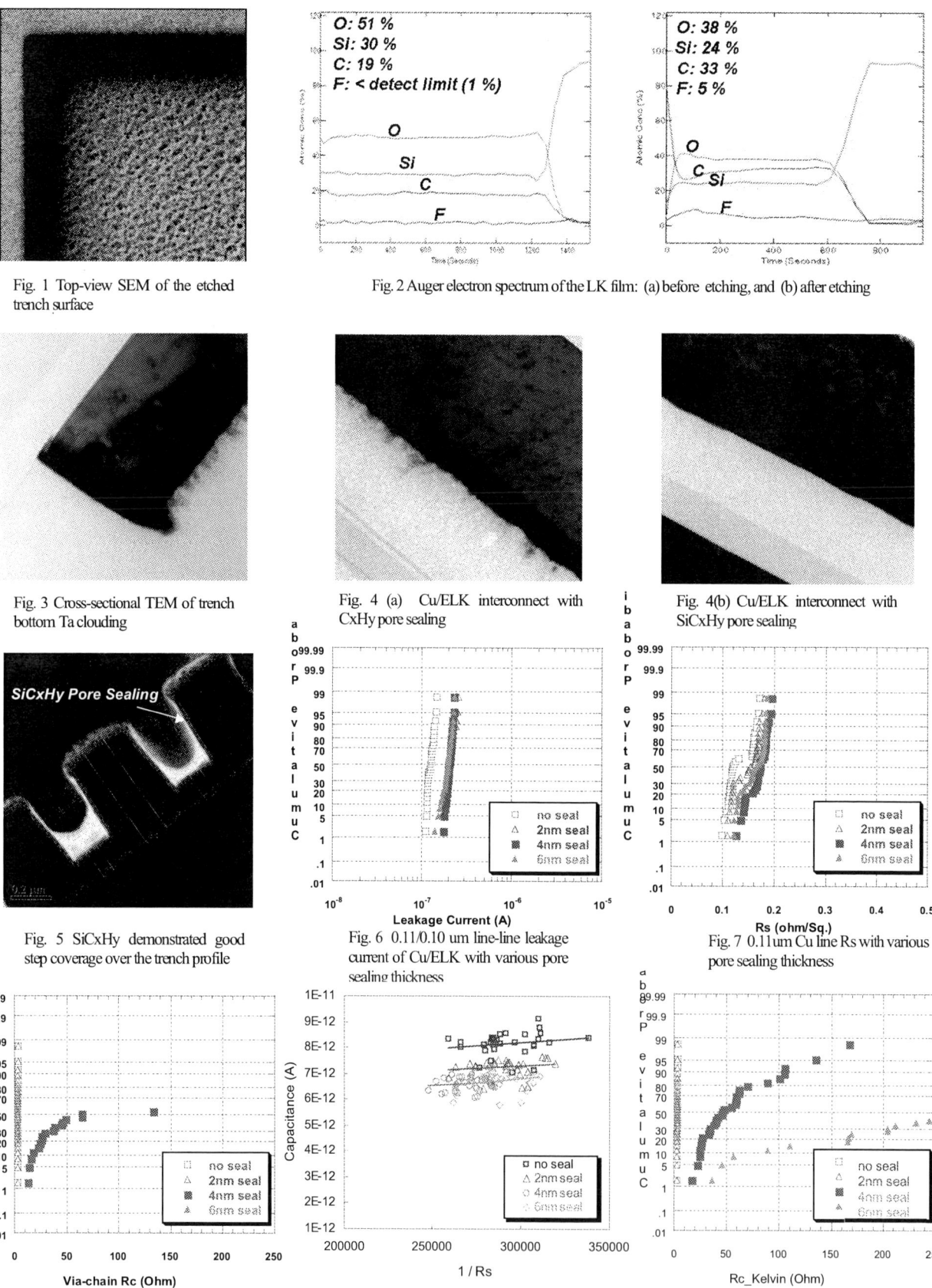

Fig. 1 Top-view SEM of the etched trench surface

Fig. 2 Auger electron spectrum of the LK film: (a) before etching, and (b) after etching

Fig. 3 Cross-sectional TEM of trench bottom Ta clouding

Fig. 4 (a) Cu/ELK interconnect with CxHy pore sealing

Fig. 4(b) Cu/ELK interconnect with SiCxHy pore sealing

Fig. 5 SiCxHy demonstrated good step coverage over the trench profile

Fig. 6 0.11/0.10 um line-line leakage current of Cu/ELK with various pore sealing thickness

Fig. 7 0.11um Cu line Rs with various pore sealing thickness

Fig. 8 0.10um via-chain Rc with various pore sealing thickness

Fig. 9 Capacitance vs 1/Rs with various pore sealing thickness

Fig. 10 Kelvin-Rc with various pore sealing thickness

2006 International Symposium on VLSI Technology, Systems, and Applications

Wafer-Level Compliant Bump for 3D Chip-Stacking

Naoya Watanabe, Takeaki Kojima, and Tanemasa Asano

Center for Microelectronic Systems, Kyushu Institute of Technology
680-4 Kawazu, Iizuka, Fukuoka 820-8502, Japan
Phone: +81-948-29-7589, Fax: +81-948-29-7586, E-mail: naoya@cms.kyutech.ac.jp

Abstract

We introduce wafer-level compliant bump for 3D chip-stacking. The inter-chip connection up to 10000 bump connections is demonstrated. It is also demonstrated that the compliant bump is very effective in minimizing strain generated in the device even when the bump bonding is performed directly on the device.

Introduction

Three-dimensional (3D) chip-stacking technology is attracting a great deal of attention for realizing advanced high-speed, compact, and highly functional electronic systems [1-4]. 3D chip stack system requires highly-reliable high-density area-bump interconnection technology. Solder bump technology encounters difficulty in shrinking bump size and its pitch. The non-melting bonding of metal bumps such as plated Au bumps, on the other hand, encounters the problems in bump height deviation which causes bonding failure and strain generation in the LSI device as the bump pitch becomes small.

To overcome these issues and meat the requirements of 3-D chip stack system, we have proposed the concept of compliant bump [5] such as the pyramid bump. The compliant bump, since it easily deforms to spread out, offers the following effects (see Fig. 1); (1) reliable bonding at low temperature (bonding even at room temperature is expected), (2) minimizing the strain generation at the device level, (3) compensation of bump height deviation, and (4) resin exclusion in chip stacking with precoated-resin. Using the pyramid bump fabricated by the bump transfer method, we have confirmed (3) and (4) of above effects [5]. However, the bump transfer method is hard to apply the wafer level process and the transfer yield is limited.

In this paper, we demonstrate a new wafer-level fabrication of compliant bump and successful bonding up to 10000 Au area bump whose pitch is 20 μm. We also prove that the compliant bump minimizes strain generation in the device by performing bonding directly on MOSFET.

Wafer-level fabrication of compliant bump

Figure 2 shows the process flow of the wafer-level compliant bump formation. First, Ti, W and Au were subsequently deposited as the seed layer for electroplating (Fig. 2(a)). Next, the resist pattern having an under-cut hole profile was formed by photolithography (Fig. 2(b)). We developed a novel process that can control the under-cut profile. Au electroplating was applied to fill the under-cut holes in the resist film. The photoresist and the seed layer were subsequently removed by a solvent and sputter etching, respectively (Fig. 2(c)).

Figure 3 shows the optical micrograph and SEM image of new compliant bumps (cone bumps) on a wafer. The number of bumps is 10000 per 2.0 × 2.0 mm^2 chip.

We can see that uniform cone bumps are formed over the chip area. The bump size is 11 μm in diameter at the basement and the bump height is 10 μm.

Effect of new compliant bump

The ability of cone bump to produce a large number of inter-chip connection is demonstrated from the results shown in Fig. 4, where results of daisy-chain connection tests were plotted. The cone bump provides successful interconnection at least up to 10000 bumps, while the conventional plated bump fails in connection at a few thousand bumps.

We investigated the strain generated in a MOSFET when a bump was directly bonded to an electrode placed just above the MOSFET (L/W= 2 μm/5 μm, t_{ox}=30 nm). Test chip is shown in Fig. 5. Change in characteristic of n-MOSFET and p-MOSFET was measured to evaluate the strain generated as the result of bump bonding. The transconductance g_m changes while threshold voltage V_{th} hardly changes. Figures 6(a) and 6(b) show the change in maximum g_m due to the direct bump bonding observed for n-MOSFET and p-MOSFET, respectively. We find that the compliant bumps (the pyramid bump and the cone bump) generate much less residual strain in the MOSFET than the conventional plated bump. From analysis of these results together with the results observed for MOSFETs placed in between bump, we can say that compliant bumps suppress chip bending and reduce the residual strain caused by deformation of the chip.

Conclusion

We have developed a new wafer-level compliant bump, which can produce high-density, large number (at least up to 10000) inter-chip connections for chip stacking. The compliant bump offers circuit design in which bump interconnection can be placed in any place over chip area. The bump size and pitch can be further shrunk because of the use of the photolithography. We have been investigating the application of the area bump technology to 3-D reconfigurable logic systems.

Acknowledgment

This work was supported in part by the Center of Excellence program of Kumamoto Prefecture, a Grant-in-Aid for Scientific Research (No. 17710116) of MEXT, the Kitakyushu knowledge-based cluster project of MEXT, and the Regional Rebirth Consortium Project ("Monodukuri-Kakushin") of METI.

References

[1] M. Koyanagi et al.: IEEE Micro **18** (1998) 17.

[2] K. Kameyama et al.: *Extended Abstracts of the 2004 International Conference on Solid State Devices and Materials* (2004) pp. 276-277.

[3] L. Schaper et al.: *Extended Abstracts of the 2004 International Conference on Solid State Devices and Materials* (2004) pp. 274-275.

[4] K. W. Guarini et al.: *International Electron Devices Meeting Technical Digest* (2002) pp. 943-945.

[5] N. Watanabe, Y. Ootani, and T. Asano: Jpn. J. Appl. Phys. **44** (2005) 2751.

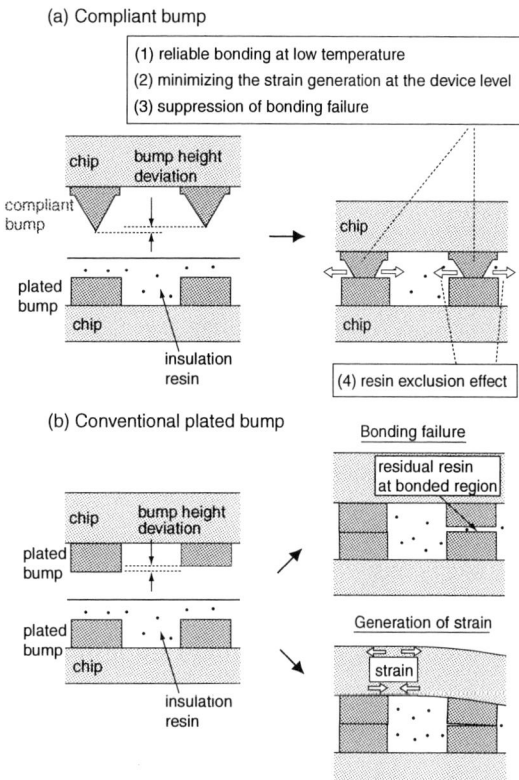

(a) Compliant bump

(1) reliable bonding at low temperature
(2) minimizing the strain generation at the device level
(3) suppression of bonding failure

(4) resin exclusion effect

(b) Conventional plated bump

Fig. 1: Advantages of the compliant bump.

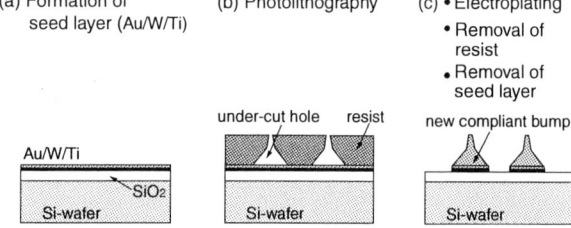

(a) Formation of seed layer (Au/W/Ti)

(b) Photolithography

(c) • Electroplating
• Removal of resist
• Removal of seed layer

Fig. 2: Process flow of new compliant bump (cone bump) using under-cut resist method.

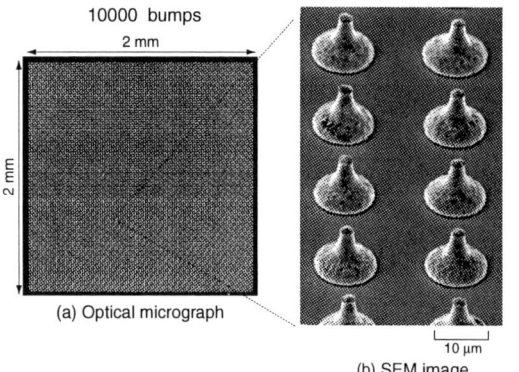

10000 bumps

(a) Optical micrograph

(b) SEM image

Fig. 3: Optical micrograph and SEM image of cone bumps on a chip in a wafer. The bump size is 11 μm in diameter at the basement and the bump height is 10 μm. The number of bump is 10000.

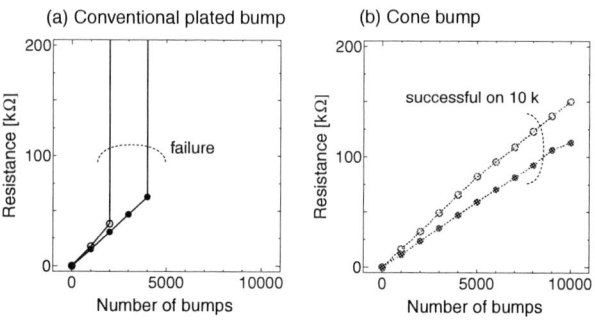

Fig. 4: Characteristic of 10000 bump connections of daisy chain after bonding of chips: (a) Plated bump. (b) Cone bump.

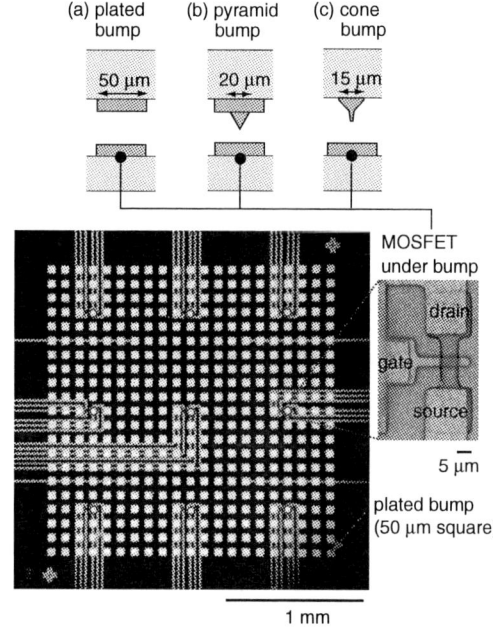

Fig. 5: Photo of the die used to measure the change in the characteristic of MOSFET due to bump bonding. MOSFETs were placed under a bump and in between bumps to evaluate strain generated by bonding.

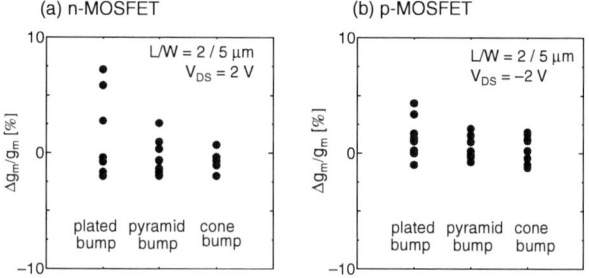

Fig. 6: Change in the maximum g_m of MOSFET after chip stacking. (a) n-MOSFET. (b) p-MOSFET. The compliant bumps (the pyramid bump and the cone bump) generate much less g_m change in the MOSFET than the conventional plated bump.

2006 International Symposium on VLSI Technology, Systems, and Applications

Strained-SOI Technology for High-Speed CMOS Operation

T. Mizuno[1,2], N. Sugiyama[3], T. Tezuka[3], Y. Moriyama[3], S. Nakaharai[3], T. Maeda[1], and S. Takagi[1,4]

[1]MIRAI-AIST, [3]MIRAI-ASET, Komukai Toshiba-cho, Saiwai-ku, Kawasaki, 212-8582, Japan
[2]Kanagawa University, 2946, Tsuchiya, Hiratsuka 259-1293, Japan
[4]The University of Tokyo, 7-3-1 Hongo, Bunkyo-ku, Tokyo 113-8656, Japan
(mizuno@info.kanagawa-u.ac.jp)

I. Introduction

The strained-Si technology, including biaxial/uniaxial strain and local stress techniques, has been widely studied for the carrier mobility enhancement using strained-Si channel layers. In order to realize high-speed CMOS operation, the SOI technology as well as the strained-Si is the key issue, because of the suppressed Coulomb scattering and the lower drain junction capacitance [1]. Using the strained-Si on (100)-surface SGOI substrate (strained-SOI) technology, we have experimentally demonstrated high performance CMOS devices [2]. However, the hole mobility of (100)-surface strained-SOIs is still lower than the electron one.

In this paper, we have discussed strained-SOI technology for further improvement of CMOS operation speed. Firstly, we have developed (110)-surface strained-SOI MOSFETs to realize much higher hole mobility [3]. We have achieved that the (110)-surface hole mobility is much higher than that of the (100)-surface strained-SOIs. Secondly, we have proposed the source-heterojunction-MOS-transistor (SHOT) for future ballistic transport transistors, using the strained-SOI substrates [4]. We have demonstrated that the transconductance of SHOTs is enhanced because of the high velocity electron injection from the source into the channel, using the excess kinetic energy due to the source heterojunction band offset.

II. Stained-SOI Technology

Fig.1 shows a schematic cross section of strained-SOI structures. Large lattice constant of relaxed-SiGe layer induces the tensile strain of the surface Si layer. As a result, both the conduction and the valence band structures are modulated, resulting in the carrier mobility enhancement. There are mainly two methods to fabricate the strained-SOI substrates. One is the SIMOX method [1], and the other is the Ge condensation technique [5], as shown in Fig.2.

As shown in the carrier mobility, μ, enhancement of the (100)-surface strained-SOIs against that of conventional SOIs versus the strain of top Si layer of Fig.3, the electron mobility enhancement amounts to larger than 1.8, but the maximum hole one is about 1.5 at maximum. As a result, the imbalance of the drain current drive between n- and p-MOS becomes larger. Therefore, we have developed (110)-surface strained-SOIs to overcome this problem, because the effective mass of (110) holes is smaller than that of (100) holes.

III. Surface Orientation Engineering: (110) Strained-SOIs

Fig.4 shows the two-dimensional schematic energy ellipses for (110) Si conduction band under tensile strain condition [3]. It is expected that the tensile strain perpendicular to the (110) surface, induces the energy splitting between the 4-fold and the 2-fold valleys, and the energy level of the 4-fold valleys becomes lower than that of the 2-fold valleys [6]. As a result, the electron mobility along the <001> direction can be enhanced, because of lower conductivity mass of the 4-fold valleys along the <001> direction and the reduced inter-valley scattering. On the other hand, it is also expected that the energy levels of the light and the heavy hole bands are split by the tensile strain, resulting in the hole mobility enhancement by the reduction of the interband scattering. Especially, the hole mobility enhancement has the peak value along the <110> direction, because of the anisotropic effective mass behaviors of holes at (110) strained-Si [3].

Fig.5 shows the mobility behaviors of (110) strained-SOI MOSFETs and (110) unstrained bulk- and SOI-MOSFETs, as a function of E_{eff}. Both the electron and hole mobilities in (110) strained-SOI MOSFETs are higher in whole range of E_{eff} than those of (110) unstrained-MOSFETs. Especially, the maximum hole mobility of (110) strained-SOIs amounts to 370cm^2/Vs, and is 2 times higher than the (100) universal mobility. The maximum enhancement factors of electron and hole mobilities against that μ of the (110) unstrained-MOSFETs are 23% and 50%, respectively, although the hole mobility enhancement of (110) strained-SOI decreases with increasing E_{eff}. The slope of the μ-E_{eff} relationship of (110) strained-SOI n-MOSFETs, which is almost the same as that of (110) bulk n-MOSFETs, is larger than that of (100) devices. However, the E_{eff} dependence of hole mobility of (110) unstrained

p-MOS is smaller than that of the (100) universal mobility, which means that the hole mobility reduction at high gate drive is suppressed on (110). This is one of the advantage of MOSFETs using (110) holes.

The current flow direction dependences of the carrier mobility enhancement of n- and p-MOS are shown in Fig.6. Both the electron and the hole mobility enhancements of (110) strained-SOI structures strongly depend on the current flow direction, while the current flow direction dependence of (110) unstrained-MOSFETs is smaller. The electron mobility enhancement along the <001> direction has a maximum value, while the hole one has a maximum value along the <110> direction. These current flow direction dependences can be explained by the anisotropic effective mass behaviors of the carriers on (110) surfaces. These results mean that it is necessary to optimize the current flow direction of n- and p-MOS in (110) strained-SOI CMOS devices.

Fig.7 shows the contour map of the estimated t_{pd} improvement ratio of strained-CMOS inverters to t_{pd} of (100) unstrained-ones. The t_{pd} of both the (100) and the (110) strained-CMOS is improved by about 20% under the present experimental results. By optimizing both the surface orientation and the current flow direction of strained-CMOS with (100) n- and (110) p-MOS, which can be easily realized by strained-FinFET CMOS of n-MOS with (100)-plane and p-MOS with (110)-plane on (100)-surface [3], the improvement of t_{pd} amounts to about 30%.

IV. Source Engineering: Source Heterojunction MOS Transistors (SHOT)

In order to realize SHOT structures for high-velocity electron injection, it is necessary to make the conduction band level of the source region higher than that of the channel region, as indicated by the band diagram of Fig.8 [4]. As a result, the source band-offset structures with ΔEc allow us to inject higher-velocity electrons by the excess energy of ΔEc into the channel region. Fig.9 shows the fabrication steps of the SHOTs with the channel length L_{eff} of 0.15μm. Ge ion implantation into strained-Si layers, after pattering the poly-Si gate (Fig.9 (b)), was used to form relaxed-SiGe source layers on strained-SOI MOS structures (Fig.9 (a)). An annealing process at 900°C for 30min. was carried out to anneal out the Ge implantation damage in the substrates. As$^+$ ions were implanted after forming 35-nm-thick gate side wall (Fig.9 (c)), to fabricate the SiGe/strained-Si heterojunction source outside the n$^+$ source region.

Fig.10 shows the experimental drain bias dependence of the maximum G_m. It is found that G_m enhancement of SHOTs increases with V_d. This means that high-velocity carrier injection becomes more dominant with increasing V_d. On the other hand, the small G_m at lower V_d is attributed to high energy barrier at the source heterojunction, as shown in the inset.

In order to increase the G_m enhancement of SHOTs in whole range of V_d, it is necessary to scale down the SHOT dimensions. In addition, the graded heterojunction structures should be introduced to suppress the energy barrier at the source heterojunction [7]. We have successfully optimized both ΔE_c and the length of the graded-heterojunction, using a 2-D device simulator. Fig.11 shows the contour map of G_m enhancement factors in V_d-ΔE_c plane, without considering the carrier mobility enhancement. When ΔE_c is between 0.15 and 0.25eV, the G_m enhancement value of as large as 1.2 can be achieved in the whole range of V_d, which is due to only the high velocity electron injection effects. When the channel of SHOT is composed of a strained-Si layer, the electron mobility enhancement can also be enjoyed.

V. Summary

Using the strained-SOI technology, we have developed two types of engineering for high-speed CMOS operation. One is the surface orientation engineering using (110)-surface strained-SOIs. We have demonstrated the hole mobility enhancement higher than that in (100) strained-SOIs. The other is the source engineering using source-heterojunction-MOS-transistor for larger G_m enhancement due to high velocity electron injection from the source into the channel regions. These engineering methods are the key technology for future scaled CMOS devices.

1-4244-0181-X/06/$20.00 ©2006 IEEE

Acknowledgment

This work was supported by NEDO. We would like to thank Dr. M. Hirose, and Dr. T. Kanayama for their continuous supports.

References

[1] T.Mizuno et al., IEEE, **EDL-21**, 230, 2000.[2] T.Mizuno et al., IEEE, **ED-50**, 988, 2003.[3] T.Mizuno et al., IEEE, **ED-52**, 367, 2005.[4]T.Mizuno et al., IEEE, **ED-52**, 2690, 2005.[5] T. Tezuka et al., *Symp. VLSI*, p.96, 2002.[6] H. Ezawa et al., Jpn. J. Appl. Phys., **13**, 126, 1974. [7] T.Mizuno et al., Exd. Abst. SSDM, p.262, 2005.

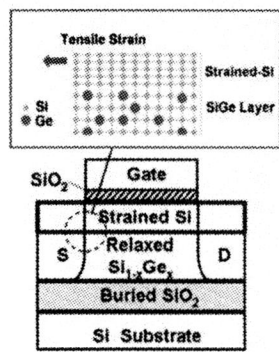

Fig.1 Schematic cross section of strained-SOI MOSFETs.

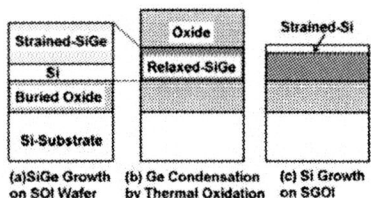

Fig.2 Process steps for strained-SOI substrates, using Ge condensation technique.

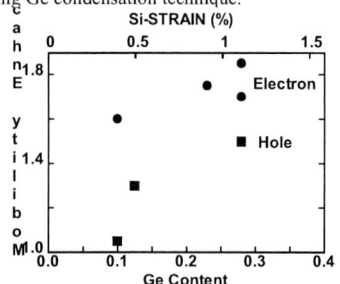

Fig.3 Electron (circles) and hole (squares) mobility enhancements versus Ge content and Si-strain of (100)-surface strained-SOI MOSFETs.

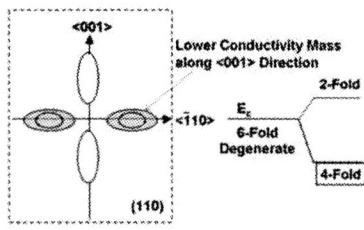

Fig.4 Two-dimensional schematic energy ellipses for (110) Si conduction band and tensile strain induced energy splitting.

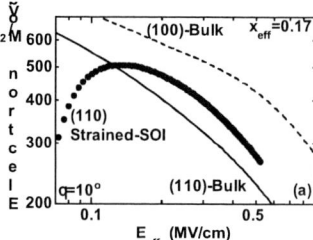

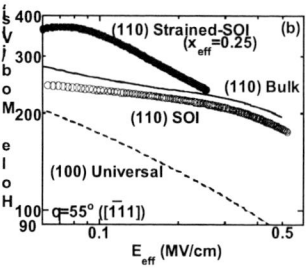

Fig.5 (a) Electron and (b) hole mobility vs. effective vertical electric field, E_{eff}, of (110) strained-SOI (closed characteristics), (110) bulk- (solid lines) and (110) SOI-MOSFETs (open circles), and the (100) universal mobility (dashed lines), where effective Ge content of n- and p-MOS are 17% and 25%, respectively. Current flow direction, θ, is defined as the angle from the [001] direction.

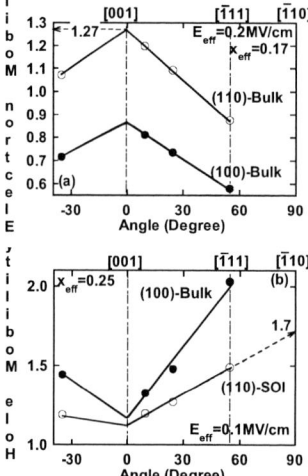

Fig.6 Current flow direction dependence of (a) electron and (b) hole mobility enhancements of (110) strained-SOI MOSFETs against those of (110) unstrained-MOSFETs (open circles) and (100) universal mobility (closed circles). The electron mobility of (110) strained-SOIs is also recovered to about 90% of (100) electron mobility.

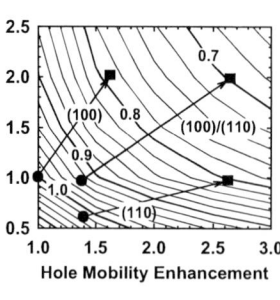

Fig.7 Estimated t_{pd} contour map of (110)/(100) strained-SOI CMOS inverter without load capacitance/resistance at sub-100-nm region.

The vertical and the lateral axes show the electron and the hole mobility enhancements in strained-CMOS. (100)/(110) indicates the data of CMOS with (100) n-MOS and (110) p-MOS. Solid arrows show the improvement of t_{pd} from unstrained- (dots) to strained-CMOS (squares) in each wafer surface orientation at the present time.

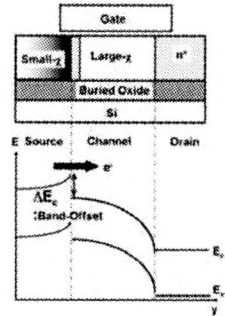

Fig.8 Schematic cross section of new source-hetero MOSFETs with SiGe (small-χ)/strained-Si (large-χ) structures and the band diagram. High-velocity electron can be injected due to excess kinetic energy of the source band-offset ΔEc.

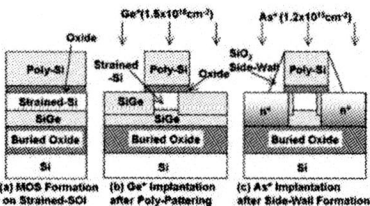

Fig.9 Process steps for fabricating the source-SiGe/channel-strained-Si hetero structures. (b) Ge ion implantation process just after the gate patterning was carried out to form the source SiGe layers on (a) strained-SOI MOS structures.

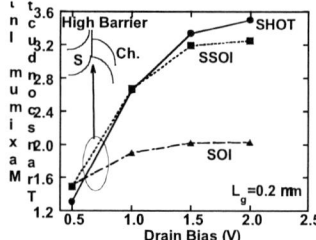

Fig.10 Intrinsic G_m as a function of the drain bias, where $Leff$=0.2µm and W=10µm. ΔEc is about 80meV obtained by the Vth shift of SHOTs.

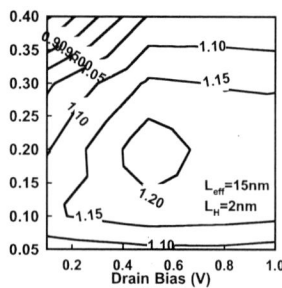

Fig.11 Contour map of simulated Gm enhancement factors in 15-nm SHOT with graded-heterojunction, compared to that of SOIs. The lateral and the vertical axes are the drain bias and ΔEc, respectively. Length of graded-heterojunction, L_H is 2nm.

138

2006 International Symposium on VLSI Technology, Systems, and Applications

RF Extrinsic Resistance Extraction Considering Neutral-Body Effect for Partially-Depleted SOI MOSFETs

Sheng-Chun Wang[1,2], Pin Su[1], Kun-Ming Chen[2], Chien-Ting Lin[3], Victor Liang[3], Guo-Wei Huang[2]

[1]Department of Electronics Engineering, National Chiao Tung University, 1001 Ta-Hsueh Rd., Hsinchu, Taiwan, R.O.C.
[2]National Nano Device Laboratories, No. 26, Prosperity Rd. 1, Science-Based Industrial Park, Hsinchu, Taiwan, R.O.C.
[3]United Microelectronics Corporation, No.3, Li-Hsin Rd. 6, Science-Based Industrial Park, Hsinchu, Taiwan, R.O.C.

INTRODUCTION

The extraction of extrinsic resistances is essential to RF CMOS modeling. With the penetration of SOI CMOS into the RF applications [1, 2], an accurate determination of gate/source/drain resistance (R_g/R_s/R_d) for these RF SOI MOSFETs becomes an important issue. References 3-5 have presented extrinsic resistance extraction methodologies for SOI MOSFETs. Among these approaches, the *zero* method [3, 6] developed under the *zero* condition (i.e., $V_{GS} = V_{DS} = 0$) is attractive because it may simplify the corresponding equivalent circuit and avoid the error caused by the NQS (non-quasi-static) effect [4]. Based on the equivalent circuit built for bulk MOSFETs under the *zero* condition [6], the following frequency-independent resistance expressions have been derived to directly determine R_s, R_d and R_g, respectively.

$$\mathrm{Re}(Z_{21}) = \mathrm{Re}(Z_{12}) = R_s \tag{1}$$

$$\mathrm{Re}(Z_{22}) - \mathrm{Re}(Z_{12}) = R_d \tag{2}$$

$$\mathrm{Re}(Z_{11}) - \mathrm{Re}(Z_{12}) = R_g \tag{3}$$

For partially-depleted (PD) SOI MOSFETs, however, Eq. (1)-(3) may not be valid due to the existence of floating body. In this work, we investigate the extrinsic resistance extraction for PD SOI MOSFETs. We will show that, for RF SOI MOSFETs, the coupling path between the source and drain terminals through the neutral-body region beneath the gate-oxide layer makes the resistance expressions behave frequency-dependently. After taking this effect into account, we develop a physical RF extrinsic resistance extraction methodology for PD SOI MOSFETs.

DEVICES AND MEASUREMENT

The PD SOI MOSFETs used in this study were fabricated using UMC 0.13 m SOI technology. The thicknesses for gate oxide, SOI layer and buried oxide are 1.4nm, 40nm and 200nm, respectively. The presence of kinks in Fig. 1 shows that the devices under test are partially depleted. These RF devices were laid out in the multi-finger and multi-group structure with various finger number NF, group number NG and finger length W_f for a given gate length L_f. On-wafer 2-port S parameters up to 20 GHz were measured, de-embedded, and then transformed to Z parameters to obtain the resistance vs. frequency characteristics.

To further minimize possible substrate resistive loss through the buried oxide layer [7], a bias-network connected to the probe station's chunk was used to provide the substrate DC ground (i.e., back-gate voltage V_E=0) with RF floating. In fact, as shown in Fig. 2, whether the substrate RF ground is provided or not, the resistance curves are almost unchanged. This indicates that the substrate effect is negligible in this experiment.

RESULTS AND DISCUSSIONS

Fig. 2 also compares the resistance vs. frequency characteristics for PD SOI MOSFET and its bulk counterpart with identical layout structure and geometry. All of these curves more or less are frequency-dependent. The poor shapes for the bulk MOSFET may result from the complicated and significant substrate resistive loss [8, 9]. For the SOI MOSFET, however, the substrate loss may not be responsible for this frequency-dependent behavior because the thick buried oxide layer in the SOI transistor may provide good isolation from the substrate.

Fig. 3 shows a cross-sectional view of the PD SOI MOSFET under the *zero* condition. The neutral-body coupling path is constituted by source- and drain-side junction capacitances ($C_{j,bs}$ and $C_{j,bd}$), and body conductances (G_{bs} and G_{bd}). Its corresponding equivalent circuit without substrate RF ground is depicted in Fig.4. Here the neutral body coupling path is represented by a lumped junction capacitance C_b ($= (1/C_{j,bs} + 1/C_{j,bd})^{-1}$) and a lumped body conductance G_b ($= (1/G_{bs} + 1/G_{bd})^{-1}$). Based on this equivalent circuit, the following resistance expressions regarding R_s, R_d and R_g can be derived.

$$\mathrm{Re}(Z_{21}) = \mathrm{Re}(Z_{12}) = R_s + A/(\omega^2 + B) \tag{4}$$

$$\mathrm{Re}(Z_{22}) - \mathrm{Re}(Z_{12}) = R_d + A/(\omega^2 + B) \tag{5}$$

$$\mathrm{Re}(Z_{11}) - \mathrm{Re}(Z_{12}) = R_g - 0.5 \times A/(\omega^2 + B) \tag{6}$$

where

$$A = 2C_b^2 G_b/D \tag{7}$$

$$B = G_b^2 \left(4C_{ds}^2 + C^2 + 4C_b^2 + 4CC_{ds} + 8C_{ds}C_b + 4CC_b\right)/D \tag{8}$$

$$D = \left(C^2 C_b^2 + 4C_{ds}CC_b^2 + 4C_{ds}^2 C_b^2\right) \tag{9}$$

Note that we have assumed $C_{gs} = C_{gd} = C$ in the derivation because of the symmetry of geometry and bias. Eq. (4)-(6) predict that the existence of neutral body may cause the resistance curves regarding R_s, R_d and R_g frequency-dependent and explains the SOI behavior in Fig. 2.

Using this new model, a physical RF extrinsic resistance extraction methodology for PD SOI MOSFETs can be developed. The model-data comparison for the extraction of R_s, R_d and R_g with various layout geometries are shown in Fig. 5 to 7, respectively. It can be seen that the measured curves are indeed frequency-dependent, and can be well fitted by our model (Eq. (4)-(9)). The extracted extrinsic resistances and fitting parameters A and B for each SOI device are listed in Table 1. Since all the involved device conductance and capacitances in Eq. (7)-(9) are proportional to the total gate electrical width W_{total} ($\approx W_f$ NF NG), the parameter A should increase as W_{total} decreases. As shown in Fig. 5 to 7, the resistance curves for the device with smaller W_{total} indeed have a larger deviation from its high-frequency asymptote in the lower frequency regime. Therefore, one may minimize the extraction error resulted from the SOI neutral body by using the wide device.

The complete flow chart for our PD-SOI extrinsic resistance extraction is summarized in Fig. 8.

CONCLUSIONS

We have investigated the extrinsic resistance extraction for PD SOI MOSFETs. Although the thick buried oxide in SOI devices can block the complicated substrate network, the SOI neutral-body coupling effect may become significant for RF applications. An equivalent circuit considering this effect has been proposed. Based

1-4244-0181-X/06/$20.00 ©2006 IEEE   139

on the equivalent circuit, a new model capturing the frequency dependence of extrinsic resistances has been derived. After considering the presence of floating body, we have developed a physically accurate RF extrinsic resistance extraction methodology for PD SOI MOSFETs.

ACKNOWLEDGEMENTS

The authors would like to thank UMC for providing the SOI devices used in this study. This work was supported in part by the National Science Council under contract NSC93-2215-E-009-042.

REFERENCES

[1] J. Kim et al., Proc. IEEE 2004 CICC, pp. 541-548.

[2] T. Douseki et al., Proc. IEEE 2003 CICC, pp. 163-168.
[3] J.-P. Raskin, et al., Proc. Electrochem. Soc., Los Angeles, CA, 1996, vol. 96-3, pp. 225–231.
[4] A. Bracale et al., Analog and Integrated Circuits and Signal Processing. Norwell, MA: Kluwer, 2000.
[5] H. Liao et al., International Conference on Solid-State and Integrated Circuit Technology (ICSSICT), p.913-915, Beijing, China, 2001.
[6] D. Lovelace et al., IEEE MTT-S International Microwave Symp. Digest, pp. 865-866, 1994.
[7] C. L. Chen et al., IEEE EDL, vol. 21, pp. 497–499, Oct. 2000.
[8] W. Liu et al., Proc. IEDM, Dec. 1997, pp. 309–312.
[9] R. T. Chang et al., IEEE TED, vol. 51, no. 4, pp. 421–426, Mar. 2004.

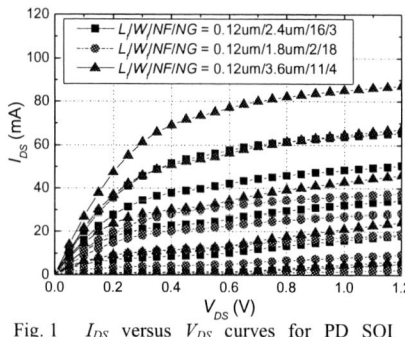

Fig. 1 I_{DS} versus V_{DS} curves for PD SOI MOSFETs used in this study. V_{GS} is from 0.4 to 1.2 V with step 0.2 V.

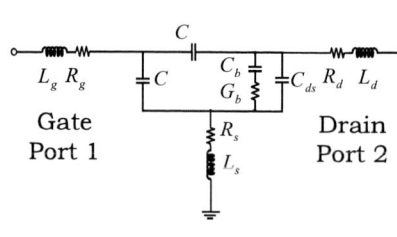

Fig. 4 Equivalent circuit for *zero* PD SOI MOSFETs

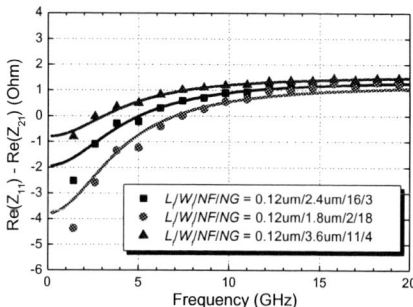

Fig. 7 Curve fitting results for R_g extraction (points: measured data. lines: model)

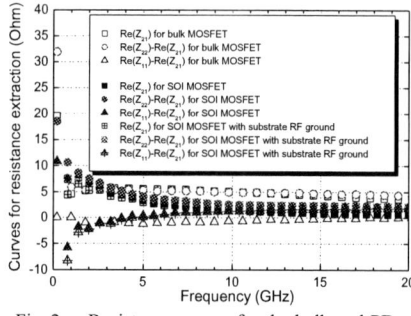

Fig. 2 Resistance curves for the bulk and PD SOI MOSFET with $L_f/W_f/NF/NG$ = 0.12 m/2.4 m/16/3.

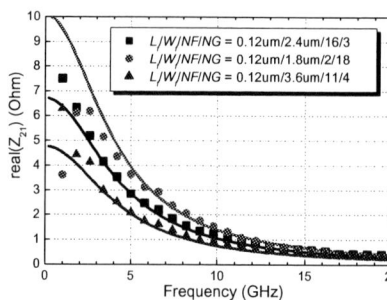

Fig. 5 Curve fitting results for R_s extraction (points: measured data. lines: model)

Table 1 Extracted values for fitting parameters A, B and extrinsic resistances (The values of $L_f/W_f/NF/NG$ for FET1, FET2 and FET3 are 0.12 m/2.4 m/16/3, 0.12 m/1.8 m/2/18, 0.12 m/3.6 m/11/4, respectively.)

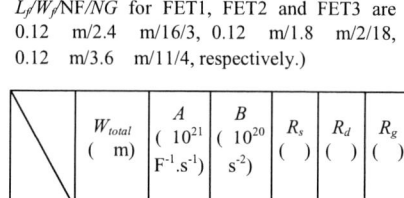
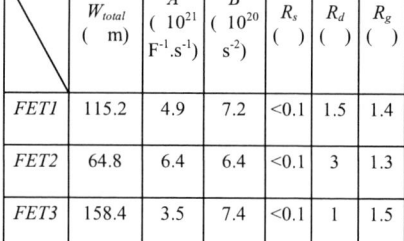

	W_{total} (m)	A (10^{21} $F^{-1}.s^{-1}$)	B (10^{20} s^{-2})	R_s ( )	R_d ( )	R_g ( )
FET1	115.2	4.9	7.2	<0.1	1.5	1.4
FET2	64.8	6.4	6.4	<0.1	3	1.3
FET3	158.4	3.5	7.4	<0.1	1	1.5

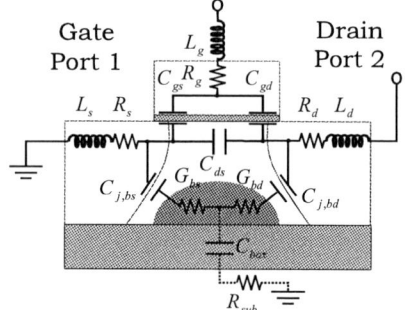

Fig. 3 Cross-sectional view of the *zero* PD SOI MOSFET

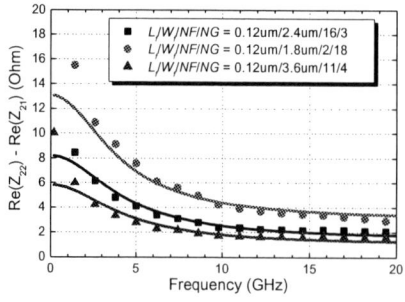

Fig. 6 Curve fitting results for R_d extraction (points: measured data. lines: model)

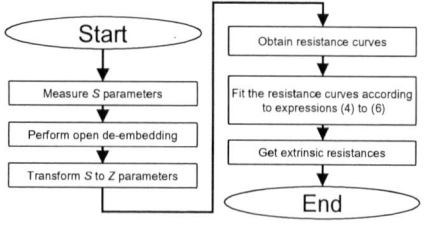

Fig. 8 Complete flow chart for the PD SOI extrinsic resistance extraction

140

2006 International Symposium on VLSI Technology, Systems, and Applications

Hot Electrons Associated with the Long-Range Coulomb Interaction under the High-Density Regime

Tadayoshi Uechi and Nobuyuki Sano
Institute of Applied Physics
University of Tsukuba
Tsukuba, Ibaraki 305-8573
JAPAN

ABSTRACT

We have carried out 1-, 2-, and 3-D Particle-Mesh simulations of the electron gas under high electron densities, in which the long-range part of the Coulomb interaction plays a dominant role in determining the transport characteristics. The simulations turn out to be extremely sensitive with the parameters such as the time step and the size of particles. It is demonstrated for the first time that, if the Coulomb potential is properly taken into account, the *bare* electrons get hotter as the electron density increases even without including the short-range scattering processes.

INTRODUCTION

The channel length in the state-of-the-art MOSFETs now reaches the order of tens of nanometers so that the highly doped source and drain regions stay close each other in nanometers [1]. Since the electron density in such high-doped regions exceeds 10^{20} cm^{-3}, the Coulomb interaction among electrons is very strong and the device properties could be greatly affected by the many-body effects associated with Coulomb interaction in high doped regions. In fact, the 2-D Monte Carlo simulations performed by Fischetti and Laux have shown that the electrons in high-doped regions become energetically hot and, therefore, the transconductance in nanoscale MOSFETs could be greatly degraded [2]. In the present study, we have carried out 1-, 2-, and 3-D particle-based simulations including the long-range part of the Coulomb interaction *as accurately as possible*. We demonstrate that the electron temperature indeed becomes hot under the high electron density regime *even without including the short-ranged scatterings such as the phonon scatterings*.

SIMULATION METHOD AND INSTABILITY

In the present study, the particle-mesh (PM) method often employed in plasma simulations is used [3], although the high electron density assumed here is seldom encountered in plasma physics. The background positive charges are assumed for charge neutrality to be either uniform (jelly) or (more realistically) randomly localized in real space, and the electrons are at first distributed in space uniformly and move in accordance with Newton's equation of motion under the self-consistent electric field. The electric potential is calculated by solving the Poisson equation with the charge distribution of electrons (and localized ions) evaluated at each mesh node via appropriate interpolation schemes. The initial electrons' velocities are prepared in accordance with the thermal equilibrium distribution at room temperature ($T = 300$ K). No external electric field is assumed so that the electron gas stays in thermal equilibrium. The periodic boundary condition is employed. Figures 1 and 2 illustrate the basic equations, a schematic drawing of electrons' trajectories, and the flowchart of the present PM simulations.

In order to check whether the current PM simulations with the long-range part of the Coulomb interaction are performed accurately, the short-range scattering processes such as the phonon scatterings are *intentionally* turned off. Hence, the physical system we consider is closed (micro-canonical) and the total energy is *strictly* conserved. As a result, the simulations become extremely difficult, compared with conventional self-consistent Monte Carlo simulations, because no energy dissipation is included; the simulations are very sensitive to the size of the mesh, the time step, and the (shape) size of the electrons employed in simulations. Without optimizing those parameters, the PM simulations easily get unstable and the total energy explodes immediately. Figure 3 shows the time evolutions of the total energy under the different interpolation schemes (NGP and CIC) and under the various sizes of simulating electrons. Notice how quickly the total energy explodes unless the simulation parameters are properly chosen. We would like to stress that such explosions seldom show up if energy dissipating processes are included. This implies that *conventional particle-based device simulations including phonon scatterings do not always guarant*ee, *in the present sense, the energy conservation during the simulations*.

HOT ELECTRONS AND RENORMALIZATION

We first demonstrate how accurately our PM simulations are carried out. Figure 4 shows the time dependence of the kinetic, potential, and total energies of electrons obtained from the present PM simulations for the two different time scales. The average electron density of 10^{20} cm^{-3} is assumed. The kinetic energy and the potential energy oscillate with the same (plasma) frequency, yet they are out of phase each other so that the total energy is conserved. Furthermore, the total energy of the entire system is well conserved for a very long time period, say, more than 100 ps. The oscillations found above are visualized in Fig. 5, in which the electron density and the corresponding electric potential in real space are plotted under the average electron density of 10^{20} cm^{-3}. The electron density profile shows rather complicated short-ranged structures, whereas the electric potential profile is much smoother than the electron density profile due to *dynamical* screening. Figure 6 shows the power spectral of density fluctuations as a function of frequency under the uniform (jelly) positive background obtained from the 1-D and 2-D PM simulations. The arrows in the figures represent the plasma frequencies corresponding to the various electron densities. Very clear peaks at the plasma frequencies are observed and, thus, the plasma waves are well excited in the electron gas. Figure 7 is our main results, which show the energy distributions as a function of the electron *kinetic* energy obtained from 1-D and 3-D PM simulations. Notice that the electron temperature is initially assumed to be 300 K. The shape of the distributions does not change much under the various electron densities in 1-D simulations, whereas the electrons get much hotter in 3-D simulations as the electron density increases. This is due to the renormalization of the kinetic energy associated with the potential fluctuation induced by the long-range part of the Coulomb interaction among electrons, as seen in Fig. 5. It should be noted that the distributions found above represent those of the *bare* electrons and the dressed (due to Coulomb interaction) electrons do show the thermal equilibrium distributions. We would also like to

1-4244-0181-X/06/$20.00 ©2006 IEEE

stress that the correct renormalization effect could be simulated *only if the Coulomb potential is accurately taken into account in 3-D particle-based simulations.*

SUMMARY

We have carried out the PM simulations for electron gas under high electron densities. It has been found that the *bare* electrons get hotter as the electron density increases due to the long-range part of the Coulomb interaction among the electrons *even without including the short-range scattering processes.*

REFERENCES

[1] N.Sano, A.Hiroki, and K.Matsuzawa, *IEEE Trans. Nanotech.*, vol.1, pp.63-71, 2002.

[2] M.V.Fischetti and S.E.Laux, *J. Appl. Phys.*, vol.89, pp.1205-1231, pp.1232-1250, 2001.

[3] C.K.Birdsall and A.B.Langdon, *Plasma Physics via Computer Simulation* (Institute of Physics, Bristol, 1991).

$$\frac{\partial f_1(\mathbf{r}_1,\mathbf{p}_1;t)}{\partial t} + \frac{\mathbf{p}_1}{m} \cdot \frac{\partial f_1(\mathbf{r}_1,\mathbf{p}_1;t)}{\partial \mathbf{r}_1} + e\frac{\partial V(\mathbf{r}_1)}{\partial \mathbf{r}_1} \cdot \frac{\partial f_1(\mathbf{r}_1,\mathbf{p}_1;t)}{\partial \mathbf{p}_1} = 0$$

$$\nabla^2 V(\mathbf{r}_1) = -\frac{e}{\varepsilon}\left\{ N_d^+(\mathbf{r}_1) - \int d\mathbf{p}_1 f_1(\mathbf{r}_1,\mathbf{p}_1;t) \right\}$$

Fig. 1 Basic (Vlasov and Poisson) equations employed for the present PM simulations. The short-ranged collision integral is intentionally set to zero throughout the present study.

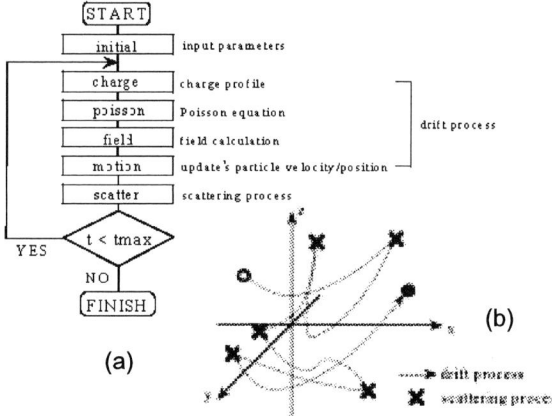

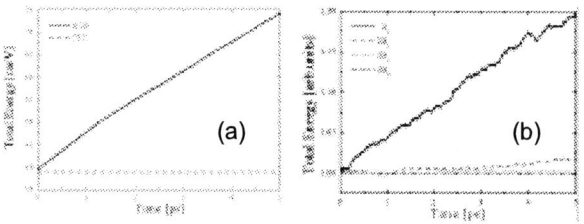

Fig. 2 (a) Flowchart of the present PM simulation code and (b) the schematic drawing of electrons' trajectories in real space (scatterings are included for clarity).

Fig. 3 Time evolutions of the total energy under (a) the NGP and CIC interpolation schemes and under (b) the various sizes of electrons. The electron density is 10^{20} cm^{-3} and the background is assumed to be jelly. These figures show how quickly the total energy explodes unless the simulation parameters are properly optimized.

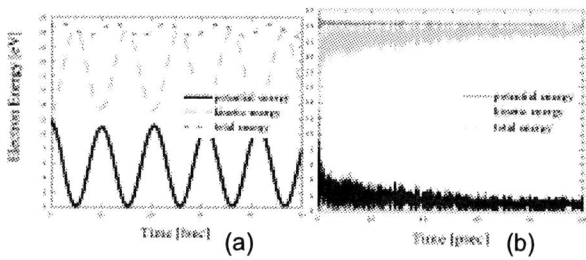

Fig. 4 Time evolutions of the kinetic, potential, and total energies of the electrons obtained from the PM simulations for the two different time scales. The kinetic and potential energies oscillate with out of phase in (a) due to the excitation of plasma oscillations. It is confirmed that the total energy is well conserved for the long time period in (b).

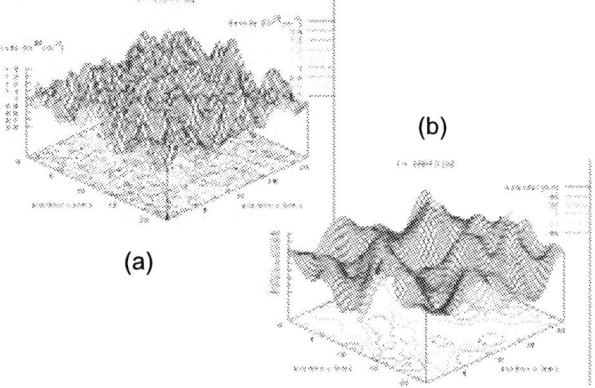

Fig. 5 (a) Electron density and (b) the corresponding electric potential in real space obtained from the PM simulations with the average electron density of 10^{20} cm^{-3}. Notice that the potential profile is much smoother than the electron density profile due to screening.

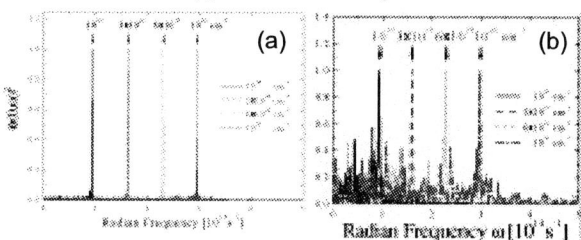

Fig. 6 Power spectral of density fluctuations as a function of frequency under the jelly background obtained from the (a) 1-D and (b) 2-D PM simulations. The arrows in the figures represent the plasma frequencies corresponding to the various electron densities.

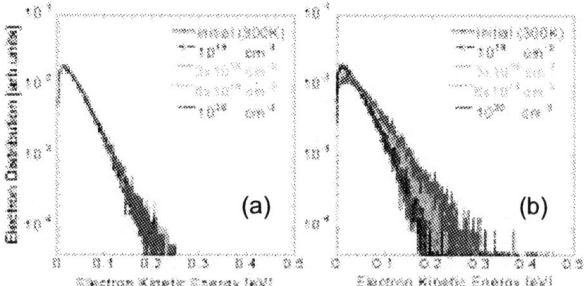

Fig. 7 Electron energy distributions as a function of the kinetic energy obtained from the (a) 1-D and (b) 3-D PM simulations for the electron densities from 10^{19} to 10^{20} cm^{-3}. Notice that the electron temperature becomes extremely hot in 3-D simulations, as the density increases.

2006 International Symposium on VLSI Technology, Systems, and Applications

A New Series Resistance and Mobility Extraction Method by BSIM Model for Nano-Scale MOSFETs

William P.N. Chen[12], P. Su[1], J.S. Wang[3], C.H. Lien[3], C.H. Chang[2], K. Goto[2], C.H. Diaz[2]

[1]Department of Electronics Engineering, National Chiao Tung University, Hsin-Chu, Taiwan, R.O.C.

[2]Advanced Device Technology Department, Taiwan Semiconductor Manufacturing Company, Hsin-Chu, Taiwan, R.O.C.

[3]Department of Electrical Engineering, National Tsing Hua University, Hsin-Chu, Taiwan, R.O.C.

INTRODUCTION

As MOSFET gate length scales down to nano scale regime, parasitic source/drain series resistance (R_{sd}) becomes one of the most critical parameters impacting device performance. Many studies have focused on developing an effective R_{sd} extraction methodology. The "Channel-Resistance method" [1] provides a simple way to extract R_{sd}; however, this method is no longer adequate as pocket implants are introduced into the nano-scale MOSFET. In this case, pocket profiles overlap in the short channel region, causing higher effective bulk concentration and resulting in a total resistance which does not scale linearly with channel gate length (L_G) [2]. Another famous method called "Shift & Ratio" (S&R) also suffers the similar issue because of its basic assumption that mobility (μ_{eff}) does not change with L_G [3]. Actually, pocket implant degrades μ_{eff} in short channel region because of higher effective bulk concentration by the halo overlapping profile [4]. Therefore, the key point for extracting R_{sd} in nano-scale MOSFETs is to take the impacts of L_G dependency of μ_{eff} into account.

In this work, a simplified BSIM-based model has been proposed to solve the above issues contributed by halo implants [5]. In this new methodology, R_{sd} and μ_{eff} can be uniquely extracted in nano-scale devices. Furthermore, the extracted L_G dependency of μ_{eff} may serve as a good indicator for monitoring the relationship between geometry and stress parameters.

DEVICES

The devices used in this experiment were fabricated by state-of-the art IC manufacturing technology [6], which provides for gate lengths ranging from 4µm down to 0.075µm. Samples with different thickness of tensile contact etch stop layer (CESL) were also adopted to verify the validity of the new extraction method.

RESULTS & DISCUSSION

Fig. 1(a) shows the procedure of our proposed BSIM-based R_{sd}/μ_{eff} extraction methodology. In the first step, careful characterization is critical for the following extraction work, especially: (i) Gate-biased drain current in the linear region (I_DV_G) with different L_G. (Note: Collected I_DV_G is extrinsic due to inseparable parasitic R_{sd}) (ii) Effective channel length (L_{eff}). Both parameters would significantly impact the accuracy of extracted R_{sd} in this methodology. Fig. 1(b) provides the related information for key parameters. In this work, we select L_G from 80nm to 120nm for R_{sd} extraction and ignore the long/medium channel region based on the following two reasons: (i) R_{sd} shows significant sensitivity in short channel devices (ii) selecting a narrower range for L_G is preferred to minimize the impact of mobility dependence on L_G. After that, we input all characterized parameters into the simplified BSIM model [7] as depicted in Eq. (1) including a temporary value for R_{sd}. Then we fit the I_DV_G characteristics by tuning mobility parameters in Eq. (2) within a limited range of L_G (80~120nm).

$$I_d = \mu_{eff} C_{ox} \frac{W}{L_{eff}} \frac{(V_g - V_{th} - V_d/2)V_d}{1 + R_{sd}\mu_{eff}C_{ox}\frac{W}{L_{eff}}(V_g - V_{th} - V_d/2)} \quad (1)$$

$$\mu_{eff} = \frac{\mu_0}{1 + \left(\frac{E_{eff}}{E_0}\right)^\nu} \quad (2)$$

Note that the bulk charge effect for short channel devices and the saturation electric field (E_{sat}) effect in the linear region ($V_d \sim 20mV$) are neglected in the drain current model of Eq. (1).

Fig. 2 shows that an optimized R_{sd} of ~165Ω*µm for our devices is obtained by minimizing the offset between experimental data and model with several iterations. Fig. 3 shows the I_DV_G fitting results in the short channel region. If the final R_{sd} value is not correct, the drain current ratio of different L_G is not correct either. Once R_{sd} is uniquely determined, the gate length dependency of mobility ($\mu_{eff}(L_G)$) may also be obtained from intrinsic drain current via Eq. (3). Fig. 4 shows that μ_{eff} tends to degrade in the deep-submicron region for both N/PMOS due to halo implants. Fig. 5 shows that good agreement between the model and experimental data for a wide range of L_G can be obtained after considering the gate length dependency of mobility.

$$I_d(int) = \frac{I_d(ext)}{1 - I_d(ext) \times \frac{R_{sd}}{V_d}} \quad (3)$$

Fig. 6 provides the R_{sd} sensitivity with variations on different key parameters, where R_{sd} is the most sensitive to L_G but this can be overcome by careful in-line measurement. It is also worth noting that the variation of μ_{eff} has to be limited to within ± 5% if ± 4% Rsd variation is the maximum tolerance level. From our previous extraction work, it is reasonable to pick up L_G equal to 80~120nm because the variation of μ_{eff} is within ±5%. Fig. 7 shows the relationship between the normalized R_{sd} and overlap capacitance (C_{ov}), where the R_{sd} value of NMOS is ~2-3X smaller than the PMOS value. Fig. 8 (a) and (b) depict the NMOS mobility behavior at L_G equal to 4µm and 0.1µm with different tensile CESL thickness. It shows ~0% improvement in the long channel region but ~14% improvement in the short channel region due to uni-axial tensile stress along the channel direction. Fig. 9 shows further details regarding the gate length dependency of mobility in this experiment. Thicker CESL enhances NMOS mobility (max. of $\Delta\mu/\mu$ ~15%) in the submicron region (<1µm), but degrades mobility in the deep-submicron region due to the halo implant effect (<0.24µm).

CONCLUSION

In conclusion, we have proposed a BSIM-based method for the extraction of source/drain series resistance which applies to nano-scale MOSFETs with halo implants. Good agreement with experimental data has been obtained. Furthermore, mobility dependence on gate length is also uniquely extracted, which may also be used as an indicator for strain engineering techniques in the future.

REFERENCES

[1] Y. Taur and Tak H. Ning, "Fundamentals of Modern VLSI Devices," *Cambridge University Press*, 1998.

[2] D. Esseni *et al.*, *IEEE Electron Device Lett.*, vol. 19, p.131, 1998.

[3] Y. Taur, *IEEE Trans. Electron Devices*, vol.47, p.160, 2000.

[4] Hans van Meer *et al.*, *IEEE Electron Device Lett.*, vol. 21, p.133, 2000.

[5] K. Goto *et al.*, *IEDM Tech. Dig.*, p. 623, 2003.

[6] S.K.H. Fung *et al.*, *Symp. VLSI Tech.*, p. 92, 2004.

[7] Y. Cheng and C. Hu, " MOSFET Modeling & BSIM3 User's Guide," KAP 1999.

1-4244-0181-X/06/$20.00 ©2006 IEEE

(a)

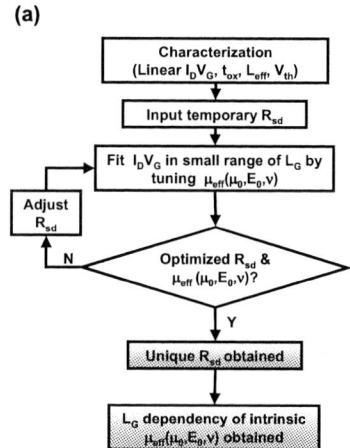

(b)

Parameter Table	Symbol	Definition
Input Parameter	I_D	*Drain current in linear region*
	μ_{eff}	*Effective mobility; Need to calibrate with R_{sd} by iterations*
	C_{ox}	*Gate oxide capacitance*
	V_{th}	*Threshold voltage from measurement*
	L_{ov}	*Gate-extension overlap distance; Calculated from C_{ox} and C_{ov}*
	L_{TEM}	*Obtained from L_{SEM} - Δ L. L_{SEM} is from in-line measurement. Δ L is a constant offset between L_{TEM} and L_{SEM}*
	L_{eff}	*L_{TEM} - L_{ov}*
	A_{bulk}	*Bulk charge parameter; A_{bulk} ~1 in short channel region*
	E_{sat}	*Saturation electrical field; Set E_{sat} ~ ∞ so that $V_{ds}/E_{sat}*L_{eff}$ <<1*
Fitting Parameter	μ_0	*Mobility fitting parameter*
	E_0	
	ν	
	R_{sd}	*Source/Drain series resistance*

Fig. 1. (a) Flow chart of BSIM-based R_{sd} extraction methodology (b) Definition table of key parameters.

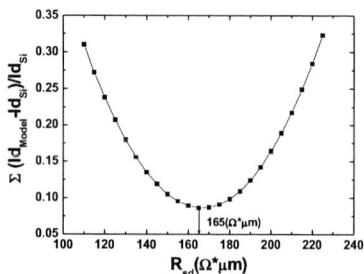

Fig. 2. Iteration work on R_{sd} & μ to extract the optimized R_{sd}.

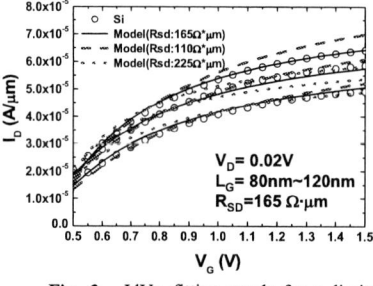

Fig. 3. IdVg fitting result for a limited range of short gate lengths.

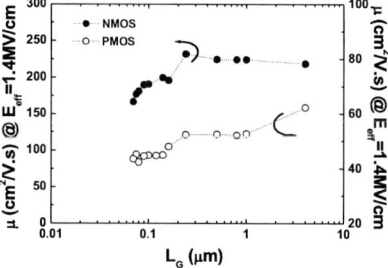

Fig. 4. Mobility dependence on gate length for NMOS/PMOS.

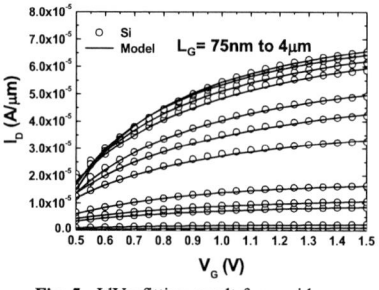

Fig. 5. IdVg fitting result for a wide range of gate lengths by considering mobility dependence on L_G.

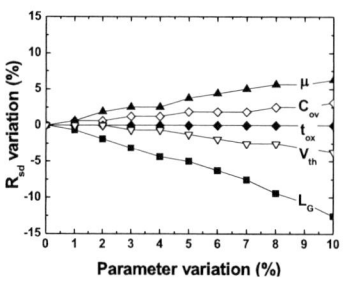

Fig. 6. R_{sd} sensitivity plot for different key parameters.

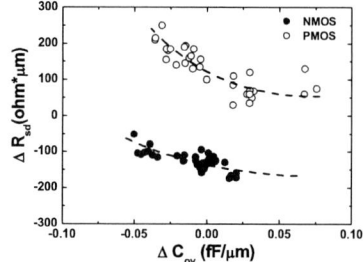

Fig. 7. Relationship between normalized Rsd and overlap capacitance.

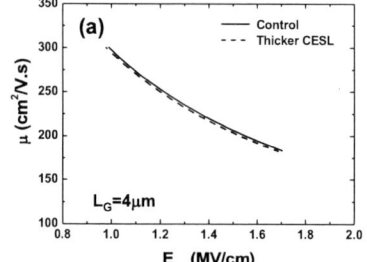

Fig. 8. NMOS mobility for different contact etch stop layer(CESL) thickness (a) L_G=4μm (b) L_G= 0.1μm.

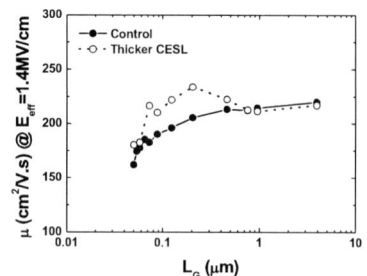

Fig. 9. Mobility dependence on L_G for different CESL thickness.

2006 International Symposium on VLSI Technology, Systems, and Applications

Coupling Advanced Atomistic Process and Device Modeling for Optimizing Future CMOS Devices

B. Colombeau[1], S.H. Yeong[1], S.M. Pandey[1], F. Benistant[1], M. Jaraiz[2] and S. Chu[1]

[1]Chartered Semiconductor Manufacturing Ltd.
60 Woodlands Industrial Park D, Street2
Singapore

[2]Department of Electronics, University of
Valladolid, 47011 Valladolid,
Spain

ABSTRACT

For the first time, we show the coupling between advanced atomistic process and device modeling and its applicability for 65nm PMOS and NMOS technology. This technique can be used to simulate and get some important insights to improve and optimize future CMOS devices.

INTRODUCTION

Front-end process modeling for nanometer structures is the key challenge for the prediction of result from device fabrication. It overlaps to some extent with the Difficult Challenge–Ultimate nanoscale CMOS simulation capability, which however also includes materials and device simulation [1]. Most important and challenging in the area of front-end process modeling is the modeling of ultra-shallow junction (USJ) formation, which starts from very low energy implant and especially focuses on the thermal annealing and diffusion of dopants.

In this paper, for the first time we will show how a highly predictive atomistic simulator can be coupled to continuum process and device modeling to predict the impact of process parameters on the device performances for 65nm technology node and below. A relevant application on the effect of the Ge preamorphized (Ge-PAI) Boron USJ for Source/Drain (S/D) extension on device results will be presented.

PHYSICAL MODELING AND METHODOLOGY

One of the main challenges is the accurate simulation of formation of P^+/N Ge-PAI USJ. The physical basis for our approach is shown schematically in Fig. 1. During regrowth of the PAI layer, B-rich clusters form in the region of high B concentration, and beyond the initial amorphous/crystalline interface the excess interstitials agglomerate into I-clusters. The driving force for subsequent diffusion and deactivation is the interstitial supersaturation generated by the ripening and dissolution of end-of-range (EOR) defects. In PAI USJs the EOR defect band is located beyond the high-concentration B region, so that deactivation requires transport of interstitials from the EOR band towards the surface, forming inactive Boron Interstitial Clusters (BICs) [3].

To reproduce this effect, accurate dopant-defect modeling has to be used. The atomistic kinetic Monte-Carlo simulator DADOS [2] has been proven to be able to reproduce and accurately predict the formation and eventual dissolution of extended defects and their interaction with dopants and impurities upon annealing [2].

After showing the calibration of the atomistic tool, we will use it to simulate 2D doping profiles (total and active) after crucial process step e.g. during spacer formation. These profiles will be used as inputs for the rest of the process into continuum based TSUPREM4. MEDICI software will finally be used for device simulation.

RESULTS AND DISCUSSION

First, 1D calibration has been performed using wide range of SIMS and sheet resistance data to extract the BIC's binding energies. One

typical example of the simulation of Ge PAI Boron USJ is shown in Fig. 2. The simulated EOR defect evolution presented in Figs. 3 & 4 also reproduce experimental TEM analysis (cross sectional and plan view). During annealing, EOR defect evolve from interstitial clusters to {113} defects (75s) up to dislocation loops for longer time (120s).

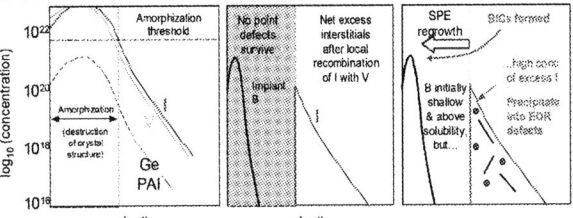

FIGURE 1. Schematic illustration of physics underlying for simulation of defect evolution and diffusion in the crystalline phase

Once calibrated this atomistic tool was used for prediction of Boron 2D profiles for full 65 nm PMOS process flow. Final process structure is shown in Fig. 5. Process and device simulation properly match SIMS, TEM and electrical data.

Same approach has been used for the 65nm NMOS case. For this, As deactivation via formation of Arsenic-Vacancy clusters had been taken into account. Figure 6 shows results after calibration for the S/D extension of 65 nm NMOS transistor.

Finally, we present one application of our accurate calibration to investigate the influence of the Ge-PAI dose/energy in the PMOS S/D extension formation on the device performances. Figure 7 shows the evolution of the predicted amorphous layer thickness for the different energy/dose considered. As expected, when the energy or dose increases, the amorphous layer increases. Device results for the different cases studied are shown in Figs. 8 & 9. One can clearly note that energy dependence has a stronger effect than the dose one on Vtsat and Idsat. Meantime, 2D simulations predict more lateral diffusion as a function of the Ge energy and dose. These trends can be explained by faster defect dissolution when the EOR are located closer to the wafer surface leading to a higher deactivation of the SD extension via BICs formation. This effect is becoming stronger when Ge-PAI energy decreases.

CONCLUSIONS

Advanced atomistic process simulator DADOS has been calibrated with wide range of physical and electrical data then used to predict the 2D doping profiles for PMOS 65 nm technology node. Based on this accurate calibration we have presented one important application to predict the impact of Ge-PAI conditions on the device performances. This technique can be used to simulate and get some important insights to improve and optimize future CMOS devices.

REFERENCES

[1]   ITRS 2003, http://public.itrs.net
[2]   M. Jaraiz, in Predictive Simulation of Semiconductor Processing, Springer-Verlag, Berlin, 2004, p.73.
[3]   B. Colombeau et al., in IEDM 2004, 971-974.

1-4244-0181-X/06/$20.00 ©2006 IEEE   145

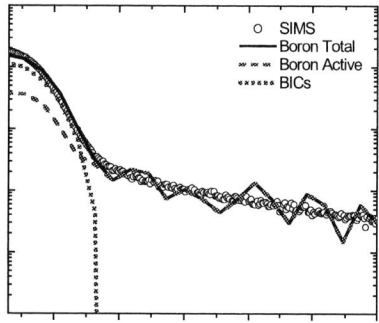

FIGURE 2. Comparison between SIMS measurements and simulation results B 500eV 1e15 Ge-PAI 30keV 1e15 annealed at 800C for 120s

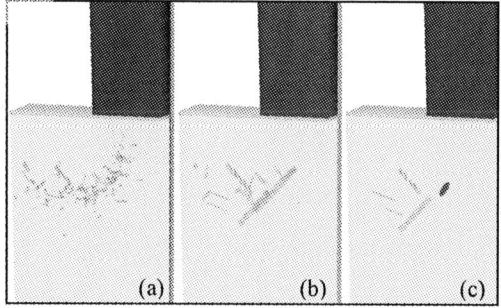

FIGURE 3. 3D simulation of EOR defect evolution formed by Ge-PAI 30keV 1e15 annealed at 800C for (a) 30s, (b) 75s, (c) 120s.

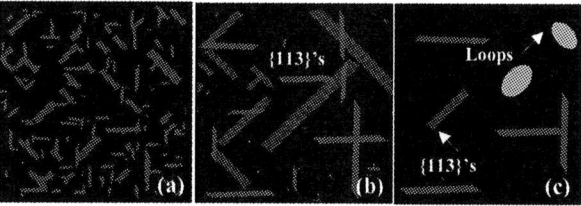

FIGURE 4. Plan view simulation of EOR defect evolution formed by Ge-PAI 30keV 1e15 annealed at 800°C for (a) 30s, (b) 75s, (c) 120s.

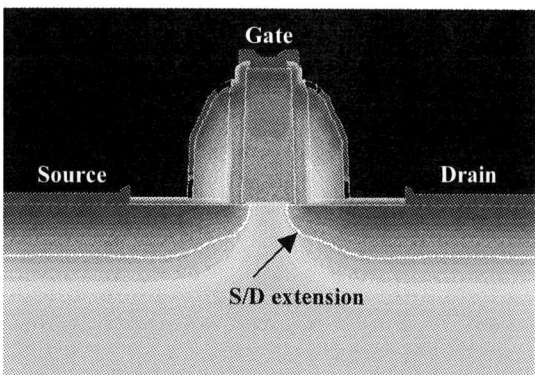

FIGURE 5. 2D simulation for 65 nm PMOS technology node. Calibration was based on SIMS, TEM and electrical data.

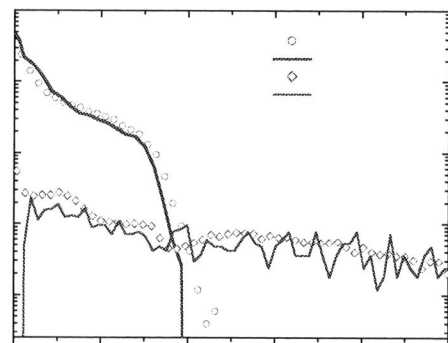

FIGURE 6. Comparison between SIMS measurements and simulation results of the arsenic S/D extension and boron halo implantation profile of a 65 nm NMOS transistor

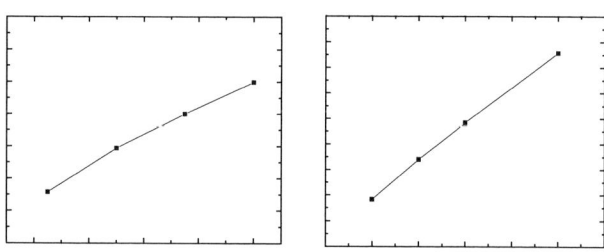

FIGURE 7. Predicted amorphous thickness as a function of Ge energy or dose.

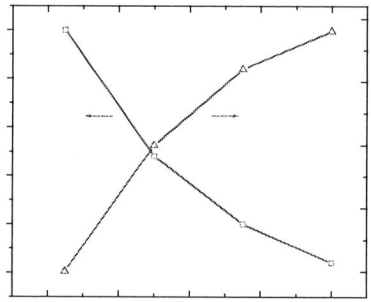

FIGURE 8. Predicted impact of Ge PAI dose in S/D extension on device performance.

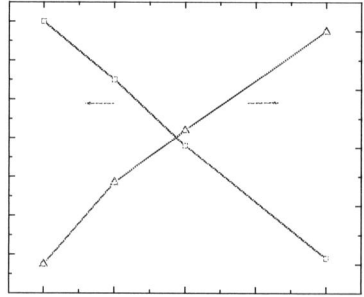

FIGURE 9. Predicted impact of Ge PAI energy in S/D extension on device performance.

2006 International Symposium on VLSI Technology, Systems, and Applications

An Asymmetrical Double-Gate VCO with Wide Frequency Range

Hung Ngo[1], Keunwoo Kim[2], Ching-Te Chuang[2], JB Kuang[1], Fadi Gebara[1], and Kevin Nowka[1]

(1) IBM Austin Research Laboratory, Austin, TX 78758
(2) IBM TJ Watson Research Center, Yorktown Heights, NY 10598

Abstract:

Independently-controlled double-gate devices using asymmetrical (n+/p+) gates are applied to the design of a novel V_T-controlled voltage controlled oscillator (VCO). The VCO supports high operating frequencies and a wide tuning range. The back gates of the asymmetrical double-gate devices are used to modulate the threshold voltages of the front gate devices. A two-fold frequency tuning range is observed without the complexity of additional devices, resulting in improved performance, noise immunity, and power efficiency over conventional well/body bias techniques. These results are validated through mixed-mode MEDICI simulations.

1. Asymmetrical Double-Gate Devices

In asymmetrical double-gate devices, the two gate electrodes consist of materials of different work functions [1]. Fig. 1(a) shows the cross-section of an asymmetrical double-gate NFET. The front and back gates, typically, consist of $n+$ and p+ polysilicon, respectively. For asymmetrical PFET, the opposite type of gate is applied, $p+$ polysilicon for the front gate and $n+$ polysilicon for the back gate. The predominant front-channel has a significantly lower V_T and much larger current drive compared with the weak back channel [2]. The front-channel V_T can be modulated by back-gate biasing through gate-to-gate coupling, as illustrated in Fig. 1(b) [3]. This V_T modulation mechanism is significantly stronger than the well/body bias in bulk and PD/SOI CMOS devices. The device performance can be maximized and is limited only by the gate RC. In addition, this modulation effect will improve with device scaling due to stronger gate-to-gate coupling. The 25nm asymmetrical double-gate device has thin physical front-gate and back-gate oxide ($t_{oxf} = t_{oxb} = 1$ nm) and un-doped ultra-thin silicon body ($t_{Si} = 5$ nm). For comparison, well-biased 25nm bulk silicon device with highly-doped halo ($N_{halo} = 5 \times 10^{18}$ cm^{-3}) to suppress short-channel effects is also studied. The gate oxide thickness is the same ($t_{ox} = 1$ nm). For a fair comparison, I_{off} is designed to be the same for double-gate and bulk silicon devices.

2. Asymmetrical Double-Gate VCO

Fig. 2 shows a five-stage double-gate VCO using the proposed scheme. The front gates are connected in a ring fashion which forms the main path of the VCO. NFET back gates are connected to the control voltage VC while the PFET back gates are connected to the control voltage VCB. In typical PLL operations, VC and VCB are usually adjusted in complementary directions, so that the VCO produces a desired frequency that is in phase with the reference clock. Fig. 1(b) illustrates the ability of asymmetrical double-gate devices to modulate the front gate V_T that is suitable for tuning applications like the VCO.

When the back gates are turned off (VC = 0 and VCB = VDD), the VCO operates at the lowest frequency. When the back gates are fully on (VC = VDD and VCB = 0), the VCO achieves its highest frequency. The output frequency can be tuned by varying the voltage control signals. Since the back gate device can modulate the front gate current by over a factor of 2, the operating frequency for this VCO is expected to exhibit a similar tuning range.

The double-gate VCO was modelled using the mixed-mode MEDICI simulator [4]. In simulation, nine double-gate inverters are connected in a fashion similar to Fig. 2. *VC* and *VCB* were swept from 0.0-1.0V and 1.0-0.0V, respectively. As the back gates are turned on, the frequency increases from approximately 12.5GHz to 25GHz. Fig. 3 shows the waveforms of the third, fourth, and fifth stage of the VCO. Fig. 4 illustrates the output frequency of the VCO as the control voltage signals vary. In comparison, applying the control voltages to the wells of a VCO of the same topology in bulk technology, an inferior operating frequency as well as narrower tuning range is observed as shown in the lower frequency curve of Fig. 4. This is because the threshold voltage change for either NFET or PFET is not sufficient to significantly improve the current drive, and hence the delay, of each stage. In addition, the respective range for each control (well) bias in bulk technology is practically limited to less than 400mV in order to avoid catastrophic junction leakage. In our simulations, as depicted in Fig. 3, *VC* and *VCB* sweeps were limited to 400mV, which abruptly cuts short the progression of the lower frequency curve of Fig. 4. Clearly, this limitation further reduces the usefulness of well bias technique in bulk technology. In contrast, the proposed scheme does not suffer junction leakage at any control bias. Although the double-gate VCO operates at a much higher frequency, at the same bias condition, than the bulk-silicon VCO, it actually consumes less power, as indicated in Fig. 5. The reason is, in bulk silicon, the forward well bias significantly increases junction capacitance causing higher dynamic power consumption.

Connecting the front gates forming the main ring and the back gates to the control voltage signals is preferred. Alternate topologies, such as swapping the front and back gate connections, might result in un-sustainable oscillations due to reduced gain. Fig. 6 shows the waveforms of the third stage of the proposed VCO and the same VCO with the connections swapped. *VC* and *VCB* are swept from t = 0.0ns to 0.5ns. The oscillation is then allowed to settle until t = 2.0ns. For the VCO that has swapped connections, the amplitude of oscillation decreases and eventually vanishes as both NFET and PFET front-gate devices are turned on, thereby, shorting the supply rail to ground. On the other hand, the back gate-biased double-gate VCO continues to ramp up in frequency w/o noticeable decrease in magnitude.

3. Conclusion

We have presented a VCO using the independent gate control of asymmetrical double-gate devices. The circuit topology requires no extra device for frequency tuning/control, thus minimizing the area/capacitance and improving performance, noise immunity, and power efficiency. The scheme, simulated in MEDICI, shows high operating frequency, wide tuning range, and superior scalability.

The authors would like to acknowledge the support of Defense Advanced Research Project Agency (DARPA) contract NBCH30390004 for this work.

[1] E. Nowak, I. Aller, T. Ludwig, K. Kim, R. V. Joshi, C. T. Chuang, K. Bernstein, and R. Puri, "Turning Silicon on its Edge," IEEE Circuits Devices Mag. pp. 20-31, Jan/Feb, 2004.

1-4244-0181-X/06/$20.00 ©2006 IEEE

[2] K. Kim and J. G. Fossum, "Double-gate CMOS: symmetrical- versus asymmetrical-gate devices," IEEE Trans. Electron Devices, vol. 48, pp. 294-299, Feb. 2001

[3] H.-K. Lim and J. G. Fossum, "Threshold Voltage of Thin-Film Silicon-on-Insulator (SOI) MOSFET's," IEEE Trans. Electron Devices, vol. ED-30, pp. 1244-1251, Oct. 1983.

[4] Taurus-MEDICI, Industry-standard device simulation tool, Mountain View, CA, synopsis, Inc., 2003.

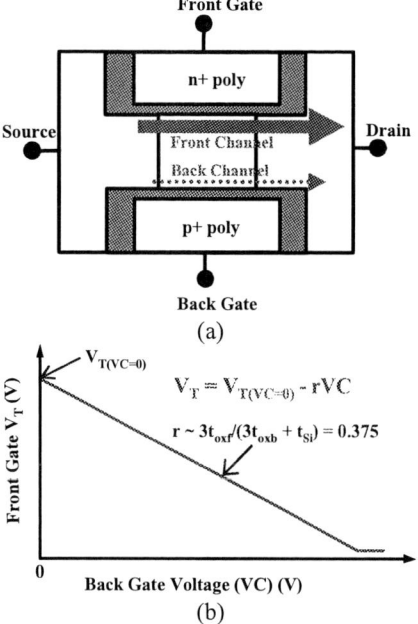

(a)

(b)

Fig. 1: (a) Schematic cross section of an asymmetrical double-gate device, and (b) front-gate V_T as a function of back-gate voltage where $t_{oxf} = 1$ nm, $t_{oxb} = 1$ nm, and $t_{Si} = 5$ nm are used for ($L_{eff} = 25$ nm) asymmetrical DG device.

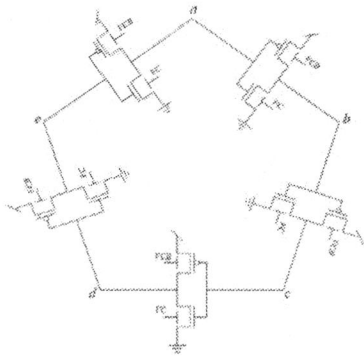

Fig. 2: Schematic diagram of the proposed asymmetrical Double-Gate VCO

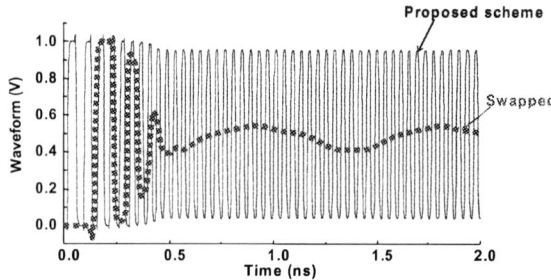

Fig. 6: Waveforms of the third stage of DG VCO. The topology with swapped connections does not sustain oscillation.

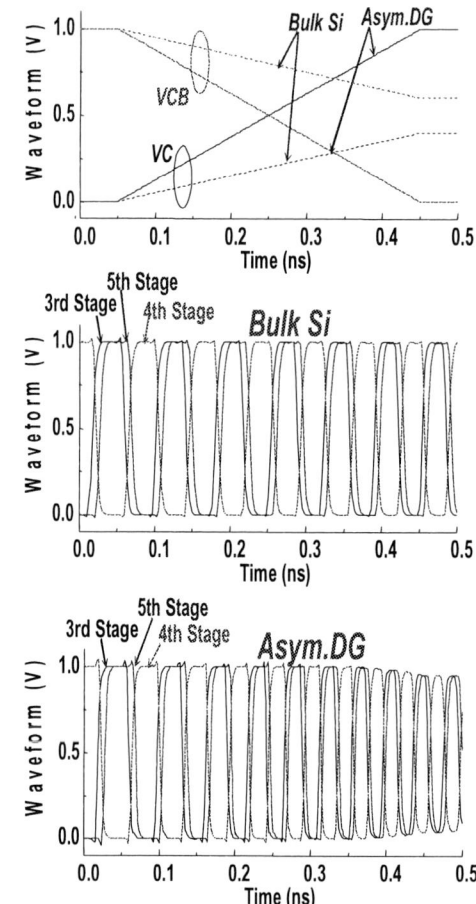

Fig. 3: Waveforms of the third, fourth, and fifth stage of the VCO utilizing bulk silicon well bias and asymmetrical double-gate back-gate bias along with their associated control voltages VC and VCB.

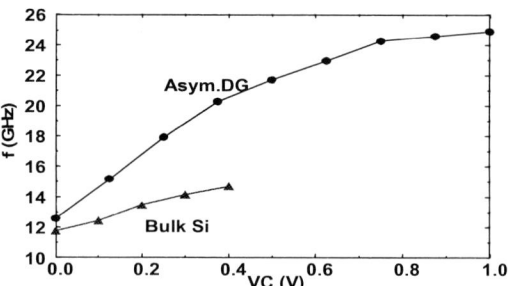

Fig. 4: Frequency range of the well biased bulk and asymmetric double-gate VCO. The well bias for bulk technology is limited to 400mV.

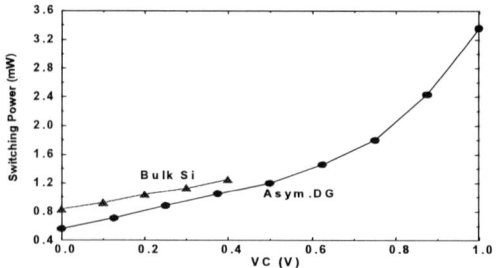

Fig. 5: Switching power of well-biased bulk silicon and back gate-biased double-gate VCO.

148

Author Index

A

Absil, P. 119
Adams, V. 58
Aichmayr, G. 92
Arnauts, S. 123
Asano, Tanemasa 135
Augendre, E. 119

B

Bao, T.I. 133
Barbagini, F. 123
Barnett, J. 111
Barral, V. 117
Bearda, T. 123
Benistant, F. 145
Bergner, W. 92
Bersuker, G. 107, 111
Beyne, Eric 19
Biesemans, S. 119
Black, L. 84
Brown, G.A. 111
Brown, George 48

C

Cabral, Jr., C. 121
Cassé, M. 117
Cave, N. 58
Chan, Y.J. 74
Chang, C.H. 143
Chang, C.P. 44
Chang, Chih-Lien 86
Chang, G. 101
Chang, J.J. 74
Chang, Josephine B. 68
Chang, L. 38
Chang, Mi-Chang 52
Chang, Y. W. 30
Chang-Liao, K. S. 36
Chang-Liao, Kuei-Shu 115
Chao, D. S. 40
Che, M. 101
Cheek, J. 58
Chen, C. H. 34
Chen, Gary 64
Chen, H. M. 32, 101
Chen, Hsien-Wei 129
Chen, Hsueh-Chung 129
Chen, Hung-Wei 62
Chen, I.I. 133
Chen, K. P. 30
Chen, K.H. 58
Chen, Kun-Ming 139
Chen, P. A. 42
Chen, Shang-Jr 109
Chen, Shih-Chang 109

Chen, Terry 34
Chen, W. -J. 36
Chen, W. S. 40
Chen, W.C. 44
Chen, William P.N. 143
Chen, Y. C. 40
Chen, Y.H. 44
Cheng, CC 101
Cheng, Chin-Lung 115
Cherif, K. Sidi Ali 117
Chiang, L.P. 42
Chiang, P. Y. 34
Chien, T. 101
Chin, H.W. 103
Choi, L.J. 88
Choi, R. 107, 111
Chou, C.C. 133
Chou, G. 101
Chou, George C. W. 34
Chou, L. 101
Chu, C. H. 34, 101
Chu, S. 145
Chu, Yu-Hung 52
Chuang, Che-Hao 56
Chuang, Ching-Te 147
Chung, C.A. 74
Chung, C.H. 101
Chung, Steve S. 34
Chuo, Y. 40
Claes, M. 123
Colombeau, B. 145
Curatola, C. 119

D

Das, K. K. 38
De Meyer, K. 119
Decoutere, S. 88
Degendt, S. 123
Deleonibus, S. 117
Demaco, C. 123
Devriendt, K. 88
Diaz, C.H. 143
Diaz, Carlos H. 109
Dokumaci, O. 84
Du, Pei-Ying 32
Dupont, T. 88

E

Eitoku, A. 123
Erben, E. 92
Eyben, P. 119

F

Falepin, A. 119
Faynot, O. 117

Author Index

Felch, S. 119
Frank, M. 123
Fujita, S. 60
Fung, Samuel K.H. 58
Fyen, W. 123

G

Gan, Tian-Choy 109
Garaud, S. 123
Garros, X. 117
Gavartin, J. 107
Ge, Chung-Hu 62
Gebara, Fadi 147
Gignac, L. 121
Gluschenkov, O. 84
Gnade, B. E. 113
Goto, K. 143
Gribelyuk, M. 121
Grosgeorges, P. 117
Grove, N. 58
Grudowski, P.A. 58
Guillaumot, B. 117
Guo, Jyh-Chyurn 32
Gusev, E. P. 121

H

Haensch, W. E. 38, 84
Hall, L. 123
Hamamoto, Takeshi 94
Hecht, T. 92
Heh, D. 111
Hellin, D. 123
Hergenrother, J. M. 84
Heyns, M. 123
Hino, Masaki 50
Hiyama, S. 60
Hoffmann, T. Y. 119
Holliger, P. 117
Holsteyns, F. 123
Hon, R.Y. 101
Hoyer, R. 123
Hsiao, G. 101
Hsiao, J.S. 101
Hsu, C.C.-H. 32
Hsu, C.-Y. 80
Hsu, H. H. 40
Hu, C. 38
Hu, T. 101
Huamg, Stanley H. 74
Huang, Guo-Wei 139
Huang, H.T. 58
Huang, J.F. 44
Huang, K.P. 133
Huang, K.Y. 101
Huang, Shao-Chang 52
Huang, T.Y. 52
Huang, Tiao-Yuan 62

Huff, Howard 48
Hung, C. H. 42
Hung, C.C. 44

I

Ieong, M. 84, 121
Ikeda, H. 60
Ikuta, T. 60
Imoto, T. 60
Iwai, Hiroshi 1
Iwamoto, H. 60
Iyer, Subramanian 90

J

Jackson, Thomas 66
Jaraiz, M. 145
Jeng, P. R. 36
Jeng, S.M. 133
Jeng, Shin-Puu 129
Jin, Ying 109
Jurczak, M. 119

K

Kadomura, S. 60
Kanda, S. 60
Kao, M. J. 40, 44
Kashiwadate, S. 60
Kataoka, T. 60
Kenis, K. 123
Ker, Ming-Dou 56
Kerner, C. 119
Kersch, A. 92
Kesters, E. 123
Kevin Nowka 147
Kikuchi, Y. 60
Kim, Jong Woo 64
Kim, K. 123
Kim, Keunwoo 147
Kim, M 113
Kim, Y.H. 121
Kirsch, P. D. 113
Ko, Chih-Hsin 62
Kobayashi, T. 60
Kocsis, M. 123
Kojima, Takeaki 135
Kolagunta, V. 58
Kotani, T. 123
Kraus, H. 123
Kraus, P. 119
Krishnan, S. 113
Kuang, JB 147
Kudelka, S. 92
Kung, C.P. 74
Kunnen, E. 88
Kuo, Chen-Chi 52
Kuo, W.S. 101

Author Index

Kurobe, Ken-ichi 50

L

Lafond, D. 117
Lau, F. 92
Le, T.Q. 123
Lee, B. H. 107, 111, 113
Lee, C. M. 40
Lee, D.H. 58
Lee, H. Y. 36
Lee, H.M. 32
Lee, J. 84
Lee, K.-L. 84
Lee, K-T. 123
Lee, L. S. 36
Lee, M. H. 40
Lee, Roger 64
Lee, Shou - Chung 131
Lee, Wen-Chin 62
Lee, Wonshick 97
Lee, Y.J. 44
Leray, P. 88
Leung, Ying Keung 109
Li, F. 101
Li, H. J. 113
Lian, S.J. 58
Liang, M.S. 133
Liang, Mong-Song 109
Liang, Victor 139
Liang, W.J. 58
Liao, H. H. 42
Lien, C.H. 143
Lim, Peng-Soon 109
Lin, C. H. 38, 42
Lin, C.S. 44
Lin, Chia-Pin 82
Lin, Chien-Ting 139
Lin, Hong-Nien 62
Lin, Horng-Chih 62
Lin, Huan-Just 109
Lin, K.C. 133
Liu, D.-H. 80
Liu, K. C. 36
Liu, Kevin_ 103
Liu, Yung 64
Lo, P.Y. 74
Loo, R. 88
Lu, Daniel 64
Lu, M. S.-C. 80
Lu, T. C. 30
Luetzen, J. 92
Lux, M. 123

M

Maikap, S. 36
Maîtrejean, S. 117
Malhouitre, S. 123

Martin, F. 117
Matsuno, Akira 50
Mazure, Carlos 78
Mertens, Paul W. 123
Mii, Yuh-Jier 109
Min, B. 58
Minoret, S. 117
Miyanami, Y. 60
Mizuno, T. 137
Molesa, Steven E. 68
Moriyama, Y. 137
Mouis, M. 117
Mueller, W. 92
Muscat, A. 123

N

Nadkarni, S. 107
Nagashima, N. 60
Nakaharai, S. 137
Nakazato, Kazuo 28
Ngo, Hung 147
Nitayama, Akihiro 94
Noda, T. 119
Nouri, F. 119
Nuetzel, J. 92

O

Oates, Anthony S. 131
Ohno, T. 60
Ohsawa, Takashi 94
Onsia, B. 123
Ooka, Masahiro 127
Orth, A. 92
Ott, J. 84

P

Pan, J. 84
Pandey, S.M. 145
Pant, G. 113
Parachiev, V. 123
Parihar, V. 119
Park, C. S. 107
Park, Donggun 97
Pei, Z. 74
Peterson, J. J. 111, 113
Pham, Daniel 48
Piontek, A. 88
Poiroux, T. 117
Previtali, B. 117

Q

Quevedo, M. A. 113

R

Ramos, J. 119

Author Index

Redinger, David R. 68
Ren, Zhibin 84
Rip, J. 123
Rogers, John A. 72
Ronsheim, P. 84
Ryu, Byung-il 97

S

Saito, M. 60
Sakurai, Takayasu 70
Sano, K. 123
Sano, Nobuyuki 141
Schloesser, T. 92
Scholz, A. 92
Schreutelkamp, R. 119
Schroeder, U. 92
Sekitani, Tsuyoshi 70
Severi, S. 119
Shen, CC 64
Shen, Rick 32
Shi, X.P. 88
Shibahara, Kentaro 50, 105
Shih, J.R. 103
Shiu, Yu-Da 56
Shluger, A. 107
Shu, H.C. 101
Sibaja-Hernandez, A. 88
Sieck, A. 92
Sim, J. 107
Singh, D. V. 84
Sioncke, S. 123
Sleight, J. 84
Snow, J. 123
Someya, Takao 70
Song, M.H. 52
Song, S. C. 113
Spitzer, A. 92
Stork, Hans 18
Strasser, M. 92
Su, K.L. 44
Su, P. 143
Su, Pin 139
Subramanian, Vivek 68
Sugiyama, N. 137
Sun, Guangyu 46
Sun, Jack Y.-C. 129
Sun, Yugang 72
Sung, C.Y. 84

T

Tagawa, Y. 60
Takagi, S. 137
Tan, Leong 64
Tao, Hun-Jan 109
Tateshita, Y. 60
Tezuka, T. 137
Thevenod, L. 117

Thompson, Scott E. 46
Toffoli, A. 117
Tsai, Ching-Wei 109
Tsai, M.-J. 36, 40, 44
Tsai, Ming-Jinn 86
Tsai, P. H. 36
Tsai, V. 101
Tsai, Y.S. 103
Tsao, C. P. 42
Tseng, M. H. 40
Tsui, Bing-Yue 82, 86
Tsui, R.F. 103
Tuan, H.C. 58
Tzeng, P. J. 36

U

Uechi, Tadayoshi 141
Ugajin, H. 60

V

Van Hoeymissen, J. 123
Van Huylenbroeck, S. 88
Vandervost, W. 119
Vanhaelemeersch, S. 88
Vereecke, G. 123
Vinet, M. 117
Vleugels, F. 88
Volkman, Steven K. 68
Vos, R. 123

W

Wallace, R. W. 113
Wang, B. 101
Wang, C. 101
Wang, C. C. 36
Wang, Howard C.-H. 109
Wang, Howard Chih-Hao 115
Wang, J. 60
Wang, J.S. 143
Wang, P-F. 92
Wang, Sheng-Chun 139
Wang, Tien-Ko 115
Wang, Tzu-Chen 115
Wang, W. H. 40
Wang, Y.H. 44
Watanabe, Naoya 135
Weber, O. 117
Wege, S. 92
Wei, Jeng-Hua 86
Weis, R. 92
Weng, Chien-Li 86
Widiez, J. 117
Williams, R. Q. 38
Wostyn, K. 123
Wu, C.H. 58
Wu, Chii-Ming M. 129

Author Index

Wu, Kenneth 103
Wu, S. 101
Wu, Z.C. 133

X

Xiao, Deyuan 64
Xu, K. 123

Y

Yamagishi, N. 60
Yamamoto, R. 60
Yamamoto, Y. 60
Yang, C.T. 58
Yang, I. C. 30
Yang, Ji-Woon 48
Yao, D.-J. 80
Yeh, M.L. 133
Yeong, S.H. 145
Yokoyama, Shin 127
Young, C. 107
Young, C.D. 111
Yu, C.H. 133

Z

Zeitzoff, P. 111
Zeitzoff, Peter 48
Zheng, L.-S. 80
Zia, O. 58

9781424401819